Evolutionary Biology

VOLUME 5

Evolutionary Biology

VOLUME 5

EDITORS:

THEODOSIUS DOBZHANSKY
University of California, Davis

MAX K. HECHT
Queens College

WILLIAM C. STEERE
The New York Botanical Garden

APPLETON-CENTURY-CROFTS
Educational Division / MEREDITH CORPORATION
NEW YORK

Softcover reprint of the hardcover 1st edition 1972

72 73 74 75 76/10 9 8 7 6 5 4 3 2 1

Library of Congress Card Number: 67-11961

ISBN 978-1-4757-0258-3 *ISBN 978-1-4757-0256-9 (eBook)*
DOI 10.107/978-1-4757-0256-9

F-27034-2 (390)

Contributors

WALTER J. BOCK, *Department of Biological Sciences, Columbia University, New York, New York, and Department of Ornithology, The American Museum of Natural History, New York, New York*

A. D. BRADSHAW, *Department of Botany, University of Liverpool, Liverpool, England*

PETER S. DAWSON, *Department of Zoology, Oregon State University, Corvallis, Oregon*

J. C. DEFRIES, *Institute for Behavioral Genetics, University of Colorado, Boulder, Colorado*

BRUCE K. ECKLAND, *Department of Sociology, The University of North Carolina, Chapel Hill, North Carolina*

ALLEN KEAST, *Department of Biology, Queens University, Kingston, Ontario, Canada*

CHARLES E. KING, *Department of Biology, Yale University, New Haven, Connecticut*

ELMAR E. LEPPIK, *Plant Science Research Division, Agricultural Research Service, United States Department of Agriculture, Beltsville, Maryland*

G. E. McCLEARN, *Institute for Behavioral Genetics, University of Colorado, Boulder, Colorado*

J. A. McKENZIE, *Department of Genetics and Human Variation, La Trobe University, Victoria, Australia*

P. A. PARSONS, *Department of Genetics and Human Variation, La Trobe University, Victoria, Australia*

Contents

Evolutionary Biology

VOLUME 5

1

Species Interactions and Macroevolution

WALTER J. BOCK
Department of Biological Sciences, Columbia University
New York, N. Y. 10027
and
Department of Ornithology, The American Museum of Natural History, New York, N. Y. 10024

INTRODUCTION

Acceptance of Darwinian concepts of evolution by biologists in the middle of the 19th century resulted in a curious mixture of new and old ideas, a familiar pattern that followed the introduction of many radically new concepts. The development of idealistic morphology and associated concepts of major evolutionary change by comparative anatomists and paleontologists in the decades following the publication of "The Origin of Species" is an excellent example of such a blend of ideas. Idealistic morphologists accepted the basic concepts of organic evolution including the notion that members of a taxonomic group descended from a single common ancestor. Yet the basic theory of idealistic morphology stemmed directly from the pre-Darwinian concept that a fundamental *bauplan* or type existed for each natural group of organisms: definite type limits or boundaries exist for each type with a sharp and absolute gap between different types. Consequently these workers assumed that different evolutionary mechanisms were responsible for changes within and between groups characterized by different bauplans. Evolutionary change between groups–the origin of new major taxa–was believed to occur by some sort of saltatory evolution, but the exact mechanisms were left unspecified or described in vague terms.

Following the emergence of genetics in 1900, the mechanism of saltatory evolution was believed to lie in the occurrence of macromutations. These large mutations gave rise to radically different individuals that were considered to be the progenitors of a new taxon. Change resulting from a macromutation is the single type jump needed to span the gap between the old and new taxa. Macromutations were regarded as differing quantitatively from the micromutations responsible for genetical variation within species. After the progenitor of the new taxon appeared, this so-called "hopeful monster" discovered, if lucky, a habitat in which it could survive. Subsequently, the members of the new taxon underwent further adaptive perfection and adaptive radiation according to conventional microevolutionary mechanisms of genetical variation and selection. But the initial steps associated with the origin of the new taxon did not involve the action of selection and hence must be regarded as nonadaptive evolutionary modifications. Indeed a common belief was that new features or new taxa appeared and became fully developed, and only subsequently acquired their functional significance and adaptive value. Or, the progenitors undergoing a type jump were believed to be released from the restraints of selection during the origin, and perhaps even during the initial adaptive radiation of the new taxon.

The origin of new taxa by saltatory evolutionary mechanisms remained the predominant belief among students of phylogeny and classification for 75 years after the advent of Darwinism. Schindewolf (1950) presents arguments for the basic concepts of typology and of evolution of type jumps from the viewpoint of a paleontologist. Concepts of idealistic morphology and evolutionary change were discussed by Neaf and Kalin, among others, and reviewed recently by Zangerl (1948). The important notion of macromutations as the basis for saltatory evolutionary change was discussed by Goldschmidt (1940).

After 1930, ideas of saltatory mechanisms of macroevolution were replaced slowly by concepts of gradual change. The most important contribution to this modification of basic macroevolutionary concepts were Simpson's analyses (1944 and 1953) of major evolutionary events documented in the fossil record. Simpson assumed that the same mechanisms acted in microevolution and macroevolution, and demonstrated in an elegant fashion that the observed modifications in the fossil record were in full accord with this assumption. After 1940, most students of evolution accepted the hypothesis that the origin of new major taxa was by gradual change.

Although Simpson considered the origin of new taxa to be essentially adaptive, he remained somewhat equivocal on this aspect of microevolution. Indeed, his initial concept of quantum evolution (Simpson, 1944:206-217) is not far removed from earlier saltatory mechanisms. Quantum evolution involves an inadaptive phase in which the organism-environment relationship is unstable, with the population facing a threshold that must be surmounted quickly or else the population is faced with likely extinction. An inadaptive phase implies the absence of selection, a point similar to ideas of saltation. Simpson's later concept (1953:389-393) of quantum evolution changed considerably toward a notion of adaptive change throughout the threshold period with continuous action of selection. The strength and "rigidity" of selection were regarded as being much more similar to that acting during slower phases of phyletic evolution. Subsequently, the concept of quantum evolution has been modified even further away from the initial concept (see Schaeffer and Hecht, 1965, and the other contributions to that symposium). Today most workers have abandoned the concept of quantum evolution in favor of the hypothesis that both slow and rapid phyletic evolution occur by the same evolutionary mechanisms.

Mayr (1963: 586-621; 1970) emphasized the role of species and of speciation in transspecific evolution, and concluded, quite rightly, that: "The species, then, is the keystone of evolution." Mayr's conclusion must not be treated lightly even though most students of major evolutionary phenomena give only lip service to the role of species in macroevolution. Rarely are the consequences of speciation and associated evolutionary mechanisms applied to facets of major changes. The relevance of species-level evolutionary mechanisms, especially those resulting from species interactions, are of particular importance to the hypothesis to be presented in this study.

Despite other areas of disagreement, advocates of gradual macroevolutionary change are in close agreement that the shift from one adaptive zone to another, or the origin of one higher taxon from another, occurs under the control of directional selection (linear selection or asymmetrical selection; Simpson, 1953:156; Hecht, 1965). The number and magnitude of the adaptive differences between most major taxa suggest that the directional selection responsible for the origin of new taxonomic groups would have to be complex, reasonably strong, and continuously active over a period of time at least several-fold as long as that required for speciation. The "direction" of this directional selection does not have to be constant, however. It must only be active as directional selection (i.e., not as centripetal or stabilizing selection). Although general agreement exists on the necessity of directional selection in the origin of new major taxa, disagreement and/or vagueness exists on the identification of the environmental factors that could provide the necessary long term directional selection.

The goal of this study is an attempt to identify those environmental factors which may be responsible for the directional selection associated with the origin of new major taxa and to inquire into some of the ramifications of these environmental factors. Most macroevolutionary theories have been based on the assumption (often tacitedly expressed) that new adaptive systems (e.g., tetrapod limbs, flight in birds) evolved in response to directional selection arising from factors of the physical environment or food items of the new major adaptive zone. I would suggest that these environmental factors provide only a very minor contribution to the total directional selection required for a macroevolutionary shift. Further, I would like to demonstrate that the major source of directional selection associated with the origin of a new major taxon and its new adaptive complexes is species interactions of all types, of which the two most important ones are exclusionary interactions and coevolutionary interactions.

BASIC MACROEVOLUTIONARY MODEL

Before inquiring into the possible source of directional selection, it is necessary to specify the mechanisms of transspecific evolution that will be adopted for this analysis, because many important differences of opinion still exist in this area of evolutionary biology. Selection must be considered both in terms of its consequence as nonrandom differential reproduction of genotypes and of the interaction of environmental factors with differing phenotypes of individual organisms; the latter is of particular importance to this study. Moreover, the total selection acting on a species expressed as the interaction (= niche) between the organism and its environment must be subdivided into component units. The fact that these units cannot be delineated sharply or that these components are not of equal magnitude is of little significance. These individual selection forces (Bock and von Wahlert, 1965:290-291) are the action of individual environmental factors on the organism and must be correlated with individual features of the organism. Most important is that the pattern of selection forces varies considerably during the course of a major evolutionary change (Bock, 1959, 1965). Many selection forces operative at the onset of such a change disappear during its time course, and new selection forces appear.

Aspects of adaptation and preadaptation have been discussed by Bock (1959) and Bock and von Wahlert (1965). Adaptations are, by definition, features of an organism favored by selection and hence associated with survival of the species. Moreover, adaptations must be judged with reference to a particular, stated environment; ideally, this judgment should be with reference to the environment selecting for the adaptation. Appraisal of the degree of adaptation of particular features should be measured independently of survival of the individual organism and its progeny. A measure of efficiency in terms of energy required by the organism to successfully maintain the interaction (= synerg) between the feature and the selection force was suggested by Bock and von Wahlert (1965: 286-290); better adaptations require less energy. A preadaptation is a feature that can assume a new biological role and hence interact with a new selection force by virtue of its existing forms and functions (Bock, 1959; Bock and von Wahlert, 1965:292). Evolutionary changes involving preadaptations necessitate modification in the selection forces acting on the organism. Most essential is that the evolution of a feature to any particular preadapted condition is under the control of existing selection forces and hence represents adaptive change. Preadaptations are fortuitous with regard to the new selection forces that will interact with them. Yet preadaptation is an extremely common, perhaps even a ubiquitous, factor in major evolutionary changes.

The onset of modifications leading to a major change is usually a behavioral shift (von Wahlert, 1965), frequently associated with a preadapted feature. These "ethological key characters" are followed by the evolution of morphological key characters adaptive to the ethological ones (von Wahlert, 1965:289). Evolution is not directed solely by environmental modifications, but is determined by the interplay between the organism and the environmental change. This interplay governs the particular selection forces that act on the organism and shape the resulting adaptations. Selection, then, is determined via a feedback mechanism according to the reaction of the organism to environmental modifications (von Wahlert, 1965:296-297).

Analyses of principles of comparison, including homology, (Bock, 1959, 1963a, 1967, and 1969) have demonstrated the distinction between adaptive and paradaptive aspects of features; a feature will possess simultaneous adaptive and paradaptive attributes. The paradaptive aspects of a feature (Bock, 1967) are those associated with and dependent upon chance-based evolutionary mechanisms (e.g., genetical) responsible for the origin and later modification of the feature. If these paradaptive aspects are favored by selection, then they will also be adaptive. Paradaptive facets of features are best seen as differences in horizontal comparisons between several features representing multiple evolutionary pathways (Bock, 1959). Adaptive facets are readily seen as differences in vertical comparisons within a single phyletic line as differences between an ancestral and a descendent feature.

A model for major evolutionary change was suggested by Bock (1965) based upon the concept of gradual adaptive change. In this model, the evolution of a new large taxon is seen as the result of a sequential series of radiations in a transitional adaptive zone spanning the gap between the major adaptive zones of the ancestral and descendent taxa; each radiation contributes a small evolutionary change to the major difference between the old and new groups. The features of the new group evolve in a pattern of stepwise or mosaic evolution (de Beer, 1954) with each step involving preadaptation, a different pattern of selection forces, and a period of postadaptation leading to a new preadaptation setting the stage for the next radiation. Selection at the onset and the close of a macroevolutionary sequence is usually quite different. Similar features may appear in several parallel, but independent, phyletic lines evolving in the transitional zone because of the similar genetical-developmental potential of these uniform, closely related lines (Bock, 1963a). No single threshold or point of origin can be ascertained for the new group. The strength of selection and the rate of evolutionary change during the transition from the ancestral to the descendent taxa are not basically different from those occurring within the limits of either major group. The entire process is considered to be adaptive with selection acting at all times; selection differs from that acting in a major adaptive zone only in being more directional than centripetal.

The number of phyletic lines involved in the origin of a new group is a key element in any theoretical model of major evolutionary change. Most earlier analyses were based upon the assumption that only a single phyletic line, or one with very few collateral branches, is involved in the origin of new major taxa. The rarity of fossil forms intermediate between taxonomic groups supports this notion, but such data are relatively weak negative evidence. I believe that no real evidence supports this view, and consequently have advocated (Bock, 1965,

1970) that, in general, continuous adaptive radiations occur during the entire evolution of a new group. Usually these radiations would be broad, each including a number of different phyletic lines most of which become extinct relatively quickly. The evolutionary situation and mechanisms of change are radically different in the case of a solitary phyletic line without close relatives as compared to the case of a number of closely related phyletic lines. In the latter the species corresponding to each phyletic line can interact with each other (e.g., competition and interbreeding) and can exert selection forces on one another; these selection forces are over and above those imposed on the species by the physical environment and food items. I have discussed species interactions earlier (Bock, 1970) and suggested that the selection forces generated by these interactions may be largely responsible for the speed of macroevolutionary change and for the total amount of change. At that time, I realized neither the full implications of species interactions for major evolutionary shifts, nor all types of interactions that may have a role in macroevolution. Such interactions appear to possess a more general role in transspecific evolution than is generally realized.

I would like to postulate that interaction between pairs of species, be it exclusionary (e.g., competition) or coevolutionary (e.g., predator-prey relationships), is the important source of directional selection required for major evolutionary modifications. Moreover, I would suggest that the rate, total amount of change, and final diversity in transpecific evolution is closely correlated with the opportunity for species interactions of all types.

ENVIRONMENTAL CHANGE AND DIRECTIONAL SELECTION

Before attempting to document the hypothesis that species interactions provide directional selection for major evolutionary shifts, it is essential to discuss the main concept held currently; namely, that most major evolutionary changes result from the availability of a new adaptive zone. This new zone could be a novel habitat or it could be an existing zone that had become empty and hence available. Directional selection forces responsible for the origin of the new taxon are believed to arise from this new adaptive zone. Simply stated, a different and empty habitat would provide the directional selection necessary to guide evolutionary changes from the ancestral taxon (and its adaptive complex of features) to the new descendant taxon (and its new adaptive complex of features). The environmental components associated with the differences between major adaptive zones are generally regarded to be factors in the physical environment or food items. This generalization has been accepted by students of major evolutionary change with only rare exception. Yet the exact nature and magnitude of these environmental differences between the old and the new adaptive zones are rarely considered, either in theoretical or actual terms. Although these environmental differences are a necessary part of the evolutionary mechanisms involved in major shifts, I doubt that they are the source for the essential directional selection, for several important reasons.

The possible patterns of selection forces arising from the differences between adaptive zones have not been appraised by advocates of this hypothesis. It is clear, from the sequence of stepwise changes (mosaic evolution) involved in the origin of most new groups, that the pattern of selection forces responsible for these changes must alter frequently during the entire evolutionary shift. Change in the pattern of selection forces implies change in the environmental factors. Yet the environmental differences that are usually stated or implied to exist between major adaptive zones do not provide the necessary sequences of changing selection forces.

If the spectrum of physical environments occupied by members of a vertebrate order or class and the range of food items eaten by these animals are compared with the diversity of the members of the particular taxon, some strange correlations (or lack of correlations) will be found. Many species of mammals and birds for example, have invaded and successfully occupied vastly different physical habitats with little or no phenotypical change. The same species may range over one or two continents, living in habitats as varied as hot swamps, cold forests, mountains, and dry deserts with minor morphological differences. The puma (*Felis concolor*) is an excellent example. The tiger (*Panthera tigris*), wolf (*Canis lupus*), wild boar (*Sus scrofa*), raven (*Corvus corax*), barn owl (*Tyto alba*), and osprey (*Pandion haliaetus*) are additional good examples of widespread species. Widespread genera that occupy diverse habitats, but with little morphological diversity between the congeneric species are even easier to cite. Less is known about physiological modifications, but studies on desert mammals and birds have revealed some surprisingly minor physiological differences between desert and nondesert populations of the same species (Bartholomew, 1963, 1964; Bartholomew and Dawson, 1968; Dawson and Bartholomew, 1968). Moreover, shifts in food items often occur with little or no change in the morphology of the feeding apparatus. The Laysan finch (*Psittirostra cantans*) feeds on insects, bird eggs, dead birds, plant roots and hard seeds without change in the large conical finch bill found through the genus (Bock, 1970:716-717). Parrots feed on a wide range of food items from hard nuts and seeds to fruit, nectar, carrion, insects, and tree fungus (the woodpecker-like pygmy parrots, *Micropsitta*) with a remarkably uniform bill structure in the entire order. The palm-nut vulture (*Gypohierax angolensis*) of Africa feeds extensively on the fruit pericarp of the oil palm (*Elaeis guineensis*) with no basic modification in its feeding apparatus from that in other Accipitridae (Thomson and Moreau, 1957).

After an extensive survey of the morphology, distribution, environments, and food items of avian taxa (families and orders), I would conclude that radically different adaptive zones (described in terms of the physical environment and food items) have often been occupied successfully by members of an avian group with relatively little adaptive modification. By "little adaptive modification" I mean the evolution of morphological differences of the degree observed between congeneric species or closely related genera. And by radically different adaptive zones, I mean differences in the physical environment and food items of the degree that characterize different avian families or orders. These comparisons are, admittedly, difficult to describe in quantitative terms, and I will not attempt to do so. The broad overlap in food items and habitats between members of different families and orders of a group such as birds or mammals raises serious doubts about the validity of discrete major adaptive zones.

The number of phyletic lines involved in the origin of a new group has direct relevance to the question of whether directional selection stems from the difference between the old and new major adaptive zones. Most earlier discussions of macroevolutionary mechanisms (e.g., Simpson, 1953; Mayr, 1963) state or imply tacitly that a single phyletic line, with a few collateral branches at the most, is involved in the origin of a new group. Adaptive radiation occurs only subsequently to the successful occupation of the new major adaptive zone. This approach may be reasonable in a "backwards" view from the descendent group toward the ancestral stock in which side branches can be ignored, but it does not provide a valid basis on which to formulate evolutionary mechanisms.

If a single phyletic line of an ancestral taxon evolved under the control of selection forces associated with a new major adaptive zone, this species would adapt to the conditions of the new adaptive zone. It is reasonable to conclude that this species will change until it is adequately adapted to the new environmental conditions and is utilizing the new resources available to it without competition from other species of the ancestral taxon. A slow environmental shift until an extreme change is achieved or a drastic shift into an adaptive zone would not necessarily result in large adaptive modifications such as those associated with the origin of a new major taxon. If entrance into the new adaptive zone required a drastic initial change in food items or habitat, then the ancestral species might not be able to enter it at all. Once the single phyletic line had adapted to the conditions of the new adaptive zone and was no longer competing with other species of the ancestral taxon, selection forces would change from directional to stabilizing. Further modification of this phyletic line from the ancestral stock would cease. Hence, if only a single phyletic line entered a new adaptive zone, it would change, but it generally would not modify drastically from the ancestral group.

On the basis of theoretical considerations, I would conclude that the pattern of directional selection forces required for the origin of new higher taxonomic groups do not arise from the environmental factors of the physical habitat and food items of the new major adaptive zone. The available evidence from comparative studies of taxonomic groups and their ecological relationships, e.g., families and orders of birds, supports this conclusion, although much of this evidence is negative or inferential.

SPECIES INTERACTIONS AND DIRECTIONAL SELECTION

The main source of directional selection associated with the origin of new major taxa is, I believe, species interactions of all types. In the environment of any species, other sympatric species are important factors (e.g., Lack, 1944, 1947), although the evolutionary role of sympatric species interactions has never received the attention it deserves in general evolutionary theory either from students of population phenomena and speciation or from students of major changes. Sympatric species can exert powerful mutual selection on one another and this selection usually has a strong directional component. Moreover, such selection may be the most intense type existing under natural conditions. The most obvious species interactions are a consequence of the continuous adaptive radiations during the origin of a new group, but there are other types of interactions that are independent of persistent speciation. It is important to note at the onset that several different types of species interactions exist and that the evolutionary mechanisms and consequences of the several types may be quite different. The major categories of species interactions to be discussed below are: (1) *exclusionary interactions*, including ecological competition and pair-bond formation (involving intrinsic isolating mechanisms); and (2) *coevolutionary interactions*, including predator-prey relationships and all sorts of symbiotic associations.

In all cases, the resulting directional selection is dependent upon coexistence of the interacting forms and will result in modification of at least one and usually all of the involved species.

"Species interactions" is a vague term covering a number of different biological phenomena. For the purposes of this discussion, I would like to limit species interactions to those phenomena between pairs or groups of sympatric species which result in mutual selection forces on each taxon that would not exist in the absence of the sympatric interacting species. The selection forces exerted by each member of the species assemblage on the others need not be the same, either in type, direction, or magnitude. Selection will persist only so long as the species continue to interact with one another. Cessation of interaction and hence mutual selection will occur when the species are no longer strictly sympatric or when the species have modified sufficiently that they no longer exert a reciprocal affect on one another.

Exclusionary species interactions involve ecological features associated with competition and/or ethological features (and all other intrinsic isolating mechanisms) associated with reduction of reproductive wastage (Bock, 1970:706) in pairs of species that are already reproductively isolated. Each species will exert selection on the other, with possible consequences ranging from the extinction of one of them, to character divergence in both (Lack, 1944, 1947). Such species interactions will be most intense between two newly evolved daughter species during the sympatric phase of speciation when they initially overlap geographically. During their neosympatric period, species frequently compete with one another and often attempt to form pair bonds and interbreed even though they are reproductively isolated by postmating intrinsic isolating mechanisms (Mayr, 1963:91). The resulting directional selection forces are strong and may be the most intense naturally occurring selection forces.

The consequence of exclusionary species interaction that has the greatest significance to major evolutionary change is character displacement. The two species will diverge from each other under the action of the mutual selection, thereby reducing the degree of interaction and hence the strength of the directional selection. Once interaction ceases between the species pair, further divergence will be greatly reduced until another cycle of speciation occurs or another new species invades the range of either member of the original species pair.

Exclusionary interaction can also occur between pairs of more distantly related species, but the resulting selection would generally be less intense depending upon the degree of their reciprocal influence. In morphologically uniform taxa, especially those with a uniform and specialized feeding apparatus such as woodpeckers (Picidae), parrots (Psittacidae), or hummingbirds (Trochilidae), almost any pair of similar species may compete strongly even if they are distantly related within the family. In the Hawaiian honeycreepers (Drepanididae) exclusionary species interactions occur between daughter species (e.g., between *Loxops v. virens* and *Loxops sagittirostris* and between *Loxops v. stejnegeri* and *Loxops pava*, Bock, 1970:711) and between closely related genera (e.g., between *Loxops v. stejnegeri, Hemignathus lucidus hanapepe*, and *Hemignathus procerus* on the island of Kauai, Bock, 1970:714-815), but apparently not between more distantly related forms. (Nothing is known about possible competition for nectar and insects between sickle-billed species of the two subfamilies.)

Coevolutionary species interactions involve a number of different types such as predator-prey relationships, pollinating organisms, and symbioses (Ehrlich and Raven, 1965; Ehrlich, 1970). This category of interaction is characterized by a direct utilization by one member of the species pair on resources of the other. In some cases, both members of the pair use resources of the other species, but generally in quite different ways. The only restriction is that use of the one species by the other species must be advantageous to the latter. In some cases, e.g., symbiosis, the reciprocal exploitation by each species is advantageous to each. In every case, each species exerts selection upon the other no matter which one is actively utilizing the other; the details of the selection forces applied by each species on the other is almost always quite different. The results of the mutual selection is modification of one or both species, but this evolutionary change is generally not a divergence as in the case of exclusionary interaction. Occasionally, e.g. in predator-prey relationships, the result of the interaction may be the extinction of one member of the species pair, but this outcome is much rarer than in exclusionary interactions. Moreover, change in either member of the interacting pair of species usually results only in a temporary, if any, reduction of selection exerted by the other species. The general consequence of coevolutionary interactions is a continued modification of selection as each of the interacting species evolves.

The strength of selection in coevolutionary interactions is difficult to estimate and is probably much more variable than those arising in exclusionary interactions. No direct correlation exists between closeness of evolutionary relationship, or morphological similarity of the taxa involved in coevolutionary interactions, on the one hand, and the strength of selection exerted reciprocally by each species, on the other hand.

Coevolutionary interactions may operate for a longer period of time and may result in greater total evolutionary change than exclusionary interactions. Moreover, coevolutionary interactions are much less dependent and indeed may be completely independent of continued speciation and formation of neosympatric zones. This class of species interactions is of special importance in considerations of the evolution of a phyletic line that lacks close relatives and hence is not subject to exclusionary interactions between morphologically similar, closely related species. The aardwolf (*Proteles*: Hyaenidae), aardvark (*Orycteropus*: Tubulidentata) and pangolins (*Manis*: Pholidota) are highly specialized mammals which may have lacked close relatives during their evolution. Each of these taxa could have become highly modified because of coevolutionary interactions with their termite and ant prey, which included the ability of these mammals to burrow into the hard nests of their insect prey.

Coevolutionary interactions may also be responsible for change of single taxa evolving on isolated oceanic islands where opportunities for speciation are lacking.

Both exclusionary and coevolutionary interactions will provide the directional selection required for major evolutionary change, but the strength of selection and resulting rate of change varies considerably. Moreover, the types of evolutionary modifications arising from these two classes of interactions are often quite different. Coevolution may result in a greater total change requiring a longer period of time. Rapid evolution of many taxa filling the multiple niches of a new major adaptive zone would follow a series of exclusionary interactions. Presumably interactions of both types would operate during the evolutionary origin of a new major taxon with a complex pattern of directional selection controlling the rate and pattern of change. Yet directional selection in groups originating via a successive series of adaptive radiations may stem largely from exclusionary species interactions. If, as postulated above (and see Bock, 1965), the origin of most major taxa involved successive adaptive radiations, then the major source of directional selection may come from exclusionary interactions, while a smaller contribution may come from coevolutionary interactions. Distinction between these origins of directional selection is an interesting evolutionary question that deserves further analysis.

As mentioned above, generalizations about the mechanisms and consequences of species interactions for macroevolution is not feasible because of basic differences between the two major types. Nevertheless, a few generalizations should be offered about exclusionary interactions because of their correlation with adaptive radiations in the origin of many new major taxa.

Exclusionary species interaction, mutual selection, and consequential divergence or change of the species will continue so long as speciation and neosympatry continue, e.g., so long as adaptive radiation continues. The total amount of change that could result from the summation of successive species interactions would depend upon the nature of the adaptive zone into which the new group is evolving. The speed of change would depend upon the time period between successive speciations and neosympatries. Most essential is the opportunity for continued speciations and radiation with a close relationship apparently existing between the rate of speciation and the rate and degree of macroevolutionary change (Bock, 1970:719).

Establishment of neosympatry and resulting species interaction is important between any pair of species, not only between newly evolved daughter species. Any individual species will be involved in a variety of interactions with a number of different species. Moreover, the number and composition of the sympatric species and the details of the species interactions will vary considerably over the geographical range of most species; the major exception being those species with restricted geographical ranges. The resulting selection forces acting on a species will vary geographically and their effect will be transmitted differentially over the species range via gene flow. Similarly, examination of the distribution of species within a higher taxon, e.g., a family or order of birds reveals that the species composition varies from locality to locality. Clearly, then, the pattern of species interactions will differ as the species composition varies. Moreover, it is fair to assume that the species composition at any given locality will modify through time with continuing speciation, range expansion and reduction, and extinction. All of these factors will affect the possible macroevolutionary change resulting from exclusionary interactions.

The consequence of species interactions on macroevolution is seen most clearly in taxa found on oceanic islands. The likelihood of such a species evolving into a new major taxon depends largely upon the opportunities for speciation, and hence radiation, regardless of the difference of the island environment from that of the ancestral mainland stock. For avian speciation, an archipelago is required; a single small island is not sufficient. The island of Madagascar is about the minimum size for speciation and radiation of a family of passerine birds (Vangidae). A single phyletic line on a island will only rarely modify sufficiently to develop into a new high taxon, partly because of the less intense directional selection arising from coevolutionary species interactions. I know of only two examples of new avian family evolving on a single isolated island, namely the kagu (Rhynochetidae: Gruiformes) on New Caledonia and the dodos (Raphidae: Columbiformes) on the Mascarene Islands. The latter is complicated since several islands are involved and considerable change may have occurred because of past exclusionary species interactions. In both cases, the island forms became flightless and much of the morphological change appears to be correlated with loss of flight. Some additional examples of distinct avian subfamilies may be cited from isolated islands, but even these are rare.

A valid objection can be raised against the above argument and conclusion based on the high rates of extinction of organisms living on oceanic islands (MacArthur and Wilson, 1967). A high probability exists that a species invading an oceanic island will become extinct before it has existed there long enough to evolve into a new higher taxon. Thus, only relatively few endemic higher taxa would be expected on oceanic islands. Both explanations are valid and both may account for the low number of endemic higher taxa on islands; the relative importance of each explanation is difficult to evaluate.

Among birds, some of the clearest examples documenting species interactions as the source for directional selection are from archipelagos or large "continental" islands. These include the Hawaiian honeycreepers (Drepanididae; Amadon, 1950; Bock, 1970), Darwin's finches (Geospizinae; Lack, 1947), the wattlebirds of New Zealand (Callaeidae; Stonor, 1942), the vanga-shrikes of Madagascar (Vangidae; Dorst, 1960), and the birds of paradise of New Guinea (Paradisaeidae; Gilliard, 1969; Bock, 1963b). An important factor in the evolution of island groups is that the range and population size of the species pairs are not too dissimilar; hence it may be assumed that the "species pressures" or the mutual selection forces exerted by each species on the other are roughly similar. Both interacting species would probably be forced to change so that little directional modification would occur at the start of the radiation. After the initial radiation filled the habitats and subhabitats originally occupied by the progenitors of the new taxon, further speciation and exclusion species interactions could force one member of an interacting species pair into new habitats. This pattern is clearly shown by the evolution of the subfamily Psittirostrinae of the Drepanididae (Bock, 1970).

Examples illustrating the role of species interactions in the evolution of continental taxa are difficult to cite because they have not been analyzed thoroughly, at least for birds. One major problem is that the classification and phylogeny of many potentially excellent cases have not been clarified sufficiently. Woodpeckers (Picidae), hummingbirds (Trochilidae), the "New-World" nine-primaried oscines (Vireonidae, Parulidae, Icteridae, Thraupidae, Fringillidae and allies), and the "Old World" insect-eaters (Turdidae, Sylviidae, Muscicapidae and their relatives) fall into this category. A second problem is that important interactions may exist between species of distantly related taxa (different families or orders) so that a large segment of the avifauna of one or more continents may have to be surveyed in order to understand the consequences of species interactions in the evolution of one family of birds.

Nevertheless, some continental groups of birds may be good prospects for study. These include the honeyeaters (Meliphagidae) of the Australian-Papuan region, the starlings (Sturnidae), the crows and jays (Corvidae), and the American blackbirds, orioles, and relatives (Icteridae).

A common mode of species formation on continents is by change in small populations found in marginal ecological areas, usually about the periphery of the main species range (see Mayr, 1963: 386-393). When these small peripheral species established neosympatry with the large central species, the strength of the mutual selection forces will be vastly different. Divergence will occur almost exclusively by evolutionary change in the peripheral species. This form will be the one forced into a new habitat with accompanying changes in its features. After adapting to the new habitat, the former peripheral species can extend its geographical range. Subsequently satellite populations will be established at the edges of the range of the new species, after which the process will repeat itself.

A number of excellent studies are available that touch upon important aspects of the role of sepcies interactions, both exclusionary and coevolutionary, in the evolution of taxa on continents and large continental islands. MacArthur (1958) examined the details of interactions between five sympatric species of warblers (*Dendroica*: Parulidae) and described many aspects of these interactions that had operated in the evolution of these birds. Unfortunately his study was never extended to include all warblers in *Dendroica* and related genera, or to inquire into details of their evolution. Moreover, no comparative studies have been made of the interactions of these *Dendroica* species in other parts of their geographical range; we do not know of possible modifications in species interactions resulting from different species compositions. Ficken et al. (1968) described the consequences of species interactions (competition) between *Dendroica dominica* and *D. pinus.* Bill size is virtually identical in allopatric populations of these pine inhabiting species of warblers. In sympatric populations, however, bill size is quite dissimilar, with no overlap in bill lengths of the two species. *Dendroica dominica* evolved a longer bill in the area of sympatry with an associated modification in feeding habits.

Keast (1968) described the species interactions between the six species of *Melithreptus* (Meliphagidae) and compared the results of the different patterns of interactions over their entire range of Australia and Tasmania. He related these interactions to the evolution of the genus with special attention given to the double invasion of these honeycreepers on Tasmania. Moreover, Keast (1968, 1970) examined the interactions between species of *Melithreptus* and other passerine families on the mainland of Australia (Victoria) and on Tasmania, showing how the differences.in the species composition, and hence species interactions, influenced the evolution of taxa reaching the continental island of Tasmania.

Cameron (1964) described competition between species of two genera (*Microtus* and *Clethrionomys*) of microtine rodents that was sufficiently strong so that each genus completely excluded the other. A comparison of mainland habits and occurrence of the two genera on offshore islands in North America and Europe showed that species of each genus can exclude species of the other.

Complex patterns of species interactions in an avifauna along the length of a mountain slope was documented by Diamond (1969) in New Guinea and by Terborgh (1971) in the eastern Andes of Peru. Both studies illustrated the complicated interactions between congeneric species and between species of different families, but these results must still be applied to the evolution of a higher taxon involved in these interactions.

Janzen (1966) described the fascinating coevolutionary mutualism between swollen-thorn acacias and obligate acaciants (*Pseudomyrmex*). A colony of ants is established by a single queen laying her eggs in the base of a swollen thorn; the colony grows until all the thorns of the tree are occupied by ants. The ants feed on nectar produced in large foliar nectaries and on specialized Beltian bodies growing on the tips of leaflets. The ants attack and drive away any insects alighting on the acacia, thus keeping the plant free of herbivores, and clear away any plants touching the acacia leaves or growing in an area of 19 to 150 cm about the trunk of the tree. Swollen-thorn acacias have apparently lost their intrinsic defenses against insect attacks and against competitive pressure of neighboring plants. Unoccupied swollen-thorn acacias show severe defoliation and loss of growing tips; experimently "de-anted" acacia plants are usually killed within a year and before they can produce a crop of mature seeds. Many of the characteristics of the swollen-thorn acacias and of the acacia-ants are the consequence of the mutual selection pressures exerted by each member of the coevolutionary complex upon the other. Janzen (1966:271) showed that swollen-thorn acacia and ant coevolutionary systems have developed in at least five independent species of *Acacia,* and that (p. 273) several additional examples of ant-plant mutualism developed involving *Pseudomyrmex* (and related genera) and various plant groups.

Smith (1971) presented an excellent analysis of the coevolutionary interactions between pine squirrels of the genus *Tamiasciurus* and conifers, notably the lodgepole pine (*Pinus contorta*). He showed that these squirrels act as a selective agent on several species of conifers, namely on the structure of the cone, and that conifers act as a selective agent on the squirrels, namely that pines with harder cones select for squirrels with stronger jaw musculature. Smith (1971:364) points out that other organisms may also exert selection on conifer cones. Several groups of insects eat the seeds and cones of most species of conifers in western North America and several birds, such as cross-bills (*Loxia*) and corvids (*Nucifraga* and *Gymnorhina*) feed extensively on seeds of conifers. Interactions between these species and conifers must be studied before the evolution of conifer cones can be understood fully, as well as the selection exerted by the conifers on all of these phytophagous species.

Ehrlich and Raven (1965) reviewed extensively the interactions between butterflies (mainly the larval forms) and their food plants, focusing attention on the "predator-prey type" of interaction between these insects and their plant hosts. They conclude that "The evolution of secondary plant substances and the stepwise evolutionary responses to these by phytophagous organisms have clearly been the dominant factor in the evolution of butterflies and other phytophagous groups. Furthermore, these secondary plant substances have probably been critical in the evolution of angiosperm subgroups and perhaps of the angiosperms themselves."

Krieger et al. (1971) have shown that many species of caterpillars possess detoxication enzymes in their gut as a countermeasure to the natural insecticides evolved by many species of plants. They demonstrated that polyphagous species have higher activity of oxidase enzymes than monophagous species, presumably as a consequence to their exposure to a greater diversity of natural toxins in their food. By limiting their exposure to one or a few toxins, monophagous forms save energy, since they need to produce a limited countermeasure rather than a range of enzymes and may possibly recycle materials comprising their detoxication enzymes. And they may use the plant toxins for their own defense against predators (e.g., birds). Lastly, the authors suggest that the "differing strategies of adaptation to plant defenses may partly account for the great diversity of insect herbivores."

Gilbert (1971) describes a different type of plant defense against herbivorous insects. *Passiflora adenopoda* (Passifloraceae) has evolved hooklike trichomes covering the surfaces of its leaves. When heliconiine butterfly larvae crawl over the leaves of *P. adenopoda*, they are caught on these hooks; death follows as a result of starvation and the loss of hemolymph draining out of the many punctures caused by the trichomes. *Heliconiine* larvae are a major herbivore of *Passiflora*, yet these insects do not attack *P. adenopoda* in the same greenhouse containing other species of *Passiflora* which suffered heavily from these caterpillars. Gilbert points out that *Passiflora adenopoda* has "won" the coevolutionary race (at least for the present) with a major group of insect herbivores attacking this plant genus. He adds that *P. adenopoda* is one of the most widespread members of this genus and if it (and its descendants) replace other passifloras, a drastic change could occur in the heliconiine butterflies.

In a like fashion, Janzen (1970) argued that the evolution and the pattern of diversity of tree species in the tropics is largely dependent upon species-specific herbivores, mainly insects. He showed, by direct breeding experiments (1970:503), that almost without exception a given insect species reproduces only on a small number (one to three) of the hundred of species of plants found in lowland Central America. The feeding by an insect on its host food plant would have strong detrimental effects on seedlings growing beneath a mature plant of the same species.

Many of the best examples of coevolutionary species interactions involve animal pollinating systems of angiosperms. The classical symbiosis developed between yuccas and their pollinator, the yucca moth (*Tegeticula*) is well known to all evolutionists. So is the mimicry of some orchids to the female genitalia of their insect pollinator. The interrelationships between hummingbirds as pollinators and "bird" flowers has been reviewed in detail by Grant and Grant (1968). They discuss the reciprocal evolution of hummingbirds and plants, showing the advantages gained by both groups of organisms in the evolution of this pollinating system. Further, they suggest (1968:94-95) that bat pollination arose from bird pollination systems; that the course of evolution from insect to bird pollination has rerun its cycle in the evolution of "bat" flowers. A fascinating aspect of this coevolutionary system was suggested by Dressler (1971). He showed that a much higher percentage of hummingbird pollinated orchids have dark pollinia than do insect pollinated orchids. Most hummingbirds have dark-colored bills. Dark pollen stuck to these dark-colored bills would be less conspicuous than a mass of light-colored pollen, and perhaps less "annoying" to the bird. Hence, Dressler suggests, dark pollen would be less likely wiped off the bill than light pollen, favoring the evolution of dark pollinia in hummingbird-pollinated orchids.

Another complex system of coevolutionary interactions between the hummingbird-pollinated *Penstemon kunthii*, a second nectar-adapted bird *Diglossa baritula* (Passeriformes: Coerebidae) and two species of bumblebees (*Bombus*) in Oaxaca, Mexico was described by Lyon and Chadek (1971). *Penstemon kunthii* has a long corolla tube adapted to hummingbird pollination. The coerebid *Diglossa* has a short hooked bill enabling it to penetrate the base of the blossom and hence obtain nectar. The bees are unable to penetrate the flower and are able to feed only on those blossoms already punctured by *Diglossa.* The resultant selection of this complex system would, the authors argued, result in a self-reinforcing feedback system leading to additional elongation of the corolla and increased nectar production. The outcome of these selection pressures on *Penstemon kunthii* would favor continued floral specialization for hummingbird pollination over bee pollination.

Many other examples of species interactions could be cited, but these are sufficient to demonstrate the widespread occurrence and diversity of this evolutionary mechanism and its bearing upon major evolutionary change.

Caution must be urged against the belief that species interaction and associated directional selection will automatically result in major evolutionary shifts. Many patterns of speciation in radiating groups result in recurrent changes, with the direction of selection reversing frequently. Thus, isolating mechanisms and ecological character displacements often tend to involve variations on a common theme with redundant changes occurring frequently. That is, certain (limited) features are often involved over and over again in newly evolved species replacing those which had become extinct earlier. Similar habitat shifts may occur; the same species specific plumage pattern, e.g., face markings, or courtship songs may be repeated in a group of birds such as the wood warblers (Parulidae), woodpeckers (Picidae), and tyrant flycatchers (Tyrannidae). In these groups, extensive speciation and species interactions will not result in any macroevolutionary change, but will simply maintain the characteristics of the original taxon. Occasionally, the consequence of species interaction will be a shift into a new direction that may result in a major evolutionary change and the development of a new higher taxon. What must be stressed is that the evolutionary mechanisms involved in recurrent changes do not differ from those involved in a shift of adaptive features.

Thus, I would identify the source of directional selection necessary for the origin of a new major taxon as coevolutionary species interaction and exclusionary species interaction, the latter occurring after the secondary overlap of a pair of species, usually after the initial overlap (neosympatry) of newly evolved species. The physical environment, food items and similar factors of the newly available adaptive zone would, in general, not provide the necessary directional selection. The directional selection from species interactions will continue to exist as long as patterns of coevolution continue and as new combinations of species interactions appear over the geographical range of the adaptive radiation leading to the new taxon. Generally no single fixed direction can be assigned to the directional selection over an extended period of time. This selection will force the interacting species to change (e.g., diverge via character displacement) and to occupy subdivisions of the habitat that would have remained empty in the absence of the directional selection. Moreover, because of competition between the species comprising the continuing adaptive radiations during the origin of a new taxon, modification of the phyletic lines (individual species) entering the newly available habitat does not cease when these species are just sufficiently different from the ancestral stock to avoid competition. Rather, each species is forced to compete with closely related, similar species and this will result in continual evolutionary change. These adaptive modifications will continue until the radiating phyletic lines reach the ecological limits of the new adaptive zone and will result in the large adaptive shifts that characterize new higher groups of organisms.

DISCUSSION

Recent studies of the evolutionary consequences of species densities by ecologists (e.g., MacArthur and Wilson, 1967) have come close to the concept of the role of species interactions in macroevolution. The information collected in these studies may prove to be valuable for future analyses of major evolutionary changes. Results of most investigations of species densities and diversities are statistical analyses of a large number of species in which the exact details of interactions between species are obscured. It will be necessary to gather and summarize the patterns of species composition and diversity more precisely in terms of the actual species level taxa coexisting at different localities within the geographical range of a major taxon for these studies to be of greatest value in the clarification of macroevolutionary mechanisms.

The complex pattern of competition and other interactions between the numerous species comprising a community is a central problem in the study of community evolution (McIntosh, 1970). Unfortunately, few evolutionists have attempted to analyze evolutionary mechanisms at this level of complexity and no one has applied the results of such studies to the problem of the role of species interactions in major evolutionary changes. Evolutionary mechanisms at the level of communities will have to be understood in detail, however, before all implications of species interactions for macroevolution can be fully comprehended.

The concept that directional selection for macroevolution arises from species interactions was based mainly on avian examples and some mammalian cases. It is dependent upon the operation of competition (competitive exclusion), selection of intrinsic isolating mechanisms to reduce reproductive wastage, and all types of coevolutionary interaction. These evolutionary mechanisms result in charactive displacement or in some other pattern of change. These mechanisms operate in all animal groups; hence, the basic conclusion that the source of the necessary directional selection in macroevolution is species interactions can be generalized for all animals. Moreover, I believe that this concept applies equally well to plants. Plants respond to selection arising from predator-prey interactions when they are the food source of many groups of animals (Ehrlich and Raven, 1964; Janzen, 1970; Smith, 1971). Selection for intrinsic isolating mechanisms to reduce reproductive costs operates in plant groups with species specific pollinators such as orchids, figs (Ramirez B., 1970), and gourds (*Cucurbita*; Hurd et al., 1971). Other coevolutionary species interactions involving symbioses associated with pollinating systems are common in plants that have animal pollinators. Some botanists deny or deemphasize the importance of competition in plant evolution, while others (e.g., McIntosh, 1970; Harper, 1967) suggest that competition operates in plants in a manner similar to its action in animals, and that its role in plant evolution is as important as in animals. Competition between species of plants may involve more subtle types of interactions than in many groups of animals, and its analysis may be more difficult in plants. Although much additional investigation is required before the significance of competition in plant evolution is fully clarified, I believe that future study will substantiate its importance and that competition will prove to be a major source of directional selection for macroevolutionary changes in plants.

If the conclusion that species interactions in a continuing adaptive radiation provide the directional selection necessary for macroevolution is substantiated further, then a number of previously accepted ideas must be reexamined. Most important is the concept that the basic features of a new taxon evolved first by some special mechanisms (e.g. quantum evolution) and that subsequent adaptive radiation filled the new adaptive zone with forms that are variants of the basic adaptation configuration of the group; this radiation is then said to be controlled by selection forces that are fundamentally different from those acting during the origin of the group. Closely associated with this concept is the notion that each taxon is characterized by some suite of adaptive features that are common to all members of the taxon including the initial members, and which serve to distinguish all species of that group from all others. Herein lies the idea that all members of each taxon possess a distinct set of derived features different from the sets of all other taxa. This concept may not always be valid if the directional selection controlling the origin of the taxon stemmed from species interactions. No definite set of adaptive features characterize the Drepanididae or either subfamily recognized within this taxon even though there is little doubt that these taxa are good monophyletic groups.

Concepts such as distinct evolutionary mechanisms, definite thresholds, unusual intensities of selection, rigidity of adaptation, and the existence of relatively inadaptive forms in the origin of new taxa or in the entrance of a group into a new major adaptive zone must be reconsidered before their validity can be maintained. A careful examination of the possible steps in the evolution of the subfamily Psittirostrinae of the Drepanididae (Bock, 1970) reveals no differences in the evolutionary mechanisms acting at any of the suggested stages, either on the specific or generic levels. Yet the evolution of a heavy bill in *Pseudonestor* as an adaptation for a particular mode of obtaining insects provided the basis for a major shift in food items, form insects and nectar to plant materials, in the descendant genus *Psittirostra.* These heavy-billed finches entered an adaptive zone quite different from that occupied by other members of the subfamily, but their evolution does not appear to differ from that of other genera in the Psittirostrinae.

The most important concept that must be reevaluated is the notion of major adaptive zones and the correlation of major taxa with these zones. This concept is certainly far more complex than generally realized and must be discussed in more concrete terms. The zones should be described in greater detail than usually attempted and preferably without using the contained taxa as a guide. Possibly major adaptive zones may exist but without a close or exclusive correlation between individual major zones and major taxa. Or distinct major adaptive zones in one group, e.g., birds, may overlap broadly distinct majo adaptive zones of another group, e.g., mammals.

Identification of the source of directional selection for macroevolution as species interactions provides strong support for several established concepts. It agrees with the well documented generalization that a large portion of the proliferation of subgroups occurs early in the history of a new taxon (Simpson, 1953). It provides a mechanism for rapid, continuous change, even when the differences in the physical environment are relatively minor. The high rate o evolution possible by this mechanism (Bock, 1970:718-719) helps to explain th relative scarcity in the fossil record of forms intermediate between major taxa

The wide variance in the opportunity for speciation in different taxa under different environmental and geographical conditions with resulting diversity in the intensity of directional selection may clarify the variation in evolutionary rates observed in the fossil record, e.g., bradytelic to tachytelic rates of evolution (Simpson, 1953).

The role of species interactions in major evolutionary modifications serves to underline the emphasis given by Mayr (1963, especially pp. 620-621; 1970) to the evolutionary significance of species and of speciation. Species formation is not a biological luxury and as Mayr states, "a prodigious multiplication of species is a prerequisite for evolutionary progress." Without continuous speciation and modification in species composition in biological communities, the rate of species interactions will decrease and the resulting strength of directional selection will drop. And major evolutionary changes will diminish rapidly. Clearly, no real advance in the comprehension of major evolutionary phenomena can be achieved unless students of these questions appreciate fully the implications of speciation for macroevolutionary change.

The identification of species interactions as the major environmental factor providing directional selection for macroevolution assigns this formerly vague evolutionary mechanism to a concrete aspect of the environment. It should be possible to formulate additional hypotheses and to make comparative observations to substantiate this concept. Moreover, the concept that directional selection arises from species interactions should allow additional observations and comparisons on which to test further the hypothesis that microevolution and macroevolution constitute a smooth continuum of change.

SUMMARY

Species interactions are suggested to be the environmental factor providing the bulk of the directional selection required for major evolutionary shifts. Factors of the physical environment and food items of the new major adaptive zone contribute a very minor part of this directional selection. The species interactions are of two major types—exclusionary and coevolutionary—which involve different mechanisms and have different outcomes.

Exclusionary species interactions include competition affecting ecological features and attempted pair formation, and interbreeding affecting ethological features and other intrinsic isolating mechanisms that would reduce reproductive wastage. The selection generated by this form of species interaction would result in divergence between members of the species pair. These selection forces are generally strongest between closely related species and usually reach their maximum intensity between newly evolved daughter species during their initial neosympatry. Selection decreases as the species diverge until it reaches a very low level. Moreover, exclusionary species interactions are dependent upon continued speciation and adaptive radiation of the group.

Coevolutionary species interactions include predator-prey relationships, pollinating systems and all sorts of symbioses in which one species is using the resources of the other. Both species exert selection upon each other, but the resulting evolutionary change is usually not divergent. Nor does a correlation exist between the degree of relationship and the intensity of selection. The strength of selection usually does not decrease as a result of evolution of one or both members of the interacting pair of species. And a coevolutionary species does not depend upon continued speciation and establishment of neosympatry.

Macroevolutionary change is a summation of the modifications in successive species interactions and is limited by the nature of the major adaptive zone into which the new taxon is evolving. The rate of macroevolution depends largely upon the time period between speciations and hence successive (exclusionary) interactions. Most essential is that the selection forces exerted by each species on other species with which it interacts, be it exclusionary or coevolutionary, modify constantly as the species changes in response to the reciprocal selection forces exerted upon it. This constant modification of selection forces stands in sharp contrast to the highly static selection exerted by the physical environment. Moreover, these ever-changing selection forces are essential for the pattern of mosaic evolution characteristic of the origin of new taxa. Lastly, the role of species interactions in providing directional selection stresses the central importance of species and of speciation in major evolutionary changes.

Acknowledgments

I would like to thank Drs. Ernst Mayr, Bobb Schaeffer, Lester Short, Frederick Warburton, and Max Hecht for reading and criticizing the manuscript and for numerous extensive discussions of this problem. I would also like to thank, with a special note of gratitutde, Mr. John Morony for posing a series of interesting and difficult questions which provided the impetus for this analysis, and for his assistance throughout the study. Previous papers on which much of this investigation depended were supported by several grants from The National Science Foundation; this study was supported by research grant NSF-GB-6909x.

REFERENCES

Amadon, D. 1950. The Hawaiian honeycreepers (Aves, Drepaniidae). Bull. Amer. Mus. Nat. Hist. 95(4):151-262.

Bartholomew, G. A. 1963. Behavioral adaptations of mammals to the desert environment. Proc. XIV Int. Cong. Zool., 3:49-52.

____1964. The role of physiology and behavior in the maintenance of homeostasis in the desert environment. Sympos. Soc. Exp. Biol., 18:7-29.

____and W. R. Dawson. 1968. Temperature regulation in desert mammals. *In* Brown, G. W., Jr., ed., Desert Biology, Vol. 1, pp. 395-421. New York and London, Academic Press.

Beer, G. de. 1954. *Archaeopteryx* and evolution. Advance. Sci., 11:160-170.

Bock, W. J. 1959. Preadaptation and multiple evolutionary pathways. Evolution, 13:194-211.

——1963a. Evolution and phylogeny in morphologically uniform groups. Amer. Natural., 97:265-285.

——1963b. Relationships between the birds of paradise and the bower birds. Condor, 65:91-125.

——1965. The role of adaptive mechanisms in the origin of higher levels of organization. Syst. Zool., 14:272-287.

——1967. The use of adaptive characters in avian classification. *In* Snow, D. W., ed., Proc. 14 Int. Ornith. Cong., pp. 61-74.

——1969. Comparative morphology in systematics, *In* Systematic Biology, pp. 411-448. Washington, D. C., Nat. Acad. Sci.

——1970. Microevolutionary sequences as a fundamental concept in macroevolutionary models. Evolution, 24:704-722.

——and G. von Wahlert. 1965. Adaptation and the form-function complex. Evolution, 19:269-299.

Cameron, A. W. 1964. Competitive exclusion between the rodent genera *Microtus* and *Clethrionomys*. Evolution, 18:630-634.

Dawson, W. R., and G. A. Bartholomew. 1968. Temperature regulation and water economy of desert birds. *In* Brown, G. W., Jr., ed., Desert Biology, Vol. 1, pp. 357-393. New York and London, Academic Press.

Diamond, J. M. 1969. Preliminary results of an ornithological exploration of the North Coastal Range, New Guinea. Amer. Mus. Novitates, No. 2362:1-57.

Dorst, J. 1960. Considerations sur les passereaux de la famille des Vangides. Proc. 12 Int. Ornith. Cong. pp. 173-177.

Dressler, R. L. 1971. Dark pollinia in hummingbird-pollinated orchids or did hummingbirds suffer from strabismus? Amer. Natural., 105:80-83.

Ehrlich, P. R. 1970. Coevolution and the biology of communities. *In* Chambers, K. L. ed., Biochemical Coevolution, pp. 1-11. Corvallis, Oregon State University Press.

——and P. H. Raven. 1965. Butterflies and plants: a study in coevolution. Evolution, 18:586-608.

Ficken, R. W., M. S. Ficken, and D. H. Morse. 1968. Competition and character displacement in two sympatric pine-dwelling warblers (*Dendroica*, Parulidae). Evolution, 22:307-314.

Gilbert, L. E. 1971. Butterfly-plant coevolution: Has *Passiflora adenopoda* won the selectional race with heliconiine butterflies? Science, 172:585-586.

Gilliard, E. T. 1969. Birds of Paradise and Bower Birds. Garden City, New York, Natural History Press.

Goldschmidt, R. 1940. The Material Basis of Evolution. New Haven, Yale University Press.

Grant, K. A., and V. Grant. 1968. Hummingbirds and Their Flowers. New York, Columbia University Press.

Harper, J. L. 1967. A Darwinian approach to plant ecology. J. Ecol., 55:247-270.

Hecht, M. K. 1965. The role of natural selection and evolutionary rates in the origin of higher levels of organization. Syst. Zool., 14:301-317.

Hurd, P. D. Jr., E. G. Linsley, and T. W. Whitaker. 1971. Squash and gourd bees (*Peponapis, Xenoglossa*) and the origin of the cultivated *Cucurbita*. Evolution, 25:218-234.

Janzen, D. H. 1966. Coevolution of mutualism between ants and acacias in Central America. Evolution, 20:249-275.

——1970. Herbivores and the number of tree species in tropical forests. Amer. Natural., 104:501-528.

Keast, A. 1968. Competitive interactions and the evolution of ecological niches as illustrated by the Australian honeyeater genus *Melithreptus* (Meliphagidae). Evolution, 22:762-784.

——1970. Adaptive evolution and shifts in niche occupation in island birds. Biotropica, 2:61-75.

Krieger, R. I., P. P. Feeny, and C. F. Wilkinson. 1971. Detoxication enzymes in the guts of caterpillars: an evolutionary answer to plant defenses? Science, 172:579-581.

Lack, D. 1944. Ecological aspects of species-formation in passerine birds. Ibis, 86:260-286.

——1947. Darwin's finches. Cambridge, Cambridge University Press.

Lyon, D. L., and C. Chadek. 1971. Exploitation of nectar resources by hummingbirds, bees (*Bombus*) and *Diglossa baritula* and its role in the evolution of *Penstemon kunthii*. Condor, 73:246-248.

MacArthur, R. H. 1958. Population ecology of some warblers of north-eastern coniferous forests. Ecology, 39:599-619.

——and E. W. Wilson. 1967. The Theory of Island Biogeography. Princeton, N. J., Princeton University Press.

McIntosh, R. P. 1970. Community, competition and adaptation. Quart. Rev. Biol., 45:259-280.

Mayr, E. 1963. Animal Species and Evolution. Cambridge, Mass., Harvard University Press.

——1970. Populations, Species and Evolution. Cambridge, Mass., Harvard University Press.

Ramirez, B. W. 1970. Host specificity of fig wasps (Agaonidae). Evolution, 24:680-691.

Schaeffer, B., and M. K. Hecht. 1965. Introduction and historical resume [for a symposium: The origin of higher levels of organization]. Syst. Zool., 14:245-248.

Schindewolf, O. H. 1950. Grundfragen der Paläontologie. Stuttgart, Schweizerbart.

Simpson, G. G. 1944. Tempo and Mode in Evolution. New York, Columbia University Press.

——1953. The major features of evolution. New York, Columbia University Press.

Smith, C. C. 1971. The coevolution of pine squirrels (*Tamiasciurus*) and conifers. Ecol. Monogr., 40:349-371.

Stonor, C. R. 1942. Anatomical notes on the New Zealand wattled crow (*Callaeas*) with special reference to its powers of flight. Ibis, (1942): 1-18.

Terborgh, J. 1971. Distribution of environmental gradients: Theory and a preliminary interpretation of distributional patterns in the avifauna of the Cordillera Vilcabamba, Peru. Ecology, 52:23-40.

Thomson, A. L. and R. E. Moreau. 1957. Feeding habits of the Palm-Nut Vulture *Gypohierax*. Ibis, 99:608-613.

Wahlert, G. von. 1965. The role of ecological factors in the origin of higher levels of organization. Syst. Zool., 14:288-300.

Zangerl, R. 1948. The methods of comparative anatomy and its contribution to the study of evolution. Evolution, 2:351-374.

2

Some of the Evolutionary Consequences of Being a Plant

A. D. BRADSHAW
*Department of Botany, University of Liverpool
Liverpool, England*

INTRODUCTION

A large number of biologists have examined evolutionary processes in plants and animals; considerably more have looked at animals than at plants, as two recent reviews purporting to be on general population genetics show (Lewontin, 1967; Spiess, 1968). Yet the most spectacular examples of specifically adapted populations within a species are to be found in plants, and the existence of ecological races or ecotypes was first appreciated in plants (Turesson, 1922). It is possible that the lack of examples of such spectacular differentiation in animal species, to match what can be found in *Achillea* (Clausen et al., 1948), or many other plants, is an accident of the interests of biologists. But this seems very unlikely. And now examples of highly localized differentiation of populations, occurring over only a few metres, are being found to be common in plants (Jain and Bradshaw, 1966) but have hardly been reported in animals, except in *Cepaea nemoralis*, a snail (Sheppard, 1952).

Differentiation of plant populations on such a small scale is remarkable. It is tempting to think that it is an example of sympatric speciation, which people have argued about (e.g., Mayr, 1963; Smith, 1966). For these reasons alone the processes of evolution in plants need looking at. But they need looking at so that they can be related to the evolutionary processes that occur in animals, if the exercise is to be properly rewarding. Perhaps then zoologists might look at plants, and botanists look at animals, and we may find that evolution in some animals is like that in plants, and evolution in some plants like that in animals, particularly when their properties overlap.

THE IMPORTANT CHARACTERISTICS OF PLANTS

It is necessary to begin by pointing out the obvious, because sometimes it is not. There are many ways in which plants differ from animals, and although the differences are not always consistent, some of them have important evolutionary consequences. The "plants" and "animals" considered can only be caricatures of reality. If this raises objections in the mind of the reader then it will have served a purpose to focus attention on cases in which we might look for similarities in evolutionary behaviour rather than differences.

Because a plant is dependent for its growth on light, CO_2, water and inorganic nutrients, its structure is singularly diffuse and vulnerable to the buffets of the environment both above ground and below. Some plants such as *Sequoiadendron giganteum* have elaborated massive protective coverings as if they were animals, and some animals are small, poikilothermic, and vulnerable. But if we generalize, it is the plants which are more likely to suffer from the environment. As a result if there is any genetic variability in adaptive characteristics, many characteristics of plants are likely to be subject to high and effective selection pressures. This is not to say that on other characteristics low selection pressures do not occur, and may indeed be important in long term evolution.

For the same reasons, a plant is stationary. It is difficult to recall a plant that is not, although as we shall consider later it does have a dispersal phase. By contrast it is not easy to think of animals that are stationary, although there are exceptions such as barnacles, limpets, and anemones, and small invertebrates with very small powers of movement. So a plant cannot move from one environment into another in the course of its lifetime, in the way that an animal may traverse numerous different environments in the course of a day. In this respect an environment that is fine grained to an animal will be coarse grained to a plant (in the sense of Levins and MacArthur, 1966). The selection pressures operating on a plant are likely to be continuous in their direction. This will be combined with the strong selection already discussed.

If by contrast the environment does vary, the plant cannot run away to a new environment or hide in a protected corner. Unless it can go into some resting stage it must accept temporal fluctuations of the environment, and evolve means of adapting to them by phenotypic plasticity. This is a particular problem in itself which has been considered generally by Levins (1968) and discussed for plants (Bradshaw, 1965), so it will not be raised again here. However, the phenotypic plasticity has another consequence in relation to the selective pressure of the environment. The effect of severe conditions on an organism is often considered to eliminate individuals. However, the alternative effect is to change the number of offspring produced by individuals without killing them. The plasticity of plants, and the fact that their reproductive organs are usually the terminal points of a branching structure, mean that they tend to respond to environmental stresses by variation in reproductive rate without death. This is a form of response to selection, and a very powerful one (Harper, 1964). Thus the seed produced by many annual plants can vary two hundredfold, which is an enormous variation in fitness matched only by some animals.

Another consequence of being stationary is that plant populations are very likely to have numbers regulated by their environment. Plants cannot run away in the face of high population densities. Although they may show phenotypic response that will allow some variation in numbers, the size of a population occupying a particular niche is essentially regulated by that niche.

The dispersal of plants takes many forms. Many of these are passive and involve the mere shedding of the seed at the mercy of the elements, sometimes with structures such as wings that cause the seed to be scattered more widely than would happen otherwise. Sometimes the dispersal is achieved with the cooperation of an animal; indeed it is the mobility of animals which is involved, either in some nonspecific manner, as with the acorn which the squirrel may dump anywhere, or in some very elaborate manner, as with the mistletoe, where the birds wipe the sticky seed off their beaks conveniently on suitable branches. While it is not possible to make one generalization to suit all cases, it is clear that usually this dispersal is random, certainly more random than the way in which most animals lay their eggs (where there may be enormous precision). Animals do provide interesting exceptions to this, however, as in the case of marine organisms with planktonic larvae; these must have interesting evolutionary consequences. The general result is that some of the offspring of a plant may find themselves in a wide variety of habitats dissimilar to that of the plant. If we couple this with the fact that growing plants are stationary, then it is clear that a single plant species will usually be subject to remarkable multidirectional selection, unlike its animal counterpart.

Another characteristic of plant dispersal which has been greatly overlooked is that it is extremely localized: this applies to both seed and pollen. It is true that pollen grains can be trapped in the middle of the Atlantic, and that coconuts can happily float thousands of miles. But it is the characteristics of the relationships of dispersal with distance which are important. The dispersal of the more mobile component, pollen, has been examined on several occasions (Bateman, 1947; Griffiths, 1950; Colwell, 1951). The results agree in showing that the distribution is leptokurtic and although some pollen moves a long way the vast preponderance only moves a few meters. This will be affected by density of plants, but on the average whether we are dealing with insect- or wind-pollinated plants, the amount of cross pollination that occurs over 10 meters is only 5%. There is little evidence about the movement of seed. This must depend on the dispersal mechanism. In *Phlox*, for example, the dispersal of seed is extremely restricted and of pollen astonishingly so (Levin and Kerster, 1968).

Such restrictions have two major effects. Firstly, plant species are broken up in populations whose effective breeding size (Wright, 1946) is remarkably small, perhaps often with a diameter of the order of 5 meters (Levin and Kerster,

1968). A continuously distributed species cannot be considered as though it were a single random mating population. This point has recently been stressed by Ehrlich and Raven (1969). As a result, species are broken up into vast arrays of small populations which are so small in diameter than they can match the size of extremely localized variations of the environment, variation occurring even over a distance of a few metres.

Secondly offspring will tend to occupy the same general habitats as their parents. It is true that, because of the random and leptokurtic nature of plant dispersal, some individuals will arrive in habitats dissimilar from their parents, particularly if there is very localized habitat variation; and few, if any, will be able to occupy the site already occupied by their parents. But for the vast majority of offspring there will be a strong correlation in habitat with that of their parents for spatial components of environmental heterogeneity. Effectively this is the same as the habitat selection which so often occurs in animals. It appears to be a very powerful force in plants leading to continued directional selection over many generations.

One of the oddest characteristics of some plants (looked at from the point of view of an animal) is connected with the fact that they can be extremely perennial. The first aspect of being perennial is that individuals will not be replaced annually, despite the fact that they will usually reproduce annually: at the same time they are static in position. If we are considering a community in a steady state, then no offspring will be able to develop until space is made available for them by the death of an adult. Much therefore of the reproduction is unused; and if there is any gene flow by mating between dissimilar populations it need have no effect.

The second aspect of perenniality in plants is that the individual is not constant in size during its lifetime and can even produce whole new individuals vegetatively. Even a tree, although it may be restricted to a single trunk, is continually enlarging its crown. If there are environmental influences, of what ever sort, they can have their selective effect on the size of individual genotypes and the contribution they make to the next generation. With elephants we really only have to consider whether or not the environment is lethal.

The extreme of perenniality is that exemplified by grasses, where an individual can continue growing indefinitely (Harberd, 1967). In this case replacement by seed may be almost irrelevant to evolutionary processes. But this is not to say that selection does not occur. It will take place by the relative vegetative growth of individuals, and it can accumulate logarithmically from one year to the next.

Plants are remarkable in the degree of flexibility of their breeding systems (Darlington, 1939) to an extent matched by only few animals. They can switch from obligate outbreeding with self incompatibility to total inbreeding, even with selfing occurring within the unopened flower. Such changes can occur with considerable ease (Mather and de Winton, 1941). Plants can dispense with sexual reproduction altogether and become apomictic: this ability is, however, matched by some animals such as aphids and rotifers. This sort of thing can have profound effect on the characteristics of variability of populations (Stebbins, 1957). Selfing will allow a species to develop uniform populations consisting of one or a few homozygous genotypes (Imam and Allard, 1965). But it will also allow a population to maintain its characteristics in the face of gene flow from other populations (Antonovics, 1968a). Apomixis will allow all this to an even greater degree.

There are other ways by which gene flow can be restricted. In animals mating discrimination is important. It seems to require only small changes to cause complete isolation between two animal populations, and this sort of isolation can be evolved very readily (e.g., Knight et al., 1956; Thoday and Gibson, 1962). In plants there is less evidence for the evolution of equivalent mating discrimination; it can occur by color differences (Levin and Kerster, 1967), or discrimination by the style of different pollen types out of a mixture (Lawrence, in Darlington and Mather, 1949) or in equivalent ways. Perhaps it needs looking for more intensively: there is certainly evidence of the evolution of intrinsic isolating mechanisms in natural populations (Kruckeberg, 1957; Vickery, 1959, 1964).

However a major way in which isolation can be achieved in plants is by differences in flowering time. For efficiency of pollination the flowering period in most species is short, so that small shifts in time of flowering could cause considerable isolation. This appears to occur often (McNeilly and Antonovics, 1968).

Polyploidy is, of course, a special characteristic of plants. It has a profound isolating effect and has had an enormous influence on plant evolution. But its effects have been well discussed by others (especially Stebbins, 1950; Grant, 1963) so that it will not be considered here.

These then are the salient characteristics of plants. They appear to mean that (1) plants species are subject to strong selective pressure; (2) are subject to selection pressures which are enhanced by the fact that they remain constant in direction; (3) are subject to selection in different directions in different populations; (4) are composed of populations of very small geographical size which therefore are likely to occupy quite distinct environments or niches; (5) between these populations there is little gene flow; (6) there is strong correlation between the general habitats of parents and offspring; (7) there may be little replacement by seed anyway because of perennation; (8) but selection will operate very strongly among perennating individuals; (9) the breeding system may change in many different ways in response to the exigencies of the environment.

NATURAL SELECTION IN PLANTS

If we argue theoretically that the selection pressures acting on plant species must be powerful, then we should be able to demonstrate this. A great variety of evidence has now accumulated that all points the same way. This evidence that is appropriate is observations on the relative performance of different genotypes of a single species. It is best if these are genotypes of a single population, and if the values for fitness are determined under conditions that are as similar as natural conditions as possible. If we compare different genotypes from different populations we may be dealing with individuals that differ in characteristics other than those in which we are interested (but not necessarily so) and therefore get exaggerated values of differences in fitness. If we examine material under unnatural experimental conditions, such as in artificial environments or in the absence of competition, we may grossly underestimate difference in fitness. But all the evidence is valuable.

Natural selection in mixtures of cultivated varieties

It is possible to make up populations from a number of specific separate gentoypes and follow their fate with the passage of time. This can be done in self-fertilizing crop plants or in perennials, where there will be no intercrossing and recombination of genotypes. The classic experiments of Harlan and Martini (1938) and Suneson (1949) with mixtures of barley varieties show very clearly the power of selection. The fitness of the varieties in these experiments have recently been computed (Allard et al., 1966; Workman and Allard, 1964): differences of fitness of 20% are common and very often exceeded. Similar differences in fitness are found between rice varieties grown in mixtures (Jennings and de Jesus, 1968), and in *Trifolium subterraneum* (Morley et al., 1962). In rice (Jennings and Aquino, 1968) and in barley (Lee, 1960) observations made on the yield of the varieties in pure stands at the same time showed that differences in performance found in mixtures were hardly visible in pure stands. This emphasizes the need to measure fitness under as natural conditions as possible especially under conditions involving competition.

In perennial grasses Charles (1961, 1964) has made up several different mixtures involving contrasting but closely related varieties. During a single generation over a single year he could demonstrate differences in survival and growth leading to differences in fitness of 50%. Growth in subsequent years without any new individuals entering the population gave further marked changes, not necessarily in the same direction as those in the first year. It is true that a much larger number of seeds were sown than developed into adult plants, and that the changes in frequency of the different genotypes were due to greater death of some genotypes than others. But this is exactly equivalent to what can happen in natural conditions: plants are very generous in their seed production.

Natural selection in artificial hybrid populations

Bulk hybrid populations of crop plants formed by the hybridization of a number of varieties provide excellent material for the determination of fitness of individual marker loci and of quantitative traits. In barley there are marked changes not only in the frequency of marker loci (Jain and Allard, 1960) but also in quantitative characters (Allard and Jain, 1962). The same is true in rice, the fitness of a dwarf genotype being about 0.70, although in spaced conditions its fitness was unity or above (Jennings and Herrera, 1968). Perhaps the most spectacular demonstration of the power of selection is in the experiment with rice of Akemine and Kikuchi (Allard and Hansche, 1964). Here time of flowering was changed from a mean of 90 days to means of 60 and 120 days in north and south Japan respectively in four generations. Other characters also changed considerably.

Natural selection in crop populations

For economic reasons the seed production of crop varieties is often carried out in regions away from the habitats for which they were developed. The effect of this was first laid bare by Sylven (1937). The effect of growing a variety of *Festuca pratensis*, adapted to northern Sweden, for one generation in southern Sweden was to increase its fitness in Southern Sweden by 145%. Similar single-generation effects, although of a smaller magnitude were observed in *Trifolium pratense* and *Phleum pratense.* A wide variety of similar investigations show the power of selection: *Trifolium pratense* bred for Tennessee, when grown with great care in Oregon, declines in fitness for Tennessee conditions by 5% per generation (Beard and Hollowell, 1952); changes in earliness of flowering occur in *Trifolium repens* (Stanford et al., 1962); changes in adaptation to soil factors occur in one generation in grass varieties (Crossley and Bradshaw, 1968). In some cases a crop variety has been subject to a number of different treatments in one environment and the effects on its composition determined. The often overlooked results of Kemp (1937) with various pasture plants are now matched by those of Brougham et al., (1960) for *Lolium perenne*: in both cases considerable changes occurred with the lifetime of a single generation.

Natural selection in natural populations.

There are two sorts of evidence here. The most direct evidence is change actually observed in the composition of plant populations with the passage of time. Direct observations where a population has been measured appropriately at succeeding times analagous to the work on industrial melanism in moths has never been done in plants. The nearest we can get to this are situations where a new selection pressure has been applied to a species that can be presumed to have been uniform to start with. Metal tolerance has clearly evolved in about 300 years in Swansea, and only in 25 years at the foot of a galvanized fence (Bradshaw et al., 1965). However, there is now evidence that this character can evolve in one generation when seed of a pasture population is sown on metal-contaminated soil since selection for tolerance on metal contaminated soils can be overwhelming (McNeilly and Bradshaw, 1968, Khan, 1969;

Bradshaw, 1970). The fertilizer treatments established about 50 years ago at Rothamsted have caused profound population differentiation in *Anthoxanthum odoratum* (Snaydon, 1963, 1970).

Natural hybrid populations are very suitable for observing the effects of natural selection. It is not without significance that, in the only two cases that have been reported, distinct changes have been observed in *Carduus* in only 5 years (Mulligan and Moore, 1961) and in *Helianthus* in only 8 years (Stebbins and Daly, 1961).

Species that have recently been introduced into a country provide further evidence. From their present day patterns of differentiation considerable change can be presumed to have occurred in many species introduced into Australia in the course of about a hundred years (Frankel, 1954).

An alternative method by which the power of selection can be deduced is by the comparison between adult populations and the seed that can be presumed to be giving rise to them. Metal-tolerant populations of *Agrostis tenuis*, occurring on small areas of metal-contaminated soil, show by the seed they produce that there is considerable gene flow from surrounding nontolerant populations. Similarly, adjacent nontolerant populations show evidence of gene flow from tolerant populations. The fact that the populations of adults do not reflect this is indicative of powerful selection, higher on the metal-contaminated soils (McNeilly and Bradshaw, 1968; McNeilly, 1968). Similar evidence has been produced for *Agrostis stolonifera* occurring in coastal habitats (Aston and Bradshaw, 1966) and for *Eucalyptus urnigera* (Barber and Jackson, 1957). A comparison of seed and adults of *Lolium perenne* populations growing on a variety of contrasting soils indicates similar selection, although part of it is a stablizing type (Crossley and Bradshaw, 1968). It must be remembered that all of the examples are of selection acting over one generation.

Performance of ecological races.

The products of evolutionary processes in plants are differentiated populations. These are the end points of selection and not the material on which selection is acting within populations. But the performance of such material is indicative of the sorts of differences in fitness that can occur within species, in natural situations.

The most appropriate material is that from adjacent populations since these can be considered to be closely related, and almost the material on which selection is operating (especially if there is gene flow). Jain and Bradshaw (1966) have reviewed the fitnesses of genotypes, from a number of adjacent populations between which gene flow is occurring. The values for coefficients of selection against genotypes when not in their normal habitats are from about 0.05 to about 0.95, with many values about 0.3. These values were rather crude estimates but where they have been re-examined critically they have been substantiated: the coefficients of selection for *Anthoxanthum odoratum* at

Rothamsted measured under natural conditions, including competition are higher than previously determined (Snaydon, 1970).

Material of populations from widely different habitats that has been tested in different environmental conditions reveals spectacular differences in fitness. An examination of the extensive transplant experiments in *Potentilla glandulosa* and *Achillea millefolium* (Clausen et al., 1940, 1948) shows that coefficients of selection readily approach 1.0, even in material that has been tested in the absence of competition. Material of *Agrostis tenuis* from more similar habitats, tested in different environments but again in the absence of competition, showed less differences, indicating coefficients of selection of about 0.3 (Bradshaw, 1960). The importance of competition, produced by the testing of material in conditions as similar as possible to those occurring in nature, cannot be overemphasized: populations of *Trifolium repens* that did not appear to differ in fitness, when grown as spaced plants in an upland soil, showed a difference of about 0.5 when planted into a natural sward (Snaydon and Bradshaw, 1962).

It is obviously impossible to produce any generalized observations on the power of selection in plants: different habitats and different species will give very different results. But it is clear that very powerful selection coefficients can occur in plant material growing in natural situations. Indeed, much of the evidence is insufficient because it has been produced from material grown in artificial situations: subsequent work is likely to produce even higher values for selection.

BARRIERS TO GENE FLOW

At some stage in the process of evolution of new forms some intrinsic isolating mechanism arises. Normally, different populations of a single plant species have no isolation between them except that arising by distance. However, closely related plant species often have very simple differences that do cause isolation between them (Grant, 1963). Such differences must therefore be able to arise readily.

The situation in nature suggests that they do arise. Polyploidy is a common, very obvious isolating mechanism, which is often associated with simple population differentiation. We shall not consider it here since its effects are so complete that the populations at different ploidy levels are in effect different species.

Differences in flowering time are another cause of isolation found in differentiating populations. The number of cases in which population differentiation is accompanied by genetically determined differences in flowering time are very many (McNeilly and Antonovics, 1968). It is not always clear whether these flowering time differences have arisen in response to selection for isolation, or whether they have arisen as a result of selection for the environmental conditions prevailing in each habitat. But whatever the cause, the effect is the same, an enhancement of the isolation between populations.

There are many other causes of isolation between populations, such as mechanical, ethological, or incompatibility barriers (for review see Grant, 1963). Again it is not clear how often they have arisen during the course of local differentiation of populations and have assisted in their isolation, or how far they have arisen subsequently as a by-product of adaptation to the exigencies of the environments in which the population find themselves. The fact that incompatibility between populations of *Streptanthus glandulosus* (Kruckeberg, 1957) and *Clarkia biloba* (Roberts and Lewis, 1955) is related to the geographical separation of the populations, suggests that they are a by-product. But the evolution of color differences in *Phlox philosa* argues the reverse (Levin and Kerster, 1967). In this case the efficiency of the isolation mechanism has been tested and is very definite.

Another mechanism we must consider affecting gene flow between populations is inbreeding. Apart from the effects inbreeding has on the general state of variability within a species (which is a separate issue) it is clear that self-pollination can reduce the amount of cross-pollination between populations subject to gene flow by allowing pollination by the pollen that is closely and copiously available. Under such circumstances providing that homozygosity does not have deleterious effects, genes increasing self-pollination will spread through the populations. Selfing is more frequent in small metal-tolerant populations of *Anthoxanthum odoratum* and *Agrostis tenuis* than in normal populations (Antonovics, 1968a). However, selfing can arise for other reasons than as a means of reducing gene flow: it is interesting that metal-tolerant populations of *Armeria maritima* have higher levels of self-fertility than normal populations although they are not subject to gene flow from neighboring nontolerant populations (Lefebvre, 1970): perhaps here it has arisen in response to low population density in a colonizing situation.

It is important to realize that the environment itself can directly determine the occurrence of an isolating mechanism. Flowering time in particular can be determined by the environment. So the populations of a species that spreads into a series of habitats may automatically acquire some measure of isolation between them. The same could be true in relation to insect pollination: if the pollinating insects tend to be restricted to the individual environments, considerably more restriction of gene flow between the populations inhabitating different environments would occur than that already due to the normal movement of the insects.

Enough has been shown to indicate that whatever changes affecting gene flow between plant populations do occur, they lead to increased, not decreased, isolation between populations. Often several changes may occur together as for instance between the sun and shade populations of *Gilia achilleaefolia* (Grant, 1954). Hence, whatever indications of gene flow between populations we have, they are likely to be overestimates, and not underestimates of the true situation. Since values for gene flow where there are no special restrictions are already low, in natural situations the gene flow must often be very low indeed.

PATTERNS OF DIFFERENTIATION

We are now in a position to assess the main evolutionary consequences of being a plant, at least those aspects which concern what we actually find in a species in nature.

In essence the problem is simple. We have the probability of powerful selection pressures. This being so, the effects of genetic drift will be overridden by selection. The same applies to mutation except insofar as mutation will provide material on which selection can act. We know almost nothing about this in natural populations, although it is clear that the supply of variation must usually be limiting the amount of change due to selection (e.g., Lewontin and Birch, 1966). In this connection it is important to realize that selection will be able to break down gene complexes held together by linkage despite any genetical "coherence" that appears (Clausen and Hiesey, 1960). This is clearly so from an analysis of correlations between different characters in *Anthoxanthum* at a mine boundary (Antonovics and Bradshaw, 1970).

We are left with migration as the only powerful factor likely to interact with selection. But we have seen that this migration is really very little indeed. It is true that some genes or individuals may travel long distances, as in the region round the Parys Mountain copper mine (Khan, 1969; Bradshaw, 1970). But in opposition to the power of selection, it is the total number of migrants which will be important. Long-distance migration is important only as a source of new advantageous variability.

As a result we can deduce that it is selection and particularly the pattern of selection which determines the patterns of differentiation in plant populations. If the environment varies then so will the plant populations, and the differentiation will follow the pattern of the environment. A general review is provided by Heslop-Harrison (1964). In any case where a species has been examined intensively the correlation between the characteristics of populations and their environment is remarkable, whether in morphological characters, such as leaf shape in *Geranium sanguineum* (Lewis, 1969), or in physiological characters, such as growth rate in *Lolium perenne* (Cooper, 1964).

The pattern of differentiation is the same as pattern of the environment not only when the environment varies only over large distances and populations occur many kilometres apart as in *Potentilla glandulosa* and *Achillea millefolium* (Clausen et al., 1940 and 1948), in many prairie grasses (McMillan, 1959), and *Prunella vulgaris* (Bocher, 1949), but also when the environment varies over very short distances and populations are not more than 100 meters apart as in *Galium pumilum* (Ehrendorfer, 1953), *Agrostis tenuis* (Bradshaw, 1959), and *Esch-*

scholtzia californica (Cook, 1962). Where there are gradual changes in the environment, clinal differentiation of populations is found, as in *Plantago maritima* (Gregor, 1930). Where there are sharp changes in the environment, population changes are equally sharp: sharp discontinuities in the characteristics of populations occurring over a few meters are now known in a number of species (Jain and Bradshaw, 1966). These two contrasting patterns of change may even be found together in the same species, *Agrostis stolonifera*, in sets of populations only a hundred meters apart (Aston and Bradshaw, 1966).

The crucial situation where the interplay of migration and gene flow can best be seen is where the environmental change occurs over a few meters or less. In all the cases recorded by Jain and Bradshaw the population change was as precise as the environmental change: there was no discrepancy between the population changes and the environmental changes. Further evidence for very sharp discontinuities is given by *Potentilla erecta*, an insect-pollinated plant (Watson, 1970), and by further work in *Agrostis tenuis* (McNeilly, 1968), and *Anthoxanthum odoratum* (Antonovics and Bradshaw, 1970). Similar very localized differentiation has even been found in a lichen, *Ramalina siliquosa* (Culberson and Culberson, 1967).

In all these cases it must be presumed that both selection is high and migration low, otherwise there would be blurring of population differences on the boundary between the contrasting environments. In both *Potentilla erecta* and *Anthoxanthum odoratum* evidence derived from the characteristics of seed produced by the adults in their natural habitats suggests that there was little gene flow between contrasting populations only 5 and 20 meters apart, respectively (Watson, 1970; Snaydon, 1970). However, in the *Agrostis tenuis* populations growing on metal-contaminated sites similar evidence suggest the occurrence of considerable gene flow from non tolerant populations, which is deduced to be held in check by selection (McNeilly and Bradshaw, 1968; McNeilly, 1968). In *Agrostis stolonifera* the situation is similar (Aston and Bradshaw, 1966).

Ehrlich and Raven (1969) ask whether there are any cases where gene flow is acting to keep populations of a species relatively undifferentiated. At the moment, for plants the answer is no. The only possible case, which is in relation to very high migration pressures arising from polarization of wind direction in a wind-pollinated plant, is that in *Agrostis tenuis*, where the populations on ordinary soil down wind from a metal-tolerant population on a metal-contaminated area are metal tolerant (McNeilly, 1968). There is a spread of genes out from the mine down wind. However, it is worth pointing out that the distance over which this occurred to any extent was only 100 meters out from the mine boundary, a distance which many investigators would not even have sampled. A second example, but over much smaller distances, is with those characters of *Anthoxanthum odoratum* occurring on metal-contaminated soils which seem to be little affected by selection (Antonovics and Bradshaw, 1970). Some of these characters, in contrast to the character of metal tolerance, form a gradual cline across the boundary between the two environments involved: but even so the cline for these characters extended only for about 20 meters from the boundary. Other characters do not change at all across the boundary: on them presumably no differential selection is operating. Calculations based on the steepness of the cline show that very different selection pressures must be operating on different characters.

The most localized differentiation recorded is that in *Anthoxanthum odoratum* at Rothamsted in the Park Grass plots (Snaydon, 1963, 1970): the distance separating contrasting populations is less than 150 mm. It could be suggested that these populations have developed by the vegetative spread of a few clones. But this seems unlikely because *Anthoxanthum odoratum* is not a long-lived perennial (Antonovics, in Harper, 1967). However in *Agrostis stolonifera* it is very likely that the distinct stream populations found only one meter from typical cliff populations have evolved in areas away from where they are now found; although it is possible that they have been formed by selection of odd extreme individuals from surrounding populations (Aston and Bradshaw, 1966). Whatever is the explanation it is clear that the perennial habit assists the evolution of an extremely localized pattern of differentiation. Critical examination of localized differentiation in annual plants is badly needed.

Some plants have pollination or dispersal mechanisms that appear to cause gene flow over longer distances, e.g., *Eucalyptus urnigera* (Barber and Jackson, 1957), *Pinus monticola* (Squillace and Bingham, 1958). These could be imagined to lead to differentiation only over large distances. But in both these species differentiation is found between populations 800 meters apart. In *Eucalyptus* gene flow was demonstrable. In these cases the populations examined were spread across rather gradual changes in the environment, so that even if more intensive sampling of populations had been made a cline would probably have been found, and it would not have been possible to decide whether this was due to selection or gene flow. Differentiation in other cases where a plant species with a distinctly wide dispersal mechanism (such as a tropical, bird dispersed species) occurs across a sharp ecological boundary needs examining.

At present the evidence suggests the possibility, at least in communities where plants are not widely scattered, of extremely localized evolution. The scale is not to be set in kilometers but in meters or even centimeters. The pattern of differentiation follows the pattern of the environment meticulously. It is typically a graded patchwork, since this is usually the pattern of the environment (Bradshaw, 1959; Cook, 1962).

THEORETICAL MODELS

Population genetics has been a fertile field for theoretical work and for model building. Some of this work has been enormously valuable in confirming ideas about the factors operating in natural populations, or in suggesting the importance of factors previously overlooked. But some of the work has been only partially relevant to natural populations. This is true particularly for plant populations.

We are concerned with a species in which:

1. The characteristics that we are considering are determined predominantly by genes acting additively.
2. There is spatial division into a series of populations occupying different environments or niches.
3. The populations are subject to very strong selection in different directions in different habitats.
4. The size of each population is limited in each habitat.
5. The sizes are unlikely to be so small that the populations will be subject to appreciable genetic drift.
6. Mating is predominantly within populations.
7. There is only limited migration between populations, by gametes, or by seed, and not by adults.
8. This migration is strongly leptokurtic.
9. Because of the characteristics of seed dispersal, the correlation between the environment of adults and of offspring is considerable (in effect there is high habitat selection).
10. There may be reproductive isolation produced as the direct result of the environment or as the outcome of selection.

In effect this is a multiple-niche model with considerable emphasis of those factors likely to enhance differentiation (3, 5, 6, 7, 8, 9, 10). As a result, many of the early models despite their undoubted value in demonstrating the significance of particular factors have only little relevance to patterns of evolution in plant species. It is necessary to say this not for the sake of being destructive but in order to ensure that the relevance of models to natural situations can be appreciated.

Population Divergence and Clines

The models of Wright (1931, 1943) are insufficient because of their emphasis on drift (5) and their disregard for differential selection in different populations (3). In the stepping stone model of Kimura and Weiss (1964) selection is again hardly considered and the pattern of gene flow is not really what occurs in nature (8), particularly if we assume a leptokurtic distribution. The model of Fisher (1950) considers the effect of selection but unfortunately assumes a gradient of selection.

The model of Haldane (1948) has been widely used and is appropriate except that he considers a normal distribution to gene flow (8). However the comparison made by Jain and Bradshaw (1966) shows that in practice the difference in the effect of the two distributions is not great. It is an indication of the views about the sort of selection pressures that Haldane considered might prevail in animal populations, that he used a coefficient of selection of 0.1%.

The differences that occur when high coefficients of selection of 25 or 75% are used, are very apparent in Jain and Bradshaw's model. They were able, for the most part, to find considerable agreement between the patterns produced by their model and what is found in nature when values for selection and migration were inserted corresponding to the values likely to be occurring in nature.

However, they admitted that certain of the cases did not fit the assumptions of the model: the clines observed in nature were steeper than predicted. In these cases it is interesting to speculate what was wrong: either migration must have been over estimated or selection underestimated. There is good reason to believe that both occurred. No allowance was made for the possibility of isolation by flowering time, differences which were shown to occur in both *Agrostis tenuis* and *Anthoxanthum odoratum* on mines (McNeilly and Antonovics, 1968) and in *Anthoxanthum odoratum* at Rothamsted (Snaydon, 1963). The pattern of gene flow assumed was that observed in widely spaced plants of *Lolium perenne* (Griffiths, 1950) and is almost certainly an overestimate of what occurs in natural populations either at overall high density or aggregated into clumps (T. Gleaves, personal communication, 1971). In the case of the *Agrostis tenuis* transect on a copper mine it was subsequently demonstrated very clearly that gene flow was in one direction only from normal populations into tolerant populations; and selection in the latter, against nontolerance, is very high, nearly 1.0. Indeed, in a transect taken in the other direction the blurring of the boundary between populations is exactly as predicted from the model (McNeilly, 1968). In both *Anthoxanthum odoratum* and *Agrostis tenuis* coefficients of selection were determined on material growing in competition, but under experimental conditions: material subject to the total of natural environmental pressures is likely to be subject to higher selection.

A two-dimensional model has been produced by Hanson (1966), assuming normal distribution of adult migrants. If we attempt to correlate this model with the one dimensional model of Jain and Bradshaw, we find that for the units of distance of their leptokurtic pollen distribution, the corresponding value in Hanson's model allowing for the fact that adults and not pollen are migrating is approximately 2. Allowing for this and for the fact that in Hanson's model coefficients of selection of 0.1 are assumed, the agreement is excellent. It is only a pity that this model has not been applied to situations found in natural populations. It was assumed that pollen would be distributed a mile or more and that as a result a "pocket" (distinctive environment surrounded by another environment) 25 to 35 km across would be too small to create a distinctive population. While this may be true in theory, in practice we know that it is unlikely, otherwise no small environment, such as on old mine workings would ever possess a distinctive population, which is very far from the case (e.g.,

Jowett, 1964). If very different coefficients of selection and values for migration are assumed, single populations in environmental pockets can become differentiated (Jain and Bradshaw, 1966).

Multiple-Niche Situations

A patchwork of environments gives rise to a multiple-niche situation, about which there has been considerable amount of discussion recently. Indeed a patchy environment such as we have been considering is perhaps the most likely way in which such a situation can arise. But if this is so then some of the models that have been produced are inadequate.

The pioneering model of Levene (1953) envisages a situation with much of what is necessary (2, 3, 4, 5) but where mating is at random between all individuals from all niches. This is clearly inappropriate (6, 7). As a result a stable polymorphism is achieved only under rather narrow limits, including the occurrence of heterozygote advantage. The model of Prout (1968) based on the same premises, generalizes the conditions sufficient for polymorphism but shows that they are still narrow.

However, Smith (1966) envisages a situation more appropriate to the conditions we are considering. By envisaging separate regulation of population size in each niche, strong selection, and habitat selection that finds its equivalent in plants by the tendency for seed to fall near their parents (9), he suggests that a stable polymorphism can be readily achieved. He investigates the effect of genetically, but not geographically, determined assortative mating. In plants we have high levels of nonrandom mating (6, 7), high levels of habitat selection (9), and high levels of natural selection (3). As a result a stable polymorphism should arise without difficulty, as indeed is the case if we assume our population in the sense of Smith's model to be composed of a set of adjacent subpopulations between which gene exchange is occurring only on a restricted basis.

The recent models by Deakin (1968) investigate this most directly. If two random mating populations are assumed to exchange genes then a stable equilibrium can be shown to exist in the absence of heterozygote advantage if selection is operating in different directions in the two populations and is large relative to migration. This then confirms algebraically the stability that can be seen in the computer models of Hanson, and Jain and Bradshaw.

Habitat heterogeneity is perhaps more important than we realize in determining variability within species. It not only has its effects in causing variability that is habitat based. But it also will have an important influence on the variability within a population occupying a single habitat because of the gene flow from neighboring populations. Amongst other things the latter will have important effects on heterozygosity. These have been discussed by Antonovics (1968b): even low levels of gene flow maintain relatively high levels of heterozygosity.

In a multiple-niche situation of this sort there will be a migrational load due to the selective disadvantage of migrants. As we have seen the amount of this load will be less, usually considerably less, than that which would occur in a situation where there is completely random mating such as envisaged by Levene. However, it will still be present. Plants produce an enormous amount of offspring, almost certainly in order to ensure that all appropriate niches are colonized. Because of the random appearance of niches–the lack of mechanisms for habitat selection by a seed–seed production has to be sufficient to saturate the environment. There is an enormous amount of random death. If the amount of selective death due to migration causing genotypes to end up in inappropriate habitats is low compared with random death, then the effect of the migrational load on fitness will be very little (Barber, 1965). It is perhaps for this reason that intrinsic isolating mechanisms reducing gene flow do not seem to be common in plant populations.

But there are many situations where there is either high gene flow or not a considerable excess of seed production, e.g., close to the boundary between two populations, or where seeds are destroyed by a predator before becoming established in an environment. Under these circumstances there will be powerful selection for the evolution of any sort of mechanism reducing gene flow, such as self fertility. The model of Antonovics (1968a) examining the effects of self fertility can be used to illustrate this. Although the selection in favor of self fertility is not actually given, it can be deduced from the differences between situations in which selfing is assumed to be complete and situations in which it is zero. A particular case where the balance between gene flow and seed production is upset is in the early stages of colonization of a habitat, when not only will seed production be low but gene flow high. Under these circumstances selection of self fertility is considerable.

Taken as a whole the models available confirm that the situation we find in nature is compatible theoretically. Indeed the models involving distance as a dimension affecting migration suggest that the assumptions that have been made about the magnitude of the factors involved are correct. However the multiple-niche models do not seem to represent all that occurs, particularly in plants. Not only are selection pressures much higher than what has been assumed, but also only recently have appropriate restrictions to random mating been considered. Yet they are commonplace in animals, as well as in plants, even if on different scales. With mating restriction the occurrence of multiple-niche polymorphism will occur much more easily and without heterozygote advantage. Indeed heterozygote advantage may readily evolve as a result of a multiple-niche situation (Antonovics, 1968b). In the context of the present interest in polymorphism this is important.

CONCLUSIONS

To understand what is *actually* happening in plant species, we need to assume very different premises from many of those currently exercising the minds of many population geneticists. It is appropriate to look at the effects of all possible factors whether or not they are important in natural situations. But it is also appropriate to try to obtain a perspective and evaluate the factors and processes which give rise to the patterns of evolution that actually exist.

Such an exercise has lead us to realize that the pattern of population differentiation in plants can be on a far smaller scale than that previously envisaged, and on a smaller scale than that found in many (but certainly not all) animal populations. But this pattern is perfectly explicable in relation to the factors that we know are operating, and can readily be simulated in theoretical models, if in particular we are prepared to realize the power of selection maintaining the integrity of populations subject to gene flow as suggested by Ford (1964). Perhaps in the future we shall see population models incorporating coefficients of selection of more than 0.5; at the moment almost no one has considered the possibility, although the evidence and supporting arguments have been forthcoming for some time (Morley, 1959).

At the same time our ideas on the amount of gene flow occurring between populations needs drastic revision. Effective population size in plants is to be measured in meters and not kilometers. As a result most populations, however close together, are parapatric and not sympatric; that is, although they exchange genes and are therefore sympatric in the definition of many authors, they are geographically separated and therefore subject to a considerable reduction in gene flow between them. The use of the term parapatric has been discussed by Jain and Bradshaw (1966).

In parapatric situations evolutionary divergence readily occurs, as we have seen in this paper. If we realize that nearly all divergent groups of individuals are parapatric to each other and not sympatric then a great number of the problems of sympatric evolution evaporate. Perhaps Mayr (1947) in arguing so strongly for the overriding importance of geographical separation in the process of evolutionary divergence was more right than he realized, for certainly later he (1963) expresses concern about evolutionary situations in which he envisaged sympatric speciation was occurring, but in which it is almost certainly parapatric speciation which is occurring.

Ehrlich and Raven (1969) have emphasized that all our ideas on gene flow in natural population of plants and animals must be revised. In plants this is overwhelmingly true. At the moment we often consider that gene flow by pollen and by seed causes a considerable congruence of neighboring populations. There must be species with particular modes of pollination and seed dispersal where this is true. But for the great mass of plant species, whether wind or insect pollinated, it is patently not true.

There is a profound ecological consequence of such patterns of evolution. Wright (1951) suggests that localized differentiation has great importance in improving the fitness of a species as a whole. In other words any process that leads to differentiation in relation to different habitats leads to increased fitness in those habitats and therefore to increase in the mean fitness of the whole species. That this is true can be seen from the values for $\bar{W}$ in Jain and Bradshaw's model (1966) for different types of gene flow.

But this observation needs translating into the context of the common natural situation of very contrasting habitats and high coefficients of selection. If a species cannot evolve a population specifically adapted to a particular environment, it may well not be able to exist in that environment. If for instance we were to assume that *Potentilla glandulosa* and *Achillea millefolium* were only lowland in type and unable to evolve alpine populations, then it is clear from transplant experiments (Clausen et al., 1940, 1948) that the two species would not be able to occupy alpine habitats. the same is absolutely true of metal-contaminated environments: nontolerant plants just do not survive in them, whatever the species considered. Nor will lowland populations of *Trifolium repens* survive in upland habitats (Bradshaw, 1969). In both these cases we are dealing with adaption to environments that are separated by only short distances: metal-contaminated environments are adjacent to normal ones, upland habitats of *Trifolium repens* are usually only a few kilometers from lowland habitats.

The range of distribution of a species is therefore very dependent on its capacity to evolve local distinct populations. If these are prevented by excessive gene flow, then the species will be very restricted in its distribution, unless it can evolve a level of phenotypic plasticity to cope with the variation in the environment (Levins, 1968), which indeed plants can (Bradshaw, 1965). The important conclusion is, however, that such plasticity is not necessary in plants even for very small-scale variations in space of the environment, since genetic differentiation of populations can so easily occur. The ecological amplitude of most species therefore has a strong genetical component.

Acknowledgments

This paper owes its origin to the hospitality and encouragement of R. C. Lewontin and to a Visiting Fellowship of the Ford Foundation at the Department of Biology, University of Chicago. Its content has depended on many friends whose enthusiastic work has taken the study of plant populations in new fields and who have always readily discussed their work. Its final form has depended considerably on those who have patiently read and commented upon it, particularly J. Antonovics, B. Charlesworth, D. Janzen, and R. Levins.

REFERENCES

Allard, R. W., and P. E. Hansche. 1964. Some parameters of population variability and their implications in plant breeding. Advances Agron., 16:281-325.

——J. Harding, and C. Wehrhahn. 1966. The estimation and use of selective values in predicting population change. Heredity (London), 21:547-563.

____and S. K. Jain. 1962. Population studies in predominantly self pollinating species. II. Analysis of quantiative genetic changes in bulk hybrid populations of barley. Evolution, 14:90-101.

Antonovics, J. 1968a. Evolution in closely adjacent plant populations. V. Evolution of self fertility. Heredity (London), 23:219-238.

____1968b. Evolution in closely adjacent plant populations. VI. Manifold effects of gene flow. Heredity (London), 23:507-524.

____and A. D. Bradshaw. 1970. Evolution in closely adjacent plant populations. VII. Clinal patterns at a mine boundary. Heredity (London), 24:349-362.

Aston, J. L., and A. D. Bradshaw. 1966. Evolution in closely adjacent plant populations. II. *Agrostis stolonifera* in maritime habitats. Heredity (London), 21:649-664.

Barber, H. N. 1965. Selection in natural populations. Heredity (London), 20:557-572.

____and W. D. Jackson. 1957. Natural selection in action in *Eucalyptus*. Nature (London), 179:1267-1269.

Bateman, A. J. 1947. Contamination in seed crops. III. Relation with isolation distance. Heredity (London), 1:303-36.

Beard, D. F., and E. A. Hollowell. 1952. The effect on performance when seed of forage crop varieties is grown under different environmental conditions. Proc. 6th Int. Grassl. Cong., 860-866.

Bocher, T. W. 1949. Racial divergences in *Prunella vulgaris* in relation to habitat and climate. New Phytol., 48:285-314.

Bradshaw, A. D. 1959. Population differentiation in *Agrostis tenuis* Sibth. I. Morphological differentiation. New Phytol., 58:208-227.

____1960. Population differentiation in *Agrostis tenuis* Sibth. III. Populations in varied environments. New Phytol., 59:92-103.

____1965. Evolutionary significance of phenotypic plasticity in plants. Advances Genet., 13:115-155.

____1969. An ecologists' viewpoint. Brit. Ecol. Soc. Sympos., 9:415-427.

____1970. Plants and industrial waste. Trans. Bot. Soc. Edinburgh, 41:71-84.

____T. S. McNeilly, and R. P. G. Gregory. 1965. Industrialisation, evolution and the development of heavy metal tolerance in plants. Brit. Ecol. Soc. Sympos., 5:327-343.

Brougham, R. W., A. C. Glenday, and S. O. Fejer. 1960. The effects of frequency and intensity of grazing on the genotypic structure of a ryegrass population. N. Z. J. Agr. Res., 3:442-453.

Charles, A. H. 1961. Differential survival of cultivars of *Lolium, Dactylis* and *Phleum*. J. Brit. Grassl. Soc., 16:69-75.

____1964. Differential survival of plant types in swards. J. Brit. Grassl. Soc., 19:198-204.

Clausen, J., and W. M. Hiesey. 1960. The balance between coherence and variation in evolution. Proc. Nat. Acad. Sci. U. S. A., 46:494-506.

____D. D. Keck, and W. M. Hiesey. 1940. Experimental studies on the nature of species. I. The effect of varied environments on western North American plants. Carnegie Inst. Washington, Publ. 520.

____D. D. Keck, and W. M. Hiesey. 1948. Experimental studies on the nature of species. III. Environmental responses of climatic races of *Achillea*. Carnegie Inst. Washington, Publ. 581.

Colwell, R. N. 1951. The use of radioactive isotopes in determining spore distribution patterns. Amer. J. Bot., 38:511-523.

Cook, S. A. 1962. Genetic system, variation and adaptation in *Eschscholtzia californica*. Evolution, 16:278-299.

Cooper, J. P. 1964. Climatic variation in forage grasses. I. Leaf development in climatic races of *Lolium* and *Dactylis*. J. Appl. Ecol., 1:45-61.

Crossley, G. K., and A. D. Bradshaw. 1968. Differences in response to mineral nutrients of populations of ryegrass, *Lolium perenne* L., and orchard grass, *Dactylis glomerata* L. Crop Sci., 8:383-387.

Culberson, W. L., and C. F. Culberson. 1967. Habitat selection by chemically differentiated races of lichens. Science, 158:1195-1197.

Darlington, C. D. 1939. The Evolution of Genetic Systems. Cambridge, Cambridge University Press.

____and K. Mather. 1949. The Elements of Genetics. London, Allen and Unwin.
Deakin, M. A. B. 1968. Genetic polymorphism in a subdivided population. Aust. J. Biol. Sci., 21:165-168.
Ehrendorfer, F. 1953. Ökologisch-geographische Mikro-Differenzierung einer Population von *Galium pumilum* Murr. s. str. Österreich. Bot. Z., 100:616-638.
Ehrlich, P. R., and P. H. Raven. 1969. Differentiation of populations. Science, 165:1228-1232.
Fisher, R. A. 1950. Gene frequencies in a cline determined by selection. Biometrics, 6:353-361.
Ford, E. B. 1964. Ecological Genetics. London, Methuen.
Frankel, D. H. 1954. Invasion and evolution of plants in Australia and New Zealand, Proc. 9th Int. Congr. Genet., Caryologia, 6 Suppl: 600-619.
Grant, V. 1954. Genetic and taxonomic studies in Gilia. IV. *Gilia achilleaefolia.* Aliso, 3:1-18.
____1963. The Origin of Adaptations. New York, Columbia University Press.
Gregor, J. W. 1930. Experiments on the genetics of wild populations I. *Plantago maritima.* J. Genet., 22:15-25.
Griffiths, D. J. 1950. The liability of seed crops of perennial ryegrass (*Lolium perenne*) to contamination by wind borne pollen. J. Agric. Sci., 40:19-38.
Haldane, J. B. S. 1948. The theory of a cline. J. Genet., 48:277-284.
Hanson, W. D. 1966. Effects of partial isolation (distance), migration and different fitness requirements among environmental pockets upon steady state gene frequencies. Biometrics, 22:453-468.
Harberd, D. J. 1967. Observations on natural clones in *Holcus mollis*. New Phytol., 66:401-408.
Harlan, H. V., and M. L. Martini. 1938. The effect of natural selection on a mixture of barley varieties. J. Agr. Res., 57:189-199.
Harper, J. L. 1964. The nature and consequences of interference amongst plants. Proc. 11th Int. Congr. Genet. 1963:465-482.
____1967. A Darwinian approach to plant ecology. J. Ecol., 55:247-270.
Heslop-Harrison, J. 1964. Forty years of genecology. Advances Ecol. Res., 2:159-247.
Imam, A. G., and R. W. Allard. 1965. Population studies in predominantly self pollinated species. VI. Genetic variability between and within natural populations of wild oats, *Avena fatua* L., from differing habitats in California, Genetics, 51:49-62.
Jain, S. K., and R. W. Allard. 1960. Population studies in predominantly self pollinating species. I. Evidence for heterozygote advantage in a closed population of barley. Proc. Nat. Acad. Sci. U. S. A., 46:1371-1377.
____and A. D. Bradshaw. 1966. Evolutionary divergence among adjacent plant populations. I. The evidence and its theoretical analysis. Heredity (London), 21:407-441.
Jennings, P. R., and R. C. Aquino. 1968. Studies of competition in rice. III. The mechanism of competition among phenotypes. Evolution, 22:529-542.
____and J. de Jesus. 1968. Studies on competition in rice. I. Competition in mixtures of varieties. Evolution, 22:119-124.
____and R. M. Herrera. 1968. Studies on competition in rice. II. Competition in segregating populations. Evolution, 22:332-336.
Jowett, D. 1964. Population studies on lead tolerant *Agrostis tenuis.* Evolution, 18:70-80.
Kemp, W. B. 1937. Natural selection within plant species as exemplified in a permanent pasture. J. Hered., 28:329-333.
Khan, M. S. I. 1969. The process of evolution of heavy metal tolerance in *Agrostis tenuis* and other grasses. MSc. thesis. University of Wales.
Kimura, M., and G. Weiss. 1964. The stepping stone model of population structure and the decrease of genetic correlation with distance. Genetics, 49:561-576.
Knight, G. R., A. Robertson, and C. H. Waddington. 1956. Selection for sexual isolation within a species. Evolution, 10:14-22.
Kruckeberg, A. R. 1957. Variation in fertility of hybrids between isolated populations of the serpentine species, *Streptanthus glandulosus* Hook. Evolution, 11:185-211.
Lee, A. J. 1960. A study of plant competition in relation to development. Evolution, 14:18-28.

Lefebvre, C. 1970. Self fertility in maritime and zinc mine populations of *Armeria maritima* (Mill.) Willd. Evolution, 24:571-577.
Levene, H. 1953. Genetic equilibrium when more than one ecological niche is available. Amer. Natural., 87:331-333.
Levin, D. A., and H. W. Kerster. 1967. Natural selection for reproductive isolation in *Phlox*. Evolution, 21:679-687.
——and H. W. Kerster. 1968. Local gene dispersal in *Phlox*. Evolution, 22:130-139.
Levins, R. 1968. Evolution in Changing Environments. Princeton, N. J., Princeton University Press.
——and R. Macarthur. 1966. Maintenance of genetic polymorphism in a heterogenous environment: variations on a theme by Howard Levene. Amer. Natural., 100:585-590.
Lewis, M. C. 1969. Genecological differentiation of leaf morphology in *Geranium sanguineum*. New Phytol., 68:481-503.
Lewontin, R. C. 1967. Population genetics. Ann. Rev. Genet., 1:37-70.
——and L. C. Birch. 1966. Hybridization as a source of variation for adaptation to new environments. Evolution, 20:315-336.
Mather, K., and D. De Winton. 1941. Adaptation and counter-adaptation of the breeding system in *Primula*. Ann. Bot., 5:297-311.
Mayr, E. 1947. Ecological factors in speciation. Evolution, 1:263-288.
——1963. Animal species and Evolution. Cambridge, Mass., Belknap Press of Harvard University Press.
McMillan, C. 1959. The role of ecotypic variation in the distribution of the central grassland of North America. Ecol. Monog., 29:285-308.
McNeilly, T. 1968. Evolution in closely adjacent plant populations. III. *Agrostis tenuis* on a small copper mine. Heredity (London), 23:99-108.
——and J. Antonovics. 1968. Evolution in closely adjacent plant populations. IV. Barriers to gene flow. Heredity (London), 23:205-218.
——and A. D. Bradshaw. 1968. Evolutionary processes in populations of copper tolerant *Agrostis tenuis* Sibth. Evolution, 22:108-118.
Morley, F. H. W. 1959. Natural selection and variation in plants. Sympos. Quant. Biol., 24:47-56.
——C. I. Davern, V. E. Rogers, and J. W. Peak. 1962. Natural selection among strains of *Trifolium subterraneum*. Evolution of Living Organisms, Sympos. Soc. Royal Soc. Victoria, 1959:181-190.
Mulligan, G. A., and R. J. Moore. 1961. Natural selection among hybrids between *Carduus acanthoides* and *C. nutans* in Ontario. Canad. J. Bot., 39:269-279.
Prout, T. 1968. Sufficient conditions for multiple niche polymorphism. Amer. Natural., 102:493-498.
Roberts, M. R., and H. Lewis. 1955. Subspeciation in *Clarkia biloba*. Evolution, 9:445-454.
Sheppard, P. M. 1952. Natural selection in two colonies of the polymorphic land snail *Cepaea nemoralis*. Heredity (London), 6:233-238.
Smith, J. M. 1966. Sympatric speciation. Amer. Natural., 100:637-650.
Snaydon, R. W. 1963. Morphological and physiological differentiation in *Anthoxanthum odoratum* on the Park Grass Experiment, Rothamsted. Heredity, (London) 18:382 (Abstr.).
——1970. Rapid population differentiation in a mosaic environment. I. The response of *Anthoxanthum odoratum* populations to soils. Evolution, 24:257-269.
——and A. D. Bradshaw. 1962. The performance and survival of contrasting natural populations of white clover when planted into an upland Festuca/Agrostis sward. J. Brit. Grassl. Soc., 17:113-118.
Spiess, E. B. 1968. Experimental population genetics. Ann. Rev. Genet., 2:165-208.
Squillace, A. E., and R. T. Bingham. 1958. Localised ecotypic variation in western white pine. Forest Sci., 4:20-34.
Stanford, E. H., H. M. Laude, and P. V. Booyzen. 1962. Effects of advance in generation under different harvesting regimes on the genetic composition of Pilgrim ladino clover. Crop Sci., 2:497-500.
Stebbins, G. L. 1950. Variation and Evolution in Plants. New York, Columbia University Press.

____1957. Self fertilization and population variability in higher plants. Amer. Natural., 91:337-354.

____and K. Daly. 1961. Changes in the variation pattern of a hybrid population of *Helianthus* over an eight year period. Evolution, 15:60-71.

Suneson, C. A. 1949. Survival of four barley varieties in a mixture. Agron. J., 41:459-461.

Sylven, N. 1937. The influence of climatic conditions on type composition. Imp. Bureau Plant Genetics, Herbage Bull., 21:8.

Thoday, J. M., and J. B. Gibson. 1962. Isolation by disruptive selection. Nature (London), 193:1164-1166.

Turesson, G. 1922. The genotypical response of the plant species to the habitat. Hereditas (Lund), 3:211-350.

Vickery, R. K. 1959. Barriers to gene exchange within *Mimulus gattatus* (Scrophulariaceae). Evolution, 13:300-310.

____1964. Barriers to gene exchange between members of the *Mimulus guttatus* complex (Scrophulariaeceae). Evolution, 18:52-69.

Watson, P. J. 1970. Evolution in closely adjacent plant populations. VII. An entomophilous species, *Potentilla erecta*, in two contrasting habitats. Heredity (London), 24:407-422.

Workman, P. L., and R. W. Allard. 1964. Population studies in predominantly self pollinating species. V. Analysis of differential and random viabilities in mixtures of competing pure lines. Heredity (London), 19:181-189.

Wright, S. 1931. Evolution in mendelian populations. Genetics, 16:97-159.

____1943. Isolation by distance. Genetics, 28:114-138.

____1946. Isolation by distance under diverse systems of mating. Genetics, 31:39-59.

____1951. The genetical structure of populations. Ann. Eugen., 15:323-354.

3

Origin and Evolution of Bilateral Symmetry in Flowers[1]

ELMAR E. LEPPIK[2]
Plant Science Research Division, Agricultural Research Service, United States Department of Agriculture, Beltsville, Maryland

INTRODUCTION

A most remarkable feature of many showy flowers is their well-proportioned and conspicuously perfected pattern of bilateral symmetry. Blossoms of violets, pansies, honeysuckles, columbines, veronicas, and African violets impress us by their delicate two-sided configuration, accompanied by harmonious colors and pleasing odors.

In these flowers we can recognize a peculiar deviation from the conventional, regular symmetry of the more primitive flowers, as a new trend in the evolution of angiosperms in a new form of floral beauty. Ordinarily all elementary flower types on lower evolutionary levels have a conspicuously regular and radial symmetry, in contrast to the bilateral tendencies of more advanced types. Floral bilateralism is consequently a significant progression in the evolution of flower types that bridges the wide structural gap between elementary radial symmetry in primitive flowers and the evolved zygomorphism of advanced types.

Beyond their esthetic appeal to man, bilateral flowers must necessarily have a definite biological function and purpose in nature. Of course, there is also an underlying genetic mechanism in plants that has produced and is still producing such spectacular patterns. An understanding of basic morphogenic problems may help us to comprehend better the evolution of flower types and to solve the

[1]Plant Introduction Investigation Paper No. 25.
[2]Present address: New Crops Research Branch, United States Department of Agriculture, Plant Industry Station, Beltsville, Maryland, 20705.

mystery of floral beauty. "If this key principle could be found, the interpretation of the flower might enter upon a new and more hopeful phase" (Arber, 1937, p. 182). Now the key principle is found in the regulating sensory perception mechanisms of pollinators, and reinterpretation of the flower is in full swing.

Despite the fact that bilateral and zygomorphic flower types have been long known and well described, no serious attempt has been made to explain their morphogenesis and evolutionary significance, except in some previous papers of the present writer (Leppik, 1957, 1958a). Numerous classical and modern floroecologists have studied bilateral flowers in detail and have emphasized their important function in the process of pollination and fertilization. However, no one has ever tried to relate floral bilateralism to the selective activity and progressive sensory capability of pollinating insects, birds, and bats. Similarly, no reliable evaluation has yet been made on the evolution of floral symmetry in connection with the development of human concept of beauty.

Basic information on floral evolution and corresponding genetic systems in plants are urgently needed now for modern agricultural and horticultural research that intends to improve cultivated crops, flowers, and other ornamentals. Any new idea or theory that may help explain the mechanism of evolution and speed up breeding work will have incalculable practical value for the agricultural and horticultural breeding of field crops and ornamentals.

Some practical aspects of these new findings, particularly a new discipline, "bee-botany," are reviewed and discussed by Deodikar (1961). This author purposely correlates recent discoveries in the sensory physiology of insects with the newest achievements in floral ecology, thus elaborating a new bottom-ground for the further progression of apiculture.

DEFINITIONS AND TERMINOLOGY

Bilateral symmetry appears in flowers as a two-sided arrangement of petals or other floral parts into equal right and left halves that are approximate mirror images of each other (Fig. 1). Usually the dividing plane lies epihypaxially (up and down) so that similar right and left (or upper and lower) halves are separated. *Radial symmetry*, on the other hand, is the arrangement of two or more planes of rays or other flower parts around a common axis.

Symmetry of the floral axis is described as *epiaxial* (epi, above), *hypaxial* (hypo, under), or *epihypaxial* (dorsiventral). The zoological terms dorsal, ventral, and dorsiventral (dorsum, back; venter, belly) are also used in botanical literature (Fig. 1).

All numerical combinations in bilateral flowers and pseudo-flowers can be expressed in common mathematical fractions, as shown in Table 1, by marking the number of epiaxial petals, or semaphylls, with the numerator, and the hypaxial petals with the denominator, as shown in Table 1. Noteworthily the

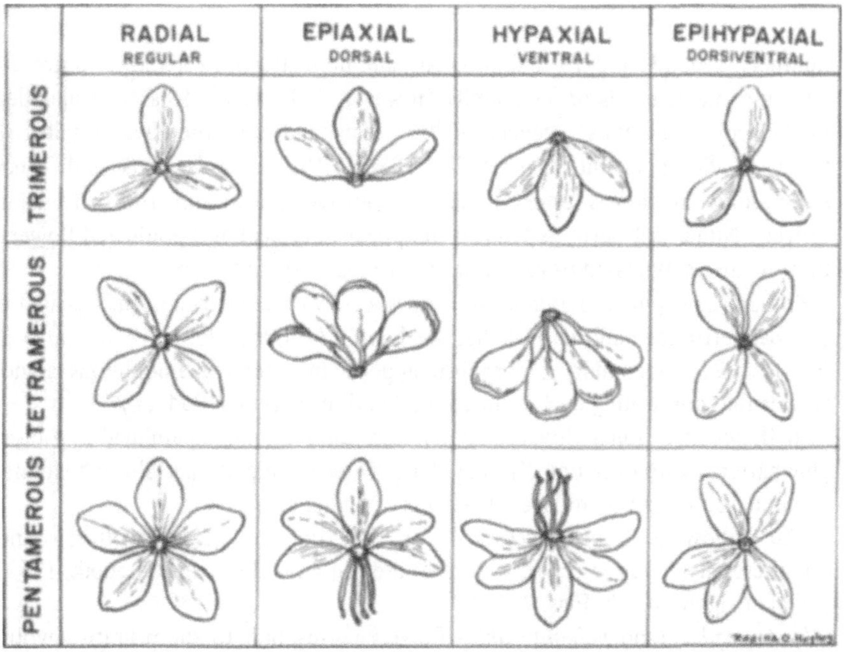

Fig. 1. Contoured types of radial (column 1) and bilateral (columns 2-4) symmetry, formed from trimerous, tetramerous, and pentamerous flowers. See Figs. 2-24 for comparisons.

Table 1. Number and arrangement of petals or semaphylls in radial and bilateral flower types and in pseudoflowers

	Radial	Epiaxial	Hypaxial	Epihypaxial
Monomerous	1	1/0[a]	0/1	–
Dimerous	2	2/0	0/2	1/1
Trimerous	3	3/0	0/3	1/2, 2/1
Tetramerous	4	4/0	0/4	2/2, 3/1, 1/3
Pentamerous	5	5/0o	0/5	2/4, 2/3, 3/2, 4/1
Hexamerous	6	6/0	0/6	1/5, 2/4, 3/3, 4/2, 5/1
Octomerous	8	–	–	–
Decamerous	10	–	–	–
Duodecamerous	12	–	–	–
Polymerous	∞	–	–	–

[a]Numerator indicates the number of epiaxial, denominator the hypaxial petals or semaphylls in a bilateral flower or pseudoflower. A dash indicates that no bilateralism is observed in octomerous, decamerous and polymerous flowers.

numbers 7, 9, 11, and 13 seldom occur, and no bilateralism is known in flowers with 8 or more petals or semaphylls. However, 7, 9, 11, and 13 are not included in the iconic numerals[3] recognized by insects, and therefore have no selective value in floral evolution (Leppik, 1956b, 1957). Birds, bats and other anthophilous mammals, unlike insects, do not respond to numerical patterns in flowers. Numerical patterns tend to disappear in typical bird-pollinated flowers; they are also suppressed or camouflaged in zygomorphic flowers.

Primary amorphic (paleomorphic) types were well developed in the Cycadatae and Bennettitatae and are described by Leppik (1960b; 1963b; 1970a, pp. 5-6, Fig. 1; 1971). Secondary amorphism appears in several anemophilous angiosperms that are turning back to insect pollination (Leppik, 1954a, b).

In three-dimensional flowers, bilateralism leads to the streamlined zygomorphic patterns with only one dividing plane. These are highly specialized bilabiate, papilionaceous, and orchidaceous flowers of higher angiosperms.

The following terms, necessary for the study of floral evolution and classification of flower types, were proposed in an earlier paper (Leppik, 1957) and are diagrammed in Figs. 2, 3, 4.

I. *Amorphic* (and *paleomorphic*) flowers correspond to the primitive evolutionary level. They have no definite shape, no iconic numerals, no particular symmetry. From green leaves they are distinguished mainly as clusters of semaphylls, characterized by white or yellowish colors. *Paleomorphic* (Fig. 4) types were developed by Cycadatae and Bennettitatae and are described elsewhere (Leppik, 1963b). In angiosperms amorphic types are formed secondarily in several anemophilous (wind-pollinated) flowers, which gradually adapt themselves to insect pollination (Leppik, 1950a).

II. *Haplomorphic* flowers have an open hemispheric shape and white or yellow color. There is no definite symmetry, no numeral patterns or variegated colors, as in *Magnolia*, water lilies, and so on.

III. *Actinomorphic* flowers typically show radial symmetry with several dividing planes around a common axis.

IV. *Pleomorphic* flowers have a reduced but definite number of petals (or semaphylls), termed iconic numerals, as indicated in Fig. 2 (IV). From the pleomorphic types bilateral flowers frequently evolved, with two dividing planes around a common axis.

V. *Stereomorphic* flowers, ordinarily with concealed nectar deposits, are stretched toward the third dimension. Such structures require that the pollinators have the ability to distinguish the three-dimensional pattern and possess a long enough proboscis to reach the concealed nectaries in flowers. Bilateralism is common among stereomorphic flowers.

VI. Zygomorphic (or "irregular") flowers are at the highest level of plant

[3] Iconic numerals (or figure numerals) are the numeral patterns of bi-, tri-, tetra-, penta-, hexa-, deca-, and polymerous flowers that enable bees, bumblebees, moths, and other pollinators with advanced senses to distinguish such flowers (Leppik, 1956b).

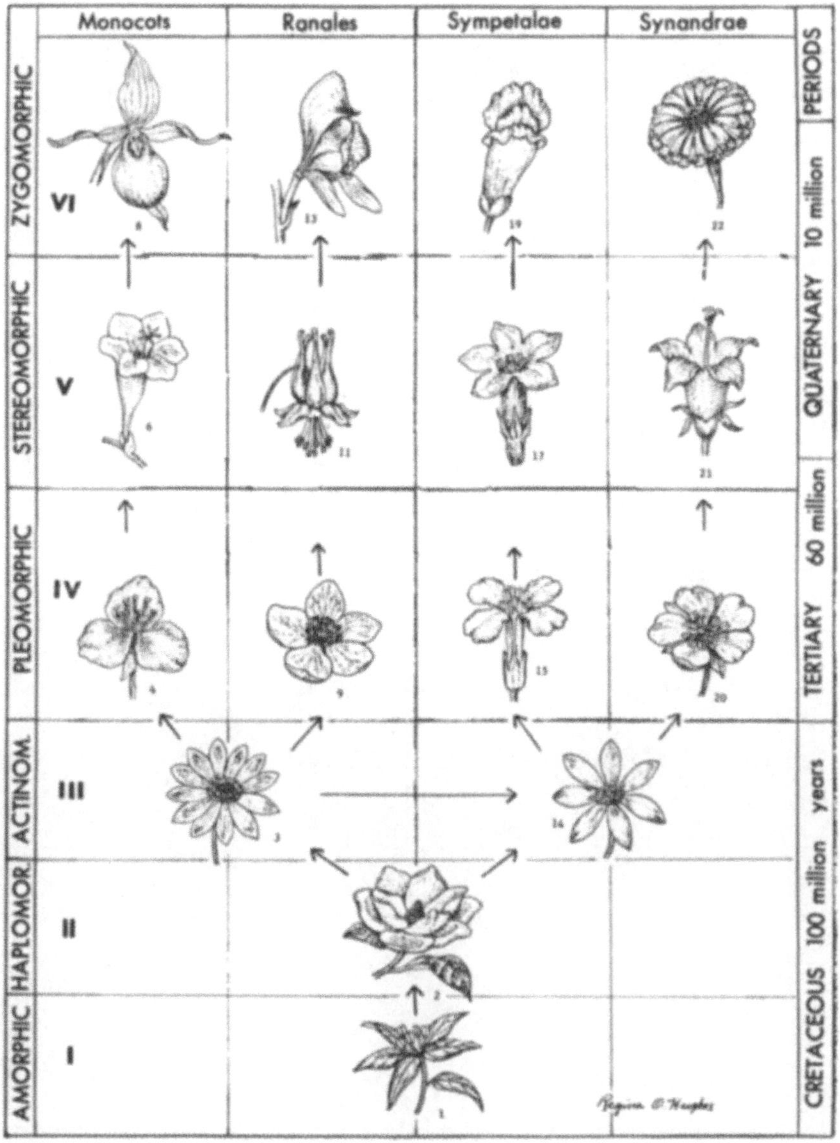

Fig. 2. Main trends of the historical development of flower types during the evolution of angiosperms in the last 100 million years. level I: amorphic flower types. level II: haplomorphic types. level III: actinomorphic. level IV: pleomorphic. level V: stereomorphic. level VI: zygomorphic flowers. Arrows indicate the successive floral evolution in main phylogenetic groups of angiosperms (monocots, Ranales, Sympetales, Synandrae). For names of plants (1-22), see Fig. 3. Rearranged scheme from Leppik (1957, p. 472).

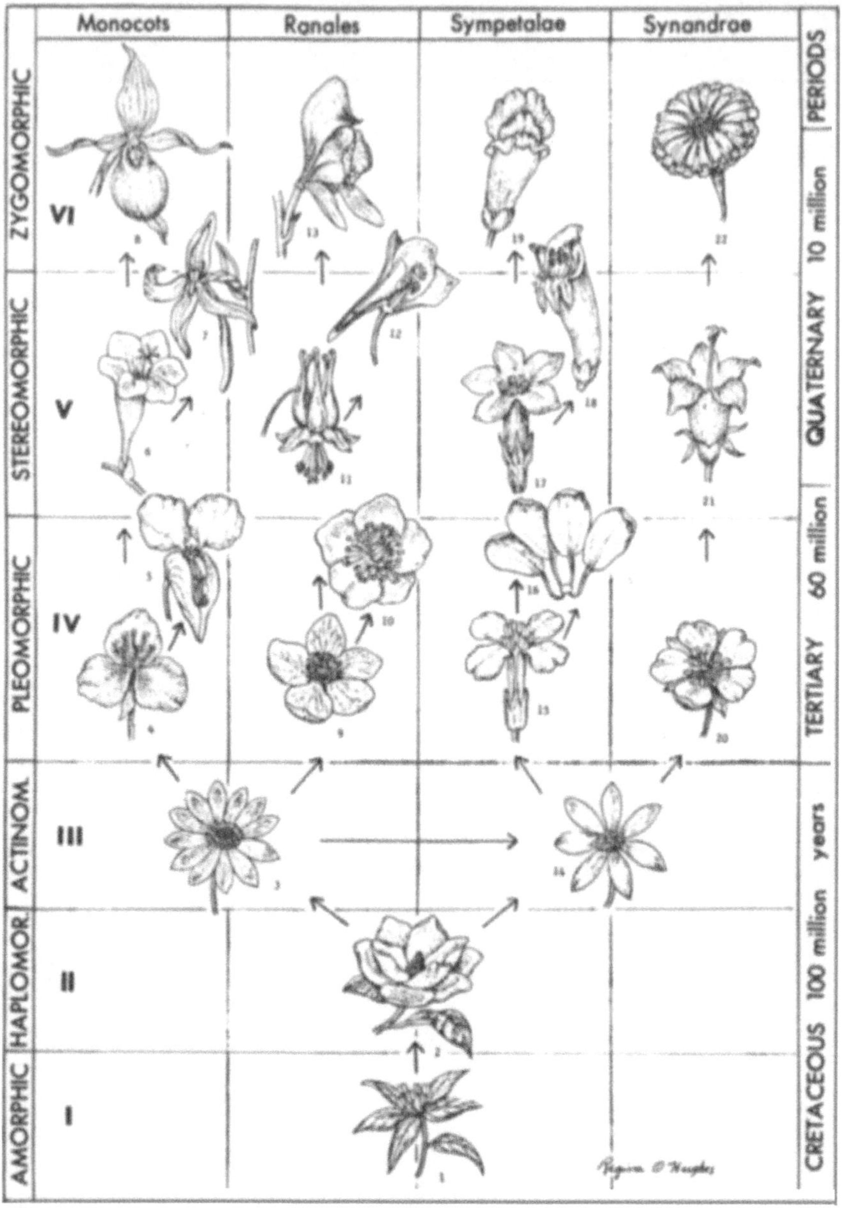

Fig. 3. Differentiation of bilateral types from flowers of radial symmetry as pictured in Fig. 1. Note the bilateral deviations from the radial symmetry on the pleomorphic (IV) and stereomorphic (V) levels (5, 7, 10, 12, 16, 18). Selected examples: 1. Hypothetical amorphic type of ancestral angiosperms. 2, Magnolia. *3,* Adonis. *4,* Tradescantia. *5,* Commelina. *6,* Freesia. *7,* Orchis. *8,* Cypripedium. *9,* Ranunculus. *10,* Helleborus. *11,* Aquilegia. *12,* Delphinium. *13,* Aconitum. *14, Oligopetalous early flowers of* Cinnamomum, Boldea, *and others. 15,* Hesperis. *16,* Cleome. *17,* Gentiana. *18,* Penstemon. *19,* Antirrhinum. *20,* Geranium. *21,* Campanula. *22,* Calendula. *(Drawn from photographs.)*

Fig. 4. Some archaic flower types of the Bennettitatae. 1, Wieladiella, *rosette-like arrangement of leaves around every single flower-like strobilus. 2,* Williamsoniella, *pedicelled strobili separated from cauline leaves. 3.* Sturianthus, *pedicelled actinomorphic pattern resembling the flower head of the Compositae. Note the exclusive radial symmetry of these Paleozoic and Mesozoic flower types, which occurred long before angiosperms. (Redrawn with permission from the models of the Chicago Natural History Museum. Adapted from Leppik, 1963b, pp. 95-98.)*

evolution. They contain many previously developed floral characteristics, such as pleomorphism, stereomorphism, bilateralism, variagated colors, and fine odors, all harmonously combined in a single flower. These streamlined, dorsiventral structures have commonly only one dividing plane of symmetry, or none. The enormously abundant array of strange and often grotesque forms makes the further morphogenic classification of zygomorphic flower types difficult to elucidate. Wettstein (1935, p. 545) subdivided them into four groups, as follows:

(1) Median zygomorphic types, if the plane of symmetry is vertical, as in the mints (Labiatae), figworts (Scrophulariaceae), peas (Papilionaceae), most orchids (Orchidaceae), and other families.
(2) Transverse zygomorphic with transverse plane of symmetry, as in Fumarioideae.
(3) Oblique zygomorphic with diagonal plane of symmetry, as in the flowers of horse chesnut (*Aesculus hippocastanum L.*), Sapindaceae, Trigoniaceae, and Vochysiaceae.
(4) Asymetric flowers without definite plane of symmetry, such as the flowers of Cannaceae and Marantaceae.

This classification system was applied also by Jaeger (1961, p. 11) and Melchior (1964, p. 4-5).

However, on the basis of the present very scanty information, Pijl and Dodson (1966) are still hesitant in attempting to organize orchid flowers into a phylogenetic sequence. Dodson termed the floral development of orchids a "leap-frog evolution" from one pollinator class to another, in a degree that any proposed historical or phylogenetic sequence could be still irrevalent.

MORPHOGENIC CLASSIFICATION OF FLOWER TYPES

The great diversity in form, symmetry, shape, size, colors, and odors of extant flowers makes their typological classification difficult. Delpino (see Knuth, 1906, vol. 1, p. 14) arranged all known floral forms into 47 types and 13 classes. Vogel (1954), Pijl (1960-1961), Faegri and Pijl (1966), Baker and Hurd (1968, p. 395), and others have used insect pollination syndromes for floral classification. Accordingly we have Melittophilae, or bee flowers; Myophilae (fly flowers); Sapromyophilae (carrion-fly flowers); Cantharophilae (beetle flowers); Psychophilae (butterfly flowers); Sphingophilae (sphinx flowers); Ornithophilae (bird flowers); Chiropterophilae (bat flowers); and so on. Unfortunately these systems are applicable only in the cases of specialized pollination mechanisms, adapted only to a few pollinators. Most flower types are accessible to and are actually pollinated by many insect groups, sometimes even by birds, bats, and other animals.

Recently a new system was proposed for the morphogenic classification of flower types (Leppik, 1956b, 1957, 1968a). This system was the first attempt to arrange existing flower types in a natural sequence of their historical development, which took place under the selective pressure of pollinators. In this sequence six successive evolutionary levels were found to concur with six type-classes that are clearly distinguishable according to their morphological characteristics and hologenic criteria. Most elementary types on amorphic and haplomorphic levels were succeeded by strictly symmetrical radiate flowers on actinomorphic and pleomorphic levels. Later, three-dimensional floral structures developed on the stereomorphic level, succeeding finally to the most complicated zygomorphic patterns at the top of the evolutionary sequence (Figs. 2, 3). Bilateral symmetry appeared in this sequence first on the pleomorphic level, intensified its stand on the stereomorphic, and integrated finally into the complicated floral structures on the uppermost zygomorphic level.

This morphogenic sequence of floral evolution is most likely correlated with the sensory development of pollinating insects, birds, and bats. Primitive insects with their elementary senses were able to distinguish and exploit amorphic and haplomorphic flowers only. With time they acquired the special ability to distinguish the radial symmetry, numeral patterns, and nectar guides, which indicated hidden nectar deposits inside showy flowers. However, there are few insect groups among hymenopterous and lepidopterous pollinators capable of distinguishing zygomorphic flowers as well as operating complicated pollination mechanisms of higher flowering plants. It follows that flowers and insects are mutually interrelated in coevolution as reciprocal selective factors, adapted to each other and to their surrounding environment. "Each pollinator is presumably programmed genetically to respond to a proper series of stimuli–an exact odor, shade of color, or shape, or a series of odors, colors or shapes. Each pollinator has adopted a strategy–in essence specialist or generalist. Similarly, each plant has adopted a strategy. As floras and faunas evolve together, the utilities of the various strategies are going to change" (Ehrlich, 1970, p. 9).

Concrete experimental findings and extensive field observations accomplished by many classical and modern workers support this theory. Corresponding data can be obtained from the papers published by Kölreuter (1761), Sprengel (1793), Müller (1883), Knuth (1906-1909), Kugler (1955), Vogel (1954), Leppik (1957, 1963a,b,c, 1964, 1966, 1968a,b), Meeuse (1961), Grant and Grant (1965), Pijl (1960-1961, 1966), Jaeger (1961), Kullenberg (1961), Barnard (1961), Macior (1964, 1965, 1966, 1967), Percival (1965), Faegri and Pijl (1966), Zimmermann (1966), Pijl and Dodson (1967), and others. Recently an extensive review of intrafloral ecology was published by Baker and Hurd (1968).

Since its first publication (Leppik, 1957), more experimental evidence and observational data have accumulated in support of this proposed classification system. After some minor changes and rearrangements, it was used as a standard scale for testing the ability grades of pollinating insects, birds, and bats (Leppik, 1956a,b, 1957, 1963a,b, 1964). The floral evolution of the Magnoliaceae, Rosaceae, Compositae, Ranunculaceae, and Leguminosae has been studied and described (Leppik, 1960a, 1961, 1964, 1966, 1970b). Fossil imprints of flowers in muds, clays, fine sands, and amber helped to connect the historical sequence of floral evolution with geological time table (Leppik, 1960b, 1963a,b, 1971). Recently, the whole system was once more revised and elaborated for practical application of floral classification (Leppik, 1968a).

It is appropriate to note here that the historical sequence of floral evolution (Figs. 2, 3) do not match with the phylogenetic development of flowering plants. Similarly, the morphogenic classification of flower types is not identical with the taxonomic classification of plants. Insects and other pollinators are not cognizant of phylogeny or taxonomy, but are searching their food plants according to typological characteristics of flowers. Accordingly, flowers of many plant families, genera, and species which are visited by the same groups of pollinators tend to develop, under the selective pressure of their pollinators, covergent flower types.

Floral evolution follows the general historical sequence from amorphic, haplomorphic, actinomorphic, pleomorphic, and stereomorphic to zygomorphic levels, as pictured in Fig. 2. This general sequence corresponds to the assumed sensory development of pollinating insects, with several deviations in flowers pollinated by birds, bats, and other animals. In this context the historical sequence of floral evolution can be as easily coordinated with the classic taxonomic orders, as with the most modern systems, developed by Cronquist (1968) and Takhtajan (1969).

IMPROPER COMMON NAMES OF FLOWERS IN OUR TEXTBOOKS

In conventional textbooks flowers are still treated and classified according to their resemblance to some simple human artifacts, such as cups, dishes, bowls, brushes, flags, gullets, trumpets, and so on (Werth, 1956; Kugler, 1955; Faegri and Pijl, 1966).

The comparison of delicate floral structures to common human tools may help us remember elementary flower types and identify them for taxonomic purposes. Yet no human tools or artifacts have even superficial resemblances to the complicated structures of the highly evolved flowers. There are no common names for numerous orchids and legume flowers, for the figwort family, or for many other zygomorphic blossoms. Moreover, human artifacts are totally unrelated to the response of insect pollinators and thus cannot provide any clue to the evolution of complicated floral patterns induced by the selective activity of pollinating insects, birds or bats. Thus all artificial systems of classification should be replaced by natural systems as soon as these can be established (Leppik, 1968a).

EARLY EVOLUTION OF FLOWER TYPES

There is considerable conclusive evidence and some paleontological proof (fossilized mouth parts of insects) that insect pollination was well established long before angiosperms, or the true flowering plants, came into existence. The strobili of some contemporary gymnosperms, such as *Encephalartos, Macrozamia, Ephedra, Gnetum,* and *Welwitschia* produce sugary fluids in their flowers and are regularly visited by insects, mainly by beetles. The extensive literature on this subject is reviewed by Baker and Hurd (1968, p. 387).

On the other hand, numerous fossil findings show that some extinct groups of gymnosperms, particularly the Bennettitatae (or Cycadeoideae) possessed strobili with all the characteristics of showy flowers. As pictured in Fig. 4, the sporophylls of these plants were assembled in symmetrical clusters, resembling showy flowers of angiosperms. In Fig. 5 the recapitulation of the archaic cycadeoid radial symmetry of *Cycadeoidea ingens* Ward.(Fig. 5:1) is compared with the modern actinomorphic patterns of *Actinotus helianthi* Labill. (Umbelliferae, Fig. 5:2) and *Helianthus grosseserratus* Mantens (Compositae, Fig. 5:3).

Functioning as true semaphylls, these flower-like strobiles must have been distinctive from vegetative organs and attractive to contemporary insects, which searched for concentrated food from fructification organs. The strobili of the Bennettitatae were even bisexual, a distinctive syndrome of entomophilous

Fig.5. Recapitulation of an archaic cycadeoid radial symmetry by modern pseudoflowers of the Umbelliferae and Compositae. 1, Cycadeoidea ingens *Ward. (Bennettitatae) from the Jurassic period. A fully expanded flower 12 cm wide (photographed from a model reconstructed in the Chicago Natural History Museum). 2,* Actinotus helianthi *Labill. (Umbelliferae) an inflorescence 5 cm wide imitating a radial (actinomorphic) solitary flower. (Collected by H. Salasoo, Lidcombe, from NSW Australia.) 3.* Helianthus grosseserratus *Mantens (Compositae) from North America (See text).*

flowers. Anemophilous flowers are unisexual.

In cauliflorous Cycadeoidea, the flower clusters were borne separately on tree trunks, easily distinguishable from vegetative leaves (Fig. 6). Bennettitatae were presumably visited and pollinated by beetles and other insects with primitive senses. The proficiency of these insects was restricted to the recognition of elementary paleomorphic patterns, such as amorphic, haplomorphic, and actinomorphic types.

Fig. 6. Cycadeoidea marshiana *Hirmer. Note the cauline radiate flowers, separated from vegetative leaves (about 1/10 natural size). (Reconstructed by Hirmer, redrawn from Wettstein, 1935, p. 489; courtesy of Franz Deuticke Verlag, Leipzig and Vienna.)*

It is highly remarkable that the primitive flower-like structures of gymnosperms in the Cretaceous period and earlier developed amorphic, haplomorphic, and actinomorphic patterns, without any trace of bilateral symmetry (Figs. 4, 6). This clearly indicates the close parallelism of the sensory evolution of pollinating insects with the morphogenic development of flower types.

Another observation that can be derived from the structure of these early flowers is that the relationship between plants and anthophilous insects obviously began long before the angiosperms or, at least, before they were abundant enough to play any significant role in pre-Cretaceous floras. Bennettitatae and some other gymnosperms, although probably not ancestral to the angiosperms, nevertheless supported an extensive fauna of anthophilous insects in the early Cretaceous period. This may have helped angiosperms in their rapid rise to domination of the flora since the upper Cretaceous period. Among angiosperm leaves found in Middle Cretaceous deposits are 26 families which might be described as typically entomophilous, and only 10 in which wind pollination (anemophily) is usual (Axelrod, 1952, p. 31). There is, of course, no indication that the Cretaceous flowers of angiosperms were bilaterally symmetrical. This type appeared in fossil records at a considerably later date.

On the other hand, the Bennettitatae, which reached their greatest abundance in the Jurassic period, declined rapidly to extinction in the upper Cretaceous period. It is tempting, therefore, to speculate that the phylogenetic youthfulness and modern floral structure of angiosperms enabled them to produce more concentrated nectar, a higher content of proteins, more intensive colors and odors for the attraction of contemporary pollinators. Biochemical evolution of the Bennettitatae did not progress fast enough and their primitive metabolism obviously could not satisfy the increasing needs of anthophilous insects. Presumably the less progressive Bennettitatae, which had already adapted themselves to insect pollination, coult not compete with modern angiosperms, and became extinct. Yet many anemophilous cycads survived until the present time in spite of their primitive floral structure and metabolism. Angiosperms, on the contrary under the selective pressure of pollinators, showed an unusual genetic elasticity and progress in their further evolution of flower types (Figs. 2, 3.)

DIFFERENTIATION OF BILATERAL SYMMETRY FROM RADIAL FLOWER PATTERNS

Observations show that all elementary flower types of phylogenetically older plant groups have a regular structure and perfect radial symmetry. Such simple floral patterns of haplomorphic, actinomorphic, and pleomorphic type classes are the most numerous in present-day floras. Fossil evidence indicates that until the Tertiary period all showy flowers were exclusively of regular shape and radial symmetry (Figs. 4, 5, 6) without any traces of bilateralism or zygomorphism (Leppik, 1960b, 1963a,b).

However, beginning with the pleomorphic level, regular flower types became subjected to a profound structural change that directed their further evolution toward bilaterialism and zygomorphism. Presumably these changes in the shape of flowers must have concurred with the corresponding changes in the sensory development of their particular pollinators. (Fig. 7). Such changes still occur in present genera, species, or even in varieties and populations of the same species. They must therefore be considered quite recent in geological terms.

At first such changes toward bilateralism were morphologically unimportant irregularities in the form or size of a single or few neighboring petals. *Veronica chamaedrys* L. (Fig. 8), for instance, has a slightly different front petal (3/1), which serves as a convenient landing place for its specialized pollinators (Fig. 8:3). Almost unnoticeable to the unskilled eye, this irregular petal is distinguishable to a syrphid fly, *Neoascia podagrica* (F.) (Fig. 8:2), and to some other small insects searching nectar from *Veronica* flowers.

Since the time of Müller (1883) it has been known that nectar is secreted in *V. chamaedrys* by a yellowish fleshy disk below the ovary, which lies in the base of the nectar tube, protected from rain by hairs on the corolla. The flowers are rendered conspicuous by their bright blue color and by clustering in racemes;

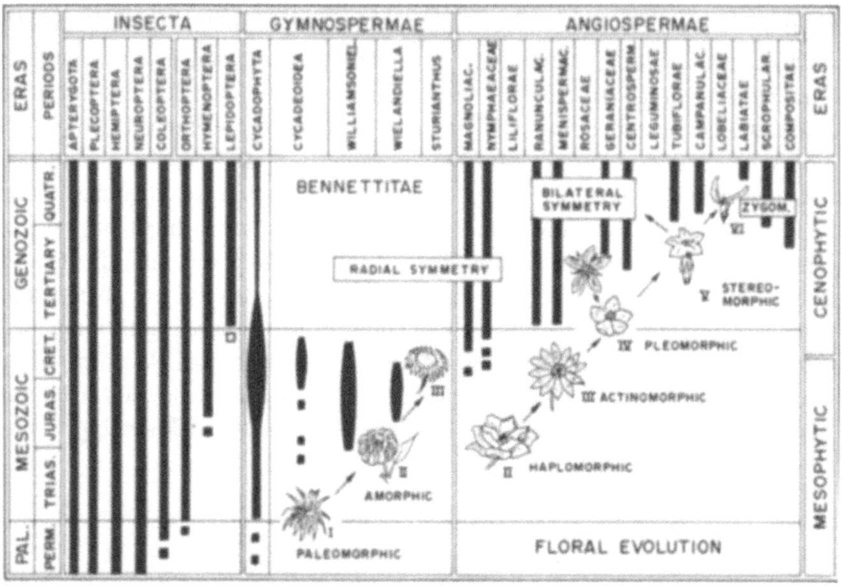

Fig. 7. Paleanthic flower types of the Bennettitatae compared with the floral differentiation of angiosperms. Roman numerals indicate the evolutionary stages I-VI and corresponding type classes as in Fig. 3. Note the early occurrence of insects, long before the appearance of flowers. Differentiation of bilateral symmetry appears relatively late, and is derived from pleomorphic (IV) and stereomorphic levels (a rearranged scheme from Leppik, 1960b).

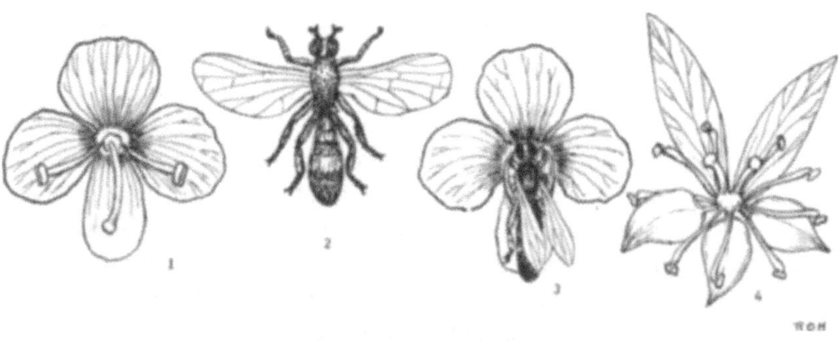

Fig. 8. An early stage of bilateral symmetry in the genus Veronica. *1,* Veronica chamaedrys *L. 2, Syrphid fly,* Neoascia podagrica *(F.). 3, Sucking fly on* Veronica. *4,* Saxifraga sarmentosa *L. (1-3 after* Muller, 1883.)

dark blue radiating lines and a central white spot on the limb of the corolla guide insects toward the nectar. Landing on the front petal an insect touches the style with its ventral surface, delivering pollen to the stigma. Creeping toward the nectar, the visitor touches the anthers and dusts its ventral surface with fresh pollen for the next flower (Fig. 8:3).

Saxifraga sarmentosa L. (Fig. 8:4) from China has two long and three shorter petals (3/2) with a single functional nectary. Cross pollination is favored by centripetal movement of stamens, which are directed towards the visitor's body.

Similar deviations from strictly radial symmetry can be seen on the corolla of *Erodium cicutarium* (L.) L'Hérit. Here the bilateralism is expressed by colored spots on somewhat smaller upper petals, as shown in Fig. 9. Two or three upper

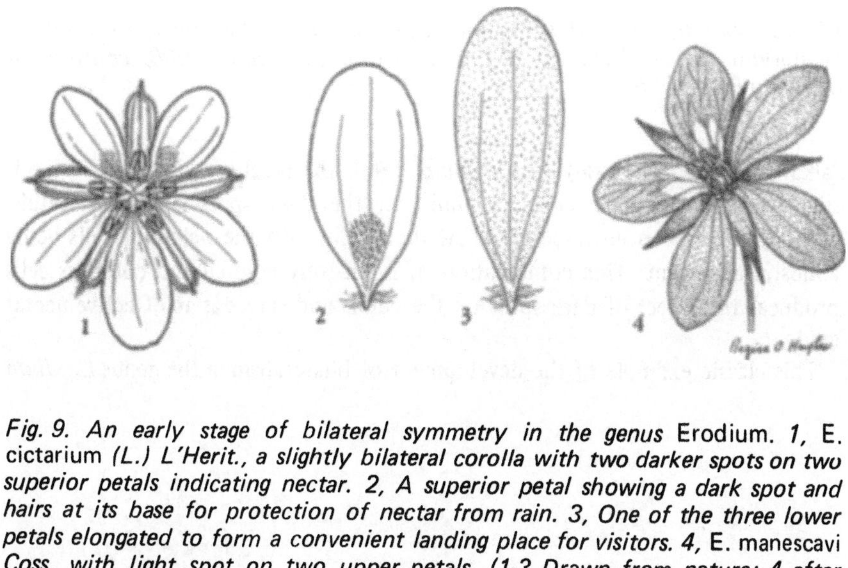

Fig. 9. An early stage of bilateral symmetry in the genus Erodium. *1,* E. cictarium *(L.) L'Herit., a slightly bilateral corolla with two darker spots on two superior petals indicating nectar. 2, A superior petal showing a dark spot and hairs at its base for protection of nectar from rain. 3, One of the three lower petals elongated to form a convenient landing place for visitors. 4,* E. manescavi *Coss. with light spot on two upper petals. (1-3 Drawn from nature; 4 after Ludwig, 1884, Plate IIIb.)*

petals are shorter and are provided with a darker red spot in the lower part of each pink petal (2/3). Marked petals are provided with large nectaries containing concentrated sugar. These flowers are usually visited by specialized pollinators that are capable of recognizing slightly bilateral flowers with spotted petals, and capable of detecting nectaries. Variation of red spots and nectar-guides on *Erodium cicutarium* was studied by Knuth (1908, vol 2:230) and is illustrated in Fig. 10.

It is a rewarding sight to observe the dark spots on the upper petals under the microscope (Fig. 9:2). The pink color of the petals is caused by the glycoside anthocyanin, which is equally distributed in the cell sap of epidermal cells. Depending on the reaction with the cell sap, the color may be red (acid), blue

Fig. 10. Variation of red spots and nectar-guides on the petals of Erodium cicutarium *(L.) L'Hérit. (Drawn from nature by Knuth, 1908; courtesy of Claredon Press, Oxford.)*

(alkali), or green (neutral). For *Erodium* petals the reaction with the cell sap is slightly acid, and the color is pink. In the dark spots of upper petals, anthocyanin is concentrated in a few dark cells with the neighbor cells being almost translucent. This combination of intensively colored and colorless cells produces the effect of dark spots on the petals and serves as an effective nectar guide.

This classic example of the development of bilateralism in the genus *Erodium*

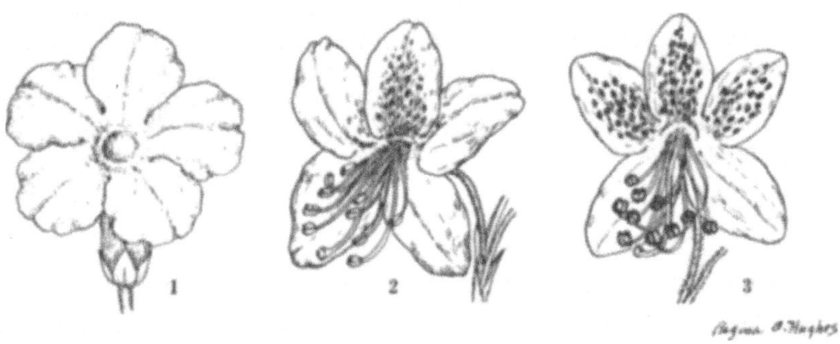

Fig. 11. Successive stages of the development of bilateral symmetry in Rhododendron *flowers. 1,* R. ledoides *Balf. & Smith. Radially symmetrical tubular corolla with pistil and stamens inside narrow corolla tube. Corolla pink. Plant introduced from Yunnan. 2,* R. yunnanense *Franch. Slightly bilateral pinkish corolla with red spotted upper petal and oriented pistil and stamens. 3,* R. camtschaticum *Pallas. Bilateral pinkish-purple corolla, three upper petals with red spots. (Drawn from photographs.)*

(Geraniaceae) was described by Ludwig (1881, 1884) but has been almost forgotten by recent students of floral ecology. Both varieties of *E. cicutarium* (with spotted and unspotted petals) are now naturalized in America and provide an interesting opportunity for floroecological observations in a new environment of plant pollinators. On the east coast, the bilateral variety appears more attractive to pollinators and has already reached a dominant position among the restricted patches of forms with radial corollas.

In several other species of the genus *Erodium*, the colored spots on the upper petals are well established and are distinctive characteristics for species identification. Examples are *E. incarnatum* L'Hérit. from South Africa, *E. guttatum* (Desf.) Willd. from North Africa, *E. hymenodes* L'Hérit., *E. macradenum* L'Hérit., *E. carvifolium* Boiss. et Reut., *E. manescavi* Coss. (Fig. 9:4), *E. moschatum* L'Hérit., *E. gussonii* Ten., and *E. ciconium* (L.) L'Hérit.

Thus, the visual effect of bilateral patterns is considerably strengthened by intensively colored spots on the upper or lower petals (Figs. 9, 11, and 12: 5). On intensively colored petals a similar effect is achieved by light spots on dark background (Fig. 9:4). The spots may occur in chromatic colors visible to both insects and man or in the ultraviolet range, perceptible to the insect only. Kugler (1963, 1966) lists several examples of zygomorphic flowers with bilateral patterns produced by ultraviolet radiation among Scrophulariaceae, Labiatae, Leguminosae, and other families.

Successive stages of the development of bilateral symmetry from radiate flowers are demonstrated in the genus *Rhododendron* (Ericaceae) (Fig. 11). Some flowers are slightly bilateral (Fig. 11:2); others are typically bilateral (3/2) with colored spots on the upper petals and with hypaxial arrangement of the stamens and pistils (Fig. 11:3).

Among Leguminosae the successive adaptation from radiate to bilateral and zygomorphic types must have started much earlier and succeeded from genus to genus as suggested in Fig. 13 (*Rosa* → *Cassia* → *Pisum*; Leppik, 1966).

Further examples of bilateral symmetry at the pleomorphic and stereomorphic levels are shown in Figs. 12 and 14.

Figure 12:

1. *Clerodendrum speciosissimum* Van Geert (Verbenaceae) has a red, pentamerous, epiaxial corolla (5/0) with nectar accessible to long-tongued visitors.
2. *Cleome spinosa* Jacq. (Capparidaceae), a white, tetramerous, epiaxial flower (4/0) with long stamens but without corolla tubes. According to Schneck (1895), Knuth et al. (1904, p. 314), and Vogel (1954, p. 144), this curious "spider-flower" is visited mainly by Lepidoptera (hawkmoths), but also by hummingbirds, wasps, bees, and bumblebees.
3. *Cassia corymbosa* Lam., a pentamerous, open flower showing a beginning stage of bilateral symmetry (3/2) with special position of stamens and pistil. Pollination biology is described by Knuth et al. (1904, pp. 361-381).

Fig. 12. Selected samples of bilateral flowers on the pleomorphic (IV) level. 1, Clerodendrum speciosissimum *Van Geert (Verbenaceae): pentamerous epiaxial corolla. 2,* Cleome spinosa *Jacq. (Capparidaceae): tetramerous epiaxial corolla. 3,* Cassia corymbosa *Lam. (Leguminosae): pentamerous epihypaxial corolla. 4,* Malpighia coccigera *L. Two pentamerous, bilateral convolute corollas with unequal, distinctly clawed, fringed or toothed long pedicelled petals. 5,* Lapeyrousia juncea *(Burm. f.) Pourr. (Iridaceae): two haxamerous corollas with bilateral dark spots on lower petals.*

4. *Malpighia coccigera* L. Two pentamerous, bilateral, convolute corollas (1/4) of unequal, distinctly claved, fringed or toothed long pedicelled petals. Known pollinators belong to Apidae (*Apis indica* F.).
5. *Lapeyrousia juncea* (Burm. f.) Pourr. (Iridaceae) from South Africa. A

hexamerous, red flower with radially arranged petals and bilateralism expressed by three dark-brown spots on the lower petals (3/3). The flowers, according to Vogel (1954, p. 99), are visited by lepidopterous pollinators (psychophile).

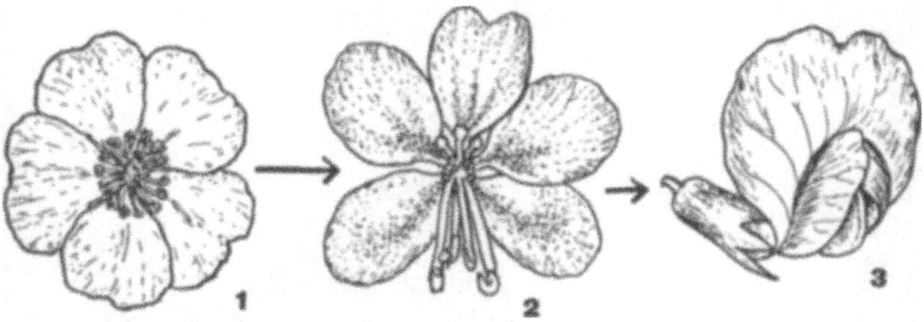

Fig. 13. Floral evolution of the Leguminosae. 1, Pentamerous strictly radially symmetrical flower of an ancestral Leguminosae. 2, Slightly bilateral floral pattern of Cassia. *3, Zygomorphic flower of* Pisum. *(After Leppik, 1966, courtesy of Soc. Bot. Zool. Fennica "Vanamo", Helsinki, Finland.)*

Figure 14:

1. *Rhododendron* hybrid (Ericaceae), a white, open, pentamerous flower with slight bilateralism (3/2), visited mainly by bumblebees.
2. *Achimenes grandiflora* (Schiede) DC. (Gesneriaceae), a pentamerous, bilateral corolla (2/3) with a long tube that is visited mainly by lepidopterous pollinators.
3. *Nemesia strumosa* Benth. (Scrophulariaceae), a pentamerous, bilateral corolla (1/4) visited mainly by hymenopterous pollinators.
4. *Proboscidea (=Martynia) lousianica* (Mill). Thell. (Martyniaceae). A bilabiate corolla (3/2) with nectar-guides, pollinated mainly by bees.

Generally, this new trend in floral evolution is not restricted to definite plant families or orders. It embraces many similarly evolving groups of flowering plants. A clear tendency to develop bilateral forms can be seen in some genera of Ranunculaceae (*Delphinium, Aconitum*), Geraniaceae (*Erodium*), Tropaeolaceae, Rutaceae, Hippocastanaceae, Capparidaceae, Cruciferae, Balsaminaceae, Scrophulariaceae, Saxifragaceae, Violaceae, Begoniaceae, Verbenaceae, Lythraceae, Bignoniaceae, Iridaceae, Gesneriaceae, Caprifoliaceae (*Lonicera*), Lentibulariaceae, Valerianaceae, and Lobeliaceae. Typically, zygomorphic flowers are found in the most advanced families, such as Orchidaceae, Scrophulariaceae, Leguminosae, Labiatae, Bignoniaceae, Balsaminaceae, and others. In addition to these families bilateralism has developed secondarily in the ray florets of the Compositae, Umbelliferae (*Daucus, Actinotus*), and Dipsacaceae (*Dipsacus,*

Fig. 14. Some examples (enlarged photographs) of bilateralism on the stereomorphic (V) level. 1, Rhododendron *hybrid (Ericaceae): a white flower with slight bilateralism. 2,* Achimenes grandiflora *(Schiede) DC. (Gesneriaceae): gamopetalous, 5-lobed, bilateral corolla. 3,* Nemesia strumosa *Benth. (Scrophulariaceae): gamopetalous, 5-lobed, bilateral corolla. 4,* Proboscidea *(=*Martynia*)* lousianica *(Mill.) Thell. (Martyniaceae).*

Scabiosa, Succisa, Morina; see below).

As seen here, bilateralism is a remarkable advantage in the evolution of floral structure because it secures cross fertilization with limited amounts of pollen between different individuals of the same species by specialized pollinators. Sensitive to the distinctive symmetry of the different bilateral flowers, these pollinators visit the different individuals of the same plant species consecutively, thus avoiding the waste of precious pollen.

ADVANTAGE OF BILATERAL STRUCTURES FOR POLLINATION AND FERTILIZATION

Bilateralism is a most drastic change in floral development and has many far-reaching effects on the evolution of flower types. With the reorientation of floral symmetry, a substantial rearrangement follows in almost all flower parts, including the petals, sepals, nectaries, pistils, stamens, and even ovaries. Gradually, complicated zygomorphic floral structures were evolved from the bilateral flowers.

Such profound changes in floral structure must be accompanied by gradual but radical changes and improvements in the process of pollination and fertilization. These changes can be outlined briefly as follows.

A radially symmetrical flower offers its visitors several easily accessible ways to reach open nectaries. Insects can enter the flower from various directions and creep around, often accomplishing only self-pollination or leaving the flower still unfertilized. Such open flowers are visited abundantly by numerous unspecialized guests that quickly exploit the available food resources, making each visit less rewarding for the specialized pollinators. As a matter of fact radial flowers are visited by beetles, flies, wasps, and many other insects with restricted sensory capabilities. Specialized pollinators, such as bees, bumblebees, and butterflies commonly avoid such less rewarding, open flowers, or visit them in the case of scarcity of their specialized food plants.

In a bilateral flower the situation is completely changed. Here the visitor can approach the flower from a certain direction only and must move in a particular way to reach the available source of food. Concealed nectaries in a bilateral flower are frequently provided with special nectar guides that conduct "literate" (specialized) visitors to a particular point inside the flower while preventing "illiterate" stragglers, which waste pollen and nectar, from entering the flowers. Hence, stamens and pistils can be arranged in such a manner that they both will touch the visitor's body on a particular spot and make the cross-pollination more effective.

Thus, the bilateral floral structure, even if quite simple and unsophisticated in the beginning, is a successful evolutionary trend that leads finally to complicated zygomorphic structures at the higher evolutionary levels. This profound morphological change in floral structure has divided flower visitors into two main groups: unspecialized pollinators, which visit the radiate flower types, and specialized visitors that visit the bilateral and zygomorphic flowers.

Bird-flowers frequently tend to become bilateral and zygomorphic. In Fig. 15, a black-chinned hummingbird, *Archilochus alexandri* (♀) Bourier & Mulsant is shown visiting *Mimulus cardinalis* Dougl. ex Benth. (Scrophulariaceae). The advantage of a bilaterally zygomorphic flower is in this case quite apparent. A hummingbird can approach this flower only from one direction and remain hovering while sucking nectar. In this way the bird can gather a maximum

Fig. 15. A hummingbird approaching the bilateral blossom of Mimulus cardinalis *Dougl. (approximately 2/3 natural size). Note the structure of the flower, favorable for bird pollination. (Drawn from photographs.)*

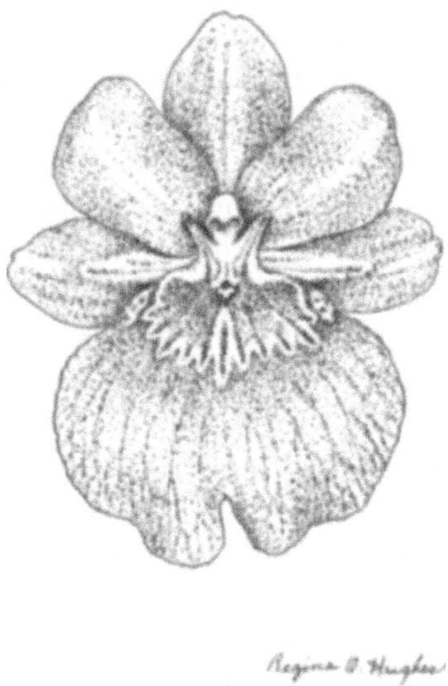

Fig. 16. Miltonia *sp. cult. grown in a greenhouse by H. F. Winters, Beltsville, Maryland. A bilaterally symmetrical purple flower with brown nectar guides and yellow inside marks makes the approch to the nectar easy and convenient (enlarged). (Drawn from a photograph.)*

amount of nectar and pollinate the greatest number of flowers during the shortest possible period of time.

A delicate bilateral symmetry is expressed in a Brazilian *Miltonia* (Orchidaceae) as shown in Fig. 16. The purple flower with brown nectar guides and yellow marks inside the flower is highly impressive with its regular bilateral symmetry. The plant pictured here was grown in a greenhouse by H. F. Winters, Beltsville, Maryland. Pollinators of this remarkable flower are still unknown.

The most complicated combinations of bilateral symmetry and delicate streamlined zygomorphism is expressed in the floral structure of the tropical *Phalaenopsis* (Orchidaceae). Several trimerous patterns are combined with variegated colors and ornamented with harmoniously arranged stripes and spots (Fig. 17). More details on floral structures and pollination mechanisms of orchids are reviewed by Pijl and Dodson (1966) and by Dodson et al. (1969).

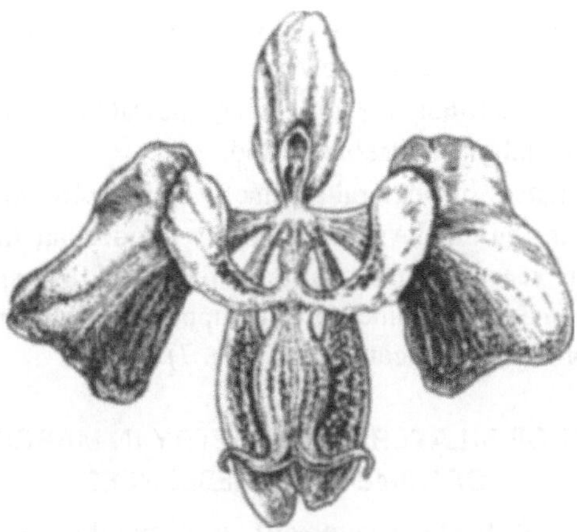

Fig. 17. A tropical Phalaenopsis *flower with several trimerous bilateral patterns harmoniously arranged in a streamlined zygomorphic structure (enlarged; Courtesy of Ginn and Company, Boston.)*

Pollination and floral evolution of Compositae, Ranunculaceae, and Leguminosae are described by Leppik (1960a, 1964, 1966, 1970b). Special patterns produced by radiation of ultraviolet rays on flowers were extensively studied by Daumer (1958) and Kugler (1963, 1966).

APPEARANCE OF BILATERAL FLOWERS IN ANGIOSPERMS

In view of the great importance of bilateralism in the evolution of modern flowers, it is interesting to determine when and where this new trend first appeared in floral history. No trace of bilateral symmetry can be seen in the Cretaceous and pre-Cretaceous flower types of the Bennettitatae and Cycadatae

(Figs. 4, 6, 7). Similarly no bilateralism can be suspected for the early angiosperms. As shown in Figs. 3 and 7 it appears that the deviation from the radial pleomorphic floral pattern began almost simultaneously in many parallel-evolving plant groups on the pleomorphic and stereomorphic levels. This is an important turning point in floral evolution, which must have coincided with a corresponding change in the sensory development of pollinating insects. One can speculate that some leading groups of anthophilous insects, among them Hymenoptera and Lepidoptera, had acquired the ability to distinguish simple bilateral forms in flowers. Because this ability was rewarded by a higher quality of food in bilateral flowers, it must have stimulated further selective activity in these pollinators. It is, therefore, highly important to establish in geological terms when this far-reaching change appeared in the sensory development of pollinating insects. Helpfully, the behavioral change of anthophilous insects is reflected in the historical sequence of floral evolution, which in turn can be connected with fossil records and geological time tables (Figs. 2, 3).

Fossilized imprints of flowers in clays, muds, fine sands, and Baltic amber were studied from this point of view and reported in several previous papers (Leppik, 1957, 1960b, 1963a,b, 1971). They indicate that radial symmetry was dominant until the end of Cretaceous period.

The first appearances of bilateral symmetry in the pleomorphic flower types might have been in the middle of the Tertiary period, but there are no certain fossil records. In recent floras, both bilateral and zygomorphic types are dominant in the modern families of the upper evolutionary levels of angiosperms, and the process is still continuing (Fig. 7).

DEVELOPMENT OF BILATERAL SYMMETRY IN MARGINAL FLOWERS OF SOME INFLORESCENCES

Frequently, perfect bilateral symmetry is distinctly expressed in marginal blossoms of compact inflorescences of some Dipsacaceae, Umbelliferae, Verbe-

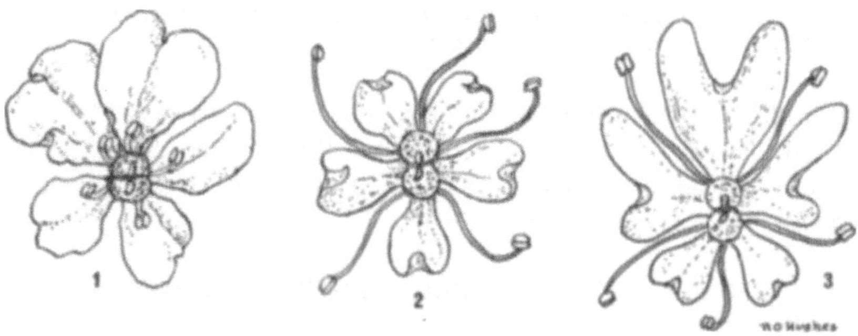

Fig. 18. Bilateral symmetry of the marginal flowers of some Umbelliferae. 1, Daucus carota *L., a bilateral marginal flower. 2, 3.* Heracleum spondylium *L., a central (2) and a marginal (3) flower (2, 3 after Kugler, 1955, p. 168; courtesy of G. Fischer Verlag, Stuttgart).*

naceae, Caprifoliaceae, Compositae, and other families (Figs. 18 to 22). In these blossoms the bilateralism cannot guide pollinators to the nectar, or favor pollination or fertilization. Marginal flowers in such inflorescences are commonly nectarless and sterile, and therefore must serve some other purpose. The marginal bilateralism heightens radial symmetry and conspicuousness of the

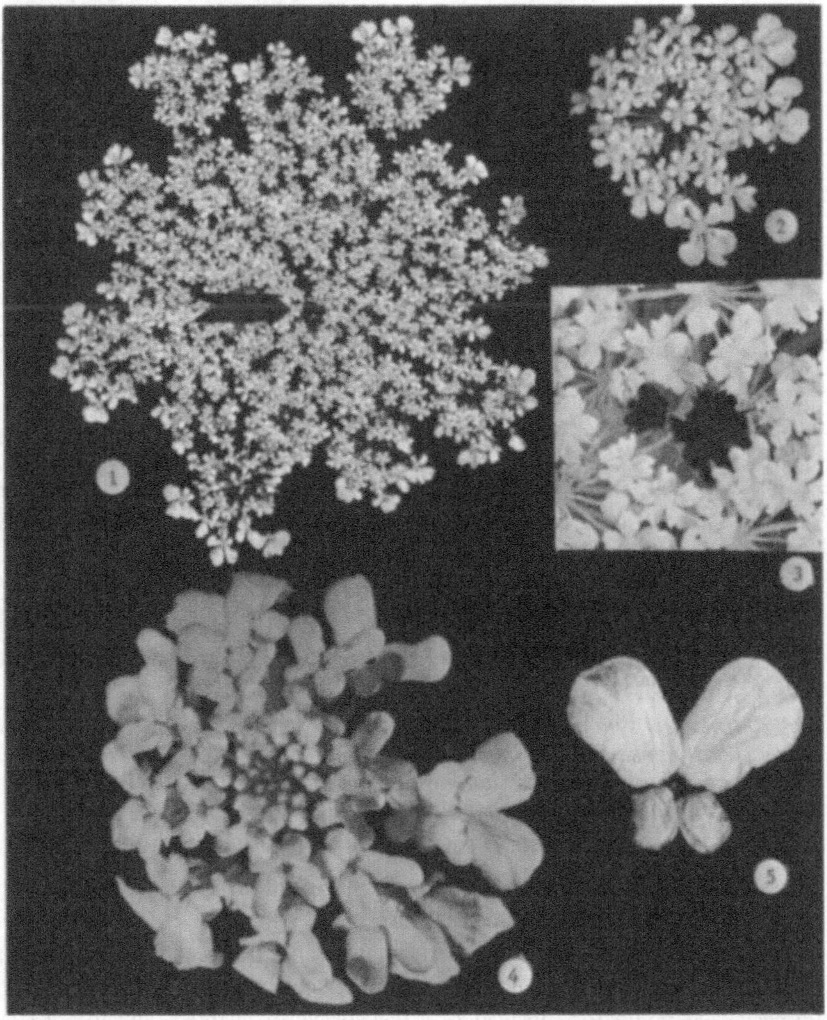

Fig. 19. Bilateral symmetry in marginal blossoms of Umbelliferae and Cruciferae. 1, Daucus carota *L., umbel about mature size. 2, Bilateral marginal blossoms enlarged. 3, Central part of an enlarged umbel with dark terminal spot. 4,* Iberis amara *L., an umbelloid inflorescence, somewhat reduced. 5, Bilateral marginal flower, enlarged.*

Fig. 20. Bilateral symmetry in the Verbenaceae. 1, Lantana camara *L., inflorescence, slightly enlarged. 2, The same two bilateral marginal flowers, enlarged. 3,* Verbena canadensis *(L.) Britt., an inflorescence about natural size. 4, A bilateral flower enlarged. (All from original photographs.)*

corresponding inflorescence as a whole. Instead of luring specialized pollinators, these inflorescences attract numerous unspecified visitors for pollination of their countless florets. The large numbers of florets in an inflorescence need individual pollination and fertilization, and therefore must attract great masses of unspecialized pollinators, instead of a few specialists adapted only to a certain pollination mechanism.

Daucus carota L. (Umbelliferae, Figs. 18:1; 19:1-3) has marginal blossoms much bigger than others, with its outer petals particularly enlarged. This increases the conspicuousness of white umbels and attracts numerous visitors. Still another remarkable feature in many (but not in all) umbels of *Daucus*

Fig. 21. Pseudoflowers of the Compositae formed by bilateral ray florets on the margin and pentamerous tubulate florets in the central part of the inflorescences. 1, Dimorphotheca sinuata *DC., white. capitulum, enlarged 2*X. *2,* Gazania rigens *(L.) Gaerin., yellow capitulum, enlarged 2*X. *3,* Cichorium intibus *L., blue capitulum, enlarged 2/1. 4,* Encelia farinosa *A. Gray., yellow capitulum, enlarged 3*X.

carota is the occurrence of one or more enlarged central flowers of a dark-purple color, indicated with a black arrow in Fig. 19:1, 3. This phenomenon cannot yet be explained. One can speculate that the dark point in the center of a larger umbel helps pollinators orient themselves during their visits. They commonly start from the margin, work to the dark central point, turn around, and fly to the next umbel. Such somewhat oriented wanderings of pollinators on floral surface enable them to better exploit certain amount of flowers during a visit and to accomplish more complete pollinations of a certain sector of the inflorescence.

Occasionally flowers of the Cruciferae are aggregated in umbellate inflorescences with bilateral marginal blossoms (2/2), such as the umbelloid corymb of *Iberis amara* L. (Fig. 19:4, 5). Similar bilateralism (3/2) appears in some Verbenaceae, such as *Lantana camara* L. (Fig. 20:1, 2) and *Verbena canadensis* (L.) Britt. (Fig. 20: 3, 4).

In the Compositae, bilateral ray florets imitate the shape of solitary flowers, producing distinctive patterns of pseudoflowers. Some selected examples of such pseudoflowers are illustrated in Fig. 21:

1. *Dimorphotheca sinuata* DC. (=*D. aurantiaca* Hort.)
2. *Gazania rigens* (L.) Gaertn.
3. *Cichorium intibus* L.
4. *Encelia farinosa* A. Gray. A perfect imitation of a flower head of the Compositae is the pseudoflower of *Actinotus helianthi* Labill., of the carrot family (Umbelliferae, Fig. 5:2).

SOME FURTHER TRENDS IN THE EVOLUTION OF BILATERAL SYMMETRY IN FLOWERS

Still further striking floral sturctures are produced from various combinations of bilateral symmetry (Fig. 22).

Lagerstroemia indica L. (Lythraceae) (Fig. 22:1), with its many epiaxial petals, resembles an inflorescence rather than a solitary flower. It is a heteranthic-pollen flower with many conspicuously yellow feeding stamens on short filaments in the center of the flower. A few dark-colored fertile stamens and a pistil, both with long filaments, are extended toward the pink petals. There is no nectar in these flowers, and visitors (mainly bees) collect sterile pollen for their own consumption, and transfer the fertile pollen from flower to flower for fertilization. The pollination ecology of *Lagerstroemia* is described by Faegri and Pijl (1966:53, 161), calling the differentiation within the androecium "heteranthy."

Crossandra infundibuliformis (L.) Nees (*C. undulifolia* Salisb.; Acanthaceae; Fig. 22:2), with its four epiaxial (5/0) gamopetalous flowers, produces a striking tetramerous pattern resembling a solitary flower. The long corolla tube of each flower is accessible to butterflies and other insects with a sufficiently long proboscis (Vogel, 1954:261; Leppik, 1961:19).

Gloriosa rothschildiana O'Brien (Liliaceae; Fig. 22:3), with a striking epiaxial (6/0) red and yellow corolla, and hypoaxial stamens (0/6) and pistils, is a typical upside-down flower. It is pollinated by butterflies (Vogel, 1954:83; Werth, 1956:98).

Lonicera japonica Thunb. (Caprifoliaceae; Fig. 22:4) strengthens the effect of its bilateral flowers (4/1) by arranging them in pairs. *L. periclymenum* L. has its bilateral flowers in parallel rows, 6 to 10 on the same twig.

Fig. 22. Some selected types of bilateralism in various solitary flowers. 1, Lagerstroemia indica *L. (Lythraceae). A flower with long pedicells and epiaxial petals resembles inflorescence, rather than solitary flower. 2,* Crossandra infundibuliformis *(L.) Nees (Acanthaceae) with four epiaxial gamopetalous blossoms imitates the tetramerous pattern of a solitary pseudoflower. 3,* Gloriosa rothschildiana *O'Brien (Liliaceae) with a striking epiaxial corolla is a typical upside-down flower. 4,* Lonicera japonica *Thunb. (Caprifoliaceae) strengthens the effect of bilateral symmetry displaying its bilateral epihypaxial flowers in pairs. (All from original photographs.)*

ABILITY OF ANTHOPHILOUS INSECTS TO DISTINGUISH BILATERAL SYMMETRY IN FLOWERS

Despite the scarcity of experimental studies concerning the sensory ability of insects to distinguish bilateral symmetry, some conclusive evidence and observational data are available. Training experiments with honey bees indicate that they can distinguish bilateral symmetry from radial patterns in natural and artificial flowers (Leppik, 1953, 1958). Further behavior tests with tropical insects were conducted by Leppik (1954, 1955a,b, 1956a, 1958), using the Petri dish method and marking insects with white or colored spots (Fig. 23) (Leppik, 1958). Small insects, such as stingless wild bees (*Trigona, Melipona*, and others) and small wasps were marked with specially colored powders which remained long on the insect body. The results of these experiments demonstrated the ability of specialized insects, and inability of unspecialized insects, to distinguish bilateral symmetry from radial flower patterns.

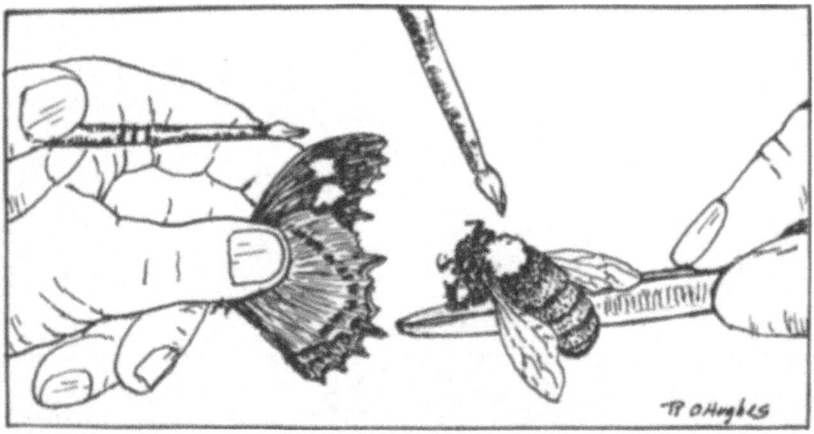

Fig. 23. Marking of butterflies and honey bees with colored spots. (Drawn from a photograph.)

There are numerous direct observations and statistical data on flower visitors. In earlier publications, all insects, occasional visitors and regular pollinators, were listed under definite plant species. In such listings it is not easy to separate occasional plant visitors from regular pollinators. It is also difficult for one to establish the exact relationship between flower types and their regular pollinators. Knuth (1906-1909) and Knuth et al. (1904) summarized contemporary floroecological literature and arranged all observed flower visitors in the taxonomic order of plants. Such an arrangement does not correspond to the sensory evolution and capabilities of insects. In recent studies more emphasis is placed on the relationship between floral structure, pollination mechanism, and sensory development of a particular pollinator (Vogel, 1954; Free, 1970).

In spite of these and other shortcomings in older literature, valuable conclusions can be deduced from the statistical material, assuming that the observations are made in sufficiently large amounts, over a long period of time. This is demonstrated in several earlier reports on the evolutionary correlation between flower types, and the sensory development of their pollinators (Leppik, 1963b, p. 108, Fig. 9; 1964, p. 52, Fig. 15).

In Fig. 24 a sphinx moth, *Macroglossum stellatarum* (L.), is shown visiting a bilateral tubiflorous blossom of *Verbena canadensis* (L.) Britt. Figure 24:4 shows a bumblebee, *Bombus hortorum* L., opening a blossom of *Antirrhinum majus* L., which is closed to other insects that are not specialized.

Fig. 24. Pollination of bilateral and zygomorphic flowers by highly specialized insects. 1, Verbena canadensis *(L.) Britt. inflorescence. 2,* Macroglossum stellatarum *(L.) observes the characteristics of* Verbena canadensis *from a short distance before stretching its proboscis into the flower tube. 3, A single flower enlarged (courtesy of the American Naturalist). 4,* Antirrhinum majus *L., zygomorphic flowers. Left: a flower closed to pollination; in the center a bumblebee (*Bombus hortorum *L.) opening the flower; right: an opened flower. (All from photographs; 1, 2 from Leppik, 1953, p. 231; 4 from Leppik, 1956b, p. 451.)*

SOME ESTHETIC ASPECTS OF FLORAL SYMMETRY

Since ancient times man has admired flowers for their mysterious beauty. Early man, living in a natural environment and being surrounded with showy wild flowers, gradually built up his own esthetic senses and concept of beauty. As a matter of fact, all present-day flower types were already perfected long before man came into existence. Flowers have evolved since about 200 million years, whereas man as an independent species has existed only about 1 to 2 million years (Leppik, 1963b, p. 112).

With the progression of civilization man began to grow flowers in his close surroundings. He planted them in gardens, flower beds, parks, greenhouses, window boxes, and brought them into his house. Man found standard elements of natural beauty in various flower types. They impressed him with the best combinations of form, symmetry, harmony, brilliancy of colors, pleasant odors, all these properties united into well-proportioned entities. Of all natural objects, none is so well adapted for the stimulation and cultivation of our esthetic senses as are flowers. The study and utilization of these living sources of beauty help us to maintain close contact and intimate communion with nature. The further development of human civilization and culture also will be equally dependent upon the utilization of these natural sources of beauty. Therefore, any new knowledge or better understanding that may help us to evaluate better and use these natural resources more properly is of great importance to the further development of civilization. We must, therefore, guard and maintain these very important natural sources for our own needs as well as for succeeding generations.

With the above described scientific knowledge at hand, we can approach now the mystery of floral beauty from a different but more objective angle. In various flower types and their evolution we have definite standard units of natural beauty, which have evolved to the highest degree of perfection and completeness during many millions of years of natural selection. There are still many types and color combinations in nature unused for human purposes.

Particularly the forms and motifs of bilateral symmetry and streamlined zygomorphic structures in flowers are seldom understood and used. The study of barely visible fine structures in flower types, such as orchids, can stimulate artists' imagination in their search for new ideas and modes of expression of beauty. Present scientific knowledge and a better understanding of these almost inexhaustible sources of natural beauty can challenge the imagination of biologists and artists, enabling them to find new concepts and ideas for their work.

MYSTERY OF FRAGRANT BLOSSOMS

There is still much mystery as to the esthetic appeal of flowers to man. Neither the antique lyricists, the classic poets, nor modern scientists, can explain why man is so dependent on flowers that do not supply him with food or anything else to support his physical well-being. Human inclination to flowers is obviously a spiritual one.

In the past the common belief was that flowers were created by God to be enjoyed by man, the king of nature. This explanation appeared so simple and acceptable to early people, that it was accepted without questioning. Then Kölreuter (1761) and Sprengel (1793) demonstrated that no flower blooms for man, but displays its colors only to attract insect and bird visitors for pollination and fertilization. This new idea was so conflicting with contemporary thoughts that the distribution of Sprengel's book long remained prohibited. It took almost a century until further study was possible and more observational and experimental evidence was gathered to support the idea of insect pollination.

However, the verification of insect pollination did not explain why flowers are so beautiful. At that time no one suspected that insects could recognize the beauty in flowers. It was argued that insects do not "enjoy" flowers for their colors and odors but for the food that they secure. Whether or not insects have an inherited sense of beauty is not easy to prove. One can assume that plant pollinators may possess a "pan-esthetic" sense or senses that are similar to but not identical with the human concept of beauty. The latter is acquired, varies from individual to individual, and is modified by experience. The pan-esthetic sense of insects is very definitely inherited, if it exists, and must be highly constant for all individuals of the same species.

Still more puzzled were plant physiologists. They considered the "fundamental" task of green plants in nature to be photosynthesis with the characteristic green color being due to chlorophyll. How then could multicolored flowers develop on green plants in a progressive sequence towards more-and-more beautiful structures? It took another century to solve these problems of coevolution of sensory abilities and selective activities of plant pollinators with the progressive development of flower types. During long evolutionary periods, pollinators have selected from available flowers the best looking types, ignoring "abominable" blossoms as less promising sources of food. Such selective activity of pollinators, guided by their inherited instincts, in the long run, must necessarily have affected the evolution of flower types. The great variety of present-day flower types, supposedly, is the outcome of this long selection, conducted by various insect groups with various plant materials in different environments. Existing flower types, therefore, represent a high degree of perfection which makes them attractive to pollinators, and beautiful to man.

In view of this development pattern, one can assume that the tropheclectic and pan-esthetic senses of insects and other pollinators, which govern their food searching activities, developed in correlation with the evolution of flower types. Later on, showy flowers became an important stimulus that inspired the development of concept of beauty in man.

BIOCHEMICAL EVOLUTION OF FLOWERING PLANTS

Biochemical evolution accompanied by many metabolic rearrangements in plants must have preceded the actual development of particular flower types. This is involved in the adaptation of plants to climatic, environmental, and various biotic factors during their vegetative growth before pollinators started to visit flowers in bloom. The selective activity of pollinators, directed towards higher content of nectar, more intense petal colors, finer odors, and more harmonious flower shapes and symmetry, caused, and is still causing, plants to produce better looking flowers with improved biochemical composition.

Stereomorphic and zygomorphic flower types represent the highest evolutionary level of flowering plants. Ordinarily, these types are associated with higher content of carbohydrates, proteins, variegated pigments, fine volatile oils, and vitamins. Thus, the biochemical evolution of plants is paralleled by their hologenetic development and with more advanced biochemical composition corresponding to the higher evolutionary levels, and vice versa.

Thus, nature has achieved the most complicated biochemical composition in carefully selected higher plants, which in their turn provide superb food for insects, higher animals, and man. It is no mere coincidence that our main highly nutritive food plants, such as cereal grains, forage grasses, and legume plants, belong to the highest evolutionary levels in the plant kingdom.

Although the grasses and grain crops are at present pollinated by wind, their floral structure reveals that the Gramineae must have reached the pleomorphic evolutionary level before they returned from entomophily to anemophily. Their ability to grow close together in monocultures, permitting high yields, and their tendency to form hybrids, enabled man to develop grain crops out of grasses as his main source of carbohydrates. But, in addition to carbohydrates, many other substances such as proteins, fats, minerals, and vitamins are necessary for the survival of insects, higher animals, and man. These are mostly produced by various plant groups of higher stereomorphic and zygomorphic evolutionary levels, such as legume plants, grasses, composites, some spurges, and so on.

Thus, the historic sequence of floral evolution outlined in Figs. 2 and 3 also reflects the biochemical evolution of entomophilous plants as expressed in their nutritive value for insects, higher animals, and man. In addition to their superior chemical components, plants of the higher evolutionary levels also possess the great advantage of having their evolution directed by agents exceptionally well adapted for this purpose. As a result, the zoophilous plants evolve rather quickly and are able to provide man with his main foods, drugs, and many raw materials for industrial use.

SUMMARY

Bilateral symmetry is a late development in floral evolution and is restricted to upper evolutionary levels. Its beginning is marked by a slight deviation from the regular radial shape, which appeared on pleomorphic and stereomorphic levels about the middle part of the Tertiary period. At that time specialized pollinators, mainly Hymenoptera and Lepidoptera and some anthophilous flies, presumably developed their sensory ability to distinguish bilateral symmetry in flowers.

With the appearance of bilateralism in flowers a new and important road was opened to the further evolution of flowering plants. Bilateral blossoms became uneasily accessible to the masses of unspecialized flower stragglers, but attracted the more specialized visitors for pollination and fertilization. The selective activity of specialized pollinators stimulated the evolution of higher angiosperms to a still more rapid progress. There is substantial evidence to show that the bilateral trend in the evolution of flower types was correlated to genetic, biochemical, and metabolic changes in evolving plants. These progressive changes are expressed in the historical sequence of floral evolution and are also reflected in the sensory development of advanced pollinators, including Hymenoptera, Lepidoptera, and some other specialized pollinators.

Specialized bird-pollinated flowers are frequently bilateral in their structure. Bilateral arrangement of flower parts favors the sucking of nectar in hovering position and secures proper fertilization in ornithophilous plants.

Acknowledgments

The author wishes to express his appreciation to Dr. G. H. M. Lawrence, Director of the Hunt Botanical Library of the Carnegie-Mellon University in Pittsburgh, Pennsylvania, for his many comments and suggestions. Dr. Hugo Salasoo of Lidcombe sent me several noteworthy flower samples from New South Wales, Australia.

REFERENCES

Arber, A. 1937. The interrelation of the flower: a study of some aspects of morphological thought. Biol. Rev., 12:157-184.

Axelrod, D. I. 1952. A theory of angiosperm evolution. Evolution, 6:29-60.

Baker, H. G., and P. D. Hurd, Jr. 1968. Intrafloral ecology. Ann. Rev. Entomol., 13:385-414.

Barnard, C. 1961. The interpretation of the angiosperm flower. Austral. J. Sci., 24(2):64-72.

Cronquist, A. 1968. The Evolution and Classification of Flowering Plants. Boston, Houghton Mifflin.

Daumer, D. 1958. Blumenfarben, wie sie die Bienen sehen. Z. Vergl. Physiol, 41:49-110.

Deodikar, G. B. 1961. Some aspects of Bee-Botany. Prof. S. P. Agharkar Commemoration, Vol. I: 61-78. Maharashtra Assoc. Cultiv. Science, Poona, India.

Dodson, C. H., R. L. Dressler, H. G. Hills, R. M. Adams, and A. H. Williams. 1969. Biologically active compounds in orchid fragrances. Science, 164:1243-1249.

Ehrlich, P. R. 1970. Coevolution and the biology of communities. Proc. 29 Ann. Biol. Colloq., 1968, edited by K. L. Chambers, Oregon State Univ. Press, Corvallis.

Faegri, K., and L. Van Der Pijl. 1966. The Principles of Pollination Ecology. Oxford and New York, Pergamon Press.
Free, J. B. 1970. Insect Pollination of Crops. London, Academic Press.
Grant, V., and K. A. Grant. 1965. Flower Pollination in the *Phlox* Family. New York, Columbia University Press.
Jaeger, P. 1961. The Wonderful Life of Flowers. London, Harrap.
Kölreuter, J. G. 1761. Vorläufige Nachricht von einigen das Geschlecht der Pflanzen betreffenden Versuchen un Beobachtungen. Leipzig. Fortsetzungen 1, 2 und 3, Leipzig, 1763, 1764, 1766. Neu herausgegeben von W. Pfeffer in Oswalds' Klassikern der exakten Wiss., 41, Leipzig, 1893.
Knuth, P. 1906-1909. Handbook of Flower Pollination. Translated by J. R. Ainsworth Davis. Vols. 1-3. Oxford, Clarendon Press.
——O. Appel and E. Loew. 1904. Handbüch der Blütenbiölogie. III.Band: Die bisher in aussereuropäischen Gebieten gemachten blütenbiologischen Beobachtungen. Leipzig, Verlag Engelmann.
Kugler, H. 1955. Einführung in die Blütenökologie. Stuttgart. G. Fischer, 2nd ed., 1970.
——1963. UV-Musterungen auf Blüten und ihr Zustandekommen. Planta, 59:296-329.
——1966. UV-Male auf Blüten. Ber. Deutsch. Bot. Ges., 79(2):57-70.
Kullenberg, B. 1961. Studies in *Ophrys* pollination. Zool. Bidr. Uppsala, 34:1-340.
Leppik, E. E. 1953. The ability of insects to distinguish number. Amer. Natural., 835:229-236.
——1954. La facultad de las mariposas para distinguir numeros figorados. Commun. Inst. Trop. Invest. Cient. Univ. El Salvador, Año III, No. 4:151-158. San Salvador.
——1955a. *Dichromena ciliata*, a noteworthy entomophilous plant among Cyperaceae. Amer. J. Bot., 42:455-458.
——1955b. La facultad de las mariposas para distinguir numeros figorados. II. Observaciones con mariposas marcados. Commun. Inst. Trop. Invest. Cient. Univ. El Salvador, Año IV, Nos. 1-2:55-59. San Salvador.
——1956a. Experimentos de amaetramiento abejas silvestres sin aguijon. Commun. Inst. Trop. Cient. Univ. El Salvador, Año V, Nos. 2-3:89-93. San Salvador.
——1956b. The form and function of numeral patterns in flowers. Amer. J. Bot., 43:445-455.
——1957. Evolutionary relationship between entomophilous plants and anthophilous insects. Evolution, 11:466-481.
——1958. A "literacy" test performed with anthophilous insects. Ann. Soc. Litt. Estonicae in Amer. (New York), 2:43-53.
——1960a. Evolutionary differentiation of the flower head of the Compositae. Arch. Soc. Zool. Bot. Fennicae 'Vanano,' 14:466-481.
——1960b. Early evolution of flower types. Lloydia, 23:72-92.
——1961. Phyllotaxis, anthotaxis, semataxis. Acta Biotheor. (Leiden), 14:1-28.
——1963a. Reconstruction of a Cretaceous *Magnolia* flower. Adv. Front. of Plant Sciences (New Delhi), 4:79-94.
——1963b. Fossil evidence for floral evolution. Lloydia, 26:91-115.
——1963c. Evolutionary correlation between plants, insects, and soils. Ann. Soc. Litt. Estonicae in Amer. (New York), 3:28-50.
——1964. Floral evolution in the Ranunculaceae. Iowa State J. Sci., 39(1):1-101.
——1966. Floral evolution and pollination of the Leguminosae. Ann. Bot. Fennici, 3:299-308.
——1968a. Morphogenetic classification of flower types. Phytomorphology (New Delhi), 18:451-466.
——1968b. Directional trend of floral evolution. Acta Biotheor. (Leiden), 18:87-102.2.
——1970a. Evolutionary correlation between plants, animals, and their environments (a review). Adv. Front. of Plant Sciences (New Delhi), 21. (In press)
——1970b. Evolutionary differentiation of the flower head of the Compositae II. Ann. Bot. Fennici, 7:325-351.
——1971. Paleontological evidence on the morphogenic development of flower types. Phytomorphology (New Delhi). (In press)

Ludwig, F. 1881 Die Anpassungen der Gattung *Erodium* an Insektenbestäubung. Kosmos (Leipzig), 4(8):357-362.

____1884. Die verschiedenen Formen des Saftmales bei *Erodium cicutarium* L'Hérit. mit Rücksicht auf die übrigen entomophilen *Erodium*-Species. Botan. Centralbl., 19:118-125.

Macior, L. W. 1964. An experimental study of the floral ecology of *Dodecatheon meadia.* Amer. J. Bot., 51:96-108.

____1965. Insect adaptation and behavior in *Asclepias* pollination. Bull. Torrey Bot. Club, 92:114-126.

____1966. Foraging behavior of *Bombus* (Hymenoptera: Apidae) in relation to *Aquilegia* pollination. Amer. J. Bot., 53:302-309.

____1967. Pollen foraging behavior of *Bombus* in relation to pollination of nototribic flowers. Amer. J. Bot., 54:359-364.

Meeuse, I. D. 1961. The Story of Pollination. New York, Ronald Press.

Melchior, H. 1964. A Englers Syllabus der Pflanzenfamilien mit besonderer Berucksichtigung der Nutzpflanzen nebst einer Übersicht über die Florenreiche und Florengebiete der Erde. H. Band, Angiospermen. Berlin, Gebrüder Borntraeger.

Müller, H. 1883. The Fertilisation of Flowers. Translated by D'Arcy W. Thompson. London, Macmillan & Co.

Percival, M. S. 1965. Floral Biology. London, Pergamon Press.

Pijl, L. Van Der. 1960-1961. Ecological aspects of flower evolution. I-II. Evolution, 14:403-416; 15:44-59.

____1966. Pollination Mechanisms in Orchids. London, Pergamon Press.

____and C. H. Dodson. 1966. Orchid flowers. Their pollination and evolution. Miami, Fla., University of Miami Press.

Schneck, J. 1895. Observations on the spider flower. Bot. Gaz., 20:168-170.

Sprengel, C. K. 1793. Das entdeckte Geheimnhis der Natur in Bau und in der Befruchtung der Blumen. Berlin. (Second edition by P. Knuth, Leipzig, 1894.)

Takhtajan, A. 1969. Flowering Plants. Origin and Dispersal. [Translated by C. Jeffrey.] Washington, D. C., Smithsonian Institution Press.

Vogel, S. 1954. Blütenbiologische Typen als Elemente der Sippengliederung, dargestellt anhand der Flora Südafrikas. Jena, G. Fischer, Verlag.

Werth, E. 1956. Bau und Leben der Blumen. Stuttgart, Verlag Enke.

Wettstein, R. 1935. Handbuch der Systematischen Botanik. Leipzig, Franz Deuticke.

Zimmermann, W. 1966. Kritische Beiträge zu einigen biologischen Problemen. VII. Die Hologenie. Z. Pflanzenphysiol., 54:125-144.

4

The Ecological Genetics of *Drosophila*

P. A. PARSONS AND J. A. MCKENZIE
Department of Genetics and Human Variation
La Trobe University, Victoria, Australia

INTRODUCTION

"It is a truism that all organisms can live under the conditions under which they do live and that they could not live under other sets of conditions that exist. To this degree, at least, and without any teleological implications, adaptation is universal. One group is better adapted to a certain environment than another if it is more successful in coping with the given conditions. From the point of evolution, the best criterion of such success is that the better-adapted group increases in numbers relative to the worse-adapted." This statement by G. G. Simpson in "Tempo and Mode in Evolution" originally published in 1944 sets the theme for the discussion to follow. It is becoming increasingly clear that to understand the evolutionary processes leading to the gene pools of species as we know them today, it is essential to consider the complex of interacting systems which comprise the ecosystem, and hence define the niche the species will occupy within that ecosystem. Inevitably, the environment will have its effect on the gene pool of the species, and hence it is necessary to study the heritable constitution of the species and how it adapts to a complex of environmental factors. At the present moment this is of peculiar significance, because of the enormous changes in the environment being created by man himself. The object of this review is to discuss the ecological genetics of the genus *Drosophila.* Ecological data are restricted in this genus, but work is being increasingly carried out in some species of *Drosophila* in relation to environment in the wild and in the laboratory. On the other hand, the genetics of some species of *Drosophila* is extremely well known, which is an important prerequisite for any study of the effect of environment on the genotype.

Most work on the genotypes found under different ecological conditions has been done on species selected for their relative ease of handling in the field, and possessing a readily and directly identifiable phenotype. Neither of these conditions so far apply to *Drosophila*, which is a reason for the lag in the study of the ecological genetics of *Drosophila*. Species such as the butterflies *Panaxia dominula* and *Maniola jurtina*, and the land snail *Cepaea nemoralis* have been studied extensively in the field (Ford, 1964), but compared with *Drosophila* in general and *D. melanogaster* in particular, the genetics of these species is not well understood. It seems important to try and reconcile the ecological difficulties of *Drosophila*, with the enormous advantages that many species of the genus have in laboratory manipulation and maintenance, since it is clear that rapid changes are possible in natural environments (Ford, 1964). The further advantage of many *Drosophila* species is that generations are short, so allowing the effects of changing environments to be observed over a short period of time. Even the question of a readily identifiable genotype in *Drosophila* has in recent years been made much easier by the large number of enzyme and protein variants now being found in natural populations. These variants have been found too recently for many associated fitness parameters to have been detected, but the ease with which certain environmental stresses can be applied in the laboratory may elucidate correlations with fitness.

The review will necessarily be restricted to a few species. The world fauna contains 1000 to 2000 identified species in the genus *Drosophila*. Up to 500 identified species occur in this and closely related genera (drosophiloids) in the Hawaiian Islands, and there are probably 200 or more species belonging to the genus *Scaptomyza* and related genera (scaptomyzoids), which also belong to the family Drosophilidae (Carson, Hardy, et al., 1970). Therefore, the Hawaiian Islands represent an adaptive radiation of this family in an analogous way to Darwin's finches on the Galapagos Islands (see Dobzhansky, 1968), probably originating from the chance arrival of one or two species of the Drosophilidae before other flies arrived. Clearly, it is to the Hawaiian species where we must look for increasing amounts of work in the future, and an excellent start has been made (see review by Carson, Hardy, et al., 1970). Of the Hawaiian species, Stone et al. (1960) estimate that about 95% are rare, and are geographically very localized and ecologically very specialized. In some of these cases, the breeding and feeding sites are relatively well known, so that interrelationships between their genetic and ecological structure can be established (Carson, Hardy, et al., 1970). This tends to be more difficult for the more widespread and cosmopolitan species, which occupy more diverse habitats.

There are two main approaches to the ecological genetics of *Drosophila*: (1) the study of flies caught in natural environments in order to correlate species, and genotypes within species, with the environment; and (2) the study of populations in artificial laboratory environments in order to try and assess what may be factors of ecological importance in nature. As will be seen, work on these two aspects has proceeded rather independently of each other and the second approach has been considered in greater depth than the first, because of the ease of breeding large numbers of *Drosophila* in the laboratory. It is, however, essential to link both approaches, and a start has been made on this. Throughout the review, examples of both approaches will be discussed, but unfortunately much of the work consists of the results of laboratory experiments being then referred back to their likely significance in wild populations.

DISTRIBUTION DATA BETWEEN SPECIES

There are a number of publications showing variations in population numbers and relative proportions of different species of *Drosophila* collected in the wild. In some cases the variations have only been marginally correlated with environmental changes. Dobzhansky and Pavan (1950) described local and seasonal variations in various *Drosophila* species in 35 localities in 17 bioclimatic regions of Brazil, and found some species to form nuclei with high and low population densities, and some were distributed relatively more uniformly. Food preference differences and seasonal differences were also found. Their work suggests many problems of an ecological-genetic nature, most of which are soluble only with detailed work. In a further experiment, several species of yeast were isolated from the crops of the sympatric species *D. persimilis* and *D. azteca*, occurring near Mather, California, and suspensions were then used as baits to attract *Drosophila* species. The different yeasts showed differential attractiveness to different species of *Drosophila*. This would be expected, since natural selection probably lessens the competition among sympatric species so that they prefer different foods normally found in their environment (da Cunha, Dobzhansky, and Sokoloff, 1951). Evidence for cyclical large population expansions of a species in a uniformly tropical wet climate (Moen Island, Truk, Eastern Caroline Islands) in *D. ananassae*, *D. hypocausta*, and *D. anuda* are believed to be explainable by a dependence upon the fruiting of certain types of fruit trees, e.g., breadfruit. *D. melanogaster*, never an abundant species on Moen Island, on the other hand did not undergo any expansions during the period of collection, and so presumably was independent of the fruiting of any major plant species in this environment (Pipkin, 1953).

Considerable work has been done on the Hawaiian species of the Drosophilidae and the accumulated data show that 32 endemic and 6 introduced genera of plants belong to 30 families serve as ovipositional sites, and of these the genera *Cheirodendron* (Araliaceae) and *Clermontia* (Lobeliaceae) serve as ovipositional and larval food sites for more than half the species that have been reared from all sources (see Carson, Hardy, et al., 1970). The natural food subtrates of the adults are less well known than are those of the larval stages, but adults of various species have been observed engaging in feeding behavior in the field, e.g., on the fermenting leaves and bark of *Cheirodendron*, the leaves, flowers, fruit, and bark of lobeliads, the leaves of *Pisonia*, various fungi, and the exuding sap from the recently cut stumps of large fern-trees (*Cibotium*). Adults do not randomly distribute themselves throughout the forest, but tend to accumulate in relatively small habitats, which suit their ecological requirements, the main controlling factors probably being wind intensity, humidity, temperature, light intensity, food sources, and acceptable ovipositional sites. Thus moderate wind currents and light intensities, humidities below 90%, and temperatures above 21.1°C seem generally to be avoided.

Detailed work on habitat has been done on the South American species *D. flavopilosa*. Females lay their eggs only in the flowers of *Cestrum parqui* (Solanaceae), and their developmental cycle takes place within these flowers (Brncic, 1966). By counting the number of flowers of *C. parqui* containing preadult forms of the fly in different geographic regions, it is possible to estimate the abundance of the species in relation to climate, available food, and other insects that inhabit the same flowers as competitors, predators, or parasites. The main factors regulating population density seem to be the climate, a species of Thysanoptera, *Thrips tabaci* (as a competitor), and two species of parasitic Hymenoptera. Brncic (1970) has reported two other species closely associated with single plant species, namely *D. allei* associated with *Datura arbustiva* (Solanaceae), and *D. appendiculata* associated with *Chusquea* (Bambuseae). It is not possible to breed these three species in the laboratory with the usual methods employed for cosmopolitan species, and this feature seems to be a product of their ecological restriction. Three or four other species of *Drosophila* have been reported which feed on a single species of flower (Pipkin et al., 1966) based on collections in Panama and Colombia in neotropical forests. In general, monophagous flower-feeding *Drosophila* occupy plants with long blossoming periods of four to nine months. Other species studied in these regions were found to be polyphagous, using a variety of host plants that generally have a short blossoming period of one to three months, the most versatile having a wide range of distribution and including the cosmopolitan *D. bucksii.* A number of polyphagous flower-feeding *Drosophila* species can be cultivated on laboratory media, but the monophagous species described by Pipkin, Rodríguez, and León with one exception refused to oviposit, or undergo a prolonged larval and pupal state and fail to eclose. In other words, the monophagous species is usually entirely adapted to its own specialized niche, whereas polyphagous species have less specialized requirements. Similarly, until recently it was almost impossible to culture most Hawaiian species, which are presumably highly specialized. A food medium has, however, been devised enabling some species to be reared (see Carson, Hardy, et al., 1970).

In a further survey (Pipkin, 1965) the influence of feeding and breeding habits upon population size was investigated in 73 species of neotropical forest-dwelling ground-feeding Drosophilidae. Some species preferred small drier fruits and blossoms for both adult and larval feeding, and entered traps to a limited extent. Because of their feeding and breeding habits, they undergo population expansions where such occur in wet seasons, when their fruits and blossoms suffer less from desiccation. Those species using fleshy fruits entered traps more readily, and some tended to undergo population expansions in the dry seasons when they could use fleshy fruits. Those species using both feeding sites also tended to undergo population expansions in the dry season. In other words, population size variations are dependent on the ecology of the species in question.

Drosophila buzzatii is of interest since it is a widespread species with niche specificity. It originated in South America in association with one or more species of the cactus genus *Opuntia.* This cactus, itself of South American origin, has become a weed in various parts of the world, and *D. buzzatii* has apparently gone along with it. There is no evidence that the species is able to breed outside its cactus niche (Carson and Wasserman, 1965). It is of interest that throughout its spread it has carried a particular inversion on chromosome 2, namely, 2j. In Argentina the population shows polymorphism on chromosome 2 for two inversions in addition to 2j. This is the presumed site of endemic origin, and during its spread it took a reduced level of chromosomal polymorphism.

So far, our discussion of niche specificity mainly involves plants, however, associations have been found between *Drosophila carcinophila* and the land crab *Gecarcinus ruricola*, such that the fly breeds as a commensal on the exterior nephric nerves of the crab (Carson, 1967). Another species, *D. endobranchia*, which is only moderately closely related to *D. carcinophila*, has been collected from *G. ruricala* and *G. lateralis* (Carson and Wheeler, 1967).

For most species of *Drosophila* our knowledge is much more restricted and this tends to apply to some of the cosmopolitan and well-studied laboratory species, e.g., the sibling species *D. pseudoobscura* and *D. persimilis*, which in nature have been found on the infected sap exudates (slime fluxes) of *Quercus kelloggii*, the Californian black oak (Carson, 1951). Occasionally *D. victoria* and/or *D. californica* were also found breeding in the same flux. A study of the organisms frequenting slime fluxes lead to the working hypothesis that of the slime flux biota, predators and associates are the main biotic factors preventing excessive oviposition by *Drosophila* at the slime flux. As a result, it is probable that competitive relationships between the various organisms associated with *Drosophila* in the exploitation of the slime flux seldom take place (Sokoloff, 1964). Other isolated observations consist of the finding of pupae of the *D. montana* complex in the rotting phloem of aspen logs (Spieth, 1951), and the finding of larvae of the *D. americana* complex in the stumps of *Salix interior* (Blight and Romano, 1953). Another commonly studied species, *D. robusta*, has a wide latitude of breeding sites, but there appears to be a distinct tendency for the American elm (*Ulmus americana*) to serve as the principal host tree, and in fact the distribution of *D. robusta* parallels *U. americana.* There are some reports of the elm fluxes as breeding sites for *D. robusta* (See Carson, 1958a).

Information on the cosmopolitan species *D. melanogaster* and its sibling species *D. simulans* is scanty, but they congregate and breed in overripe fruit. Hoenigsberg (1968) described a situation in one site in Colombia where *D. melanogaster* was 10 times more frequent than *D. simulans* in 1963, and five years later in 1967, *D. simulans* had become rather more common than *D. melanogaster.* Some ecological changes occurred in the vegetation due to the appearance of a road, but whether this was the mechanism which could account for the change is unknown. A similar observation was made by Tantawy et al. (1970), who found that *D. melanogaster* lost ground in competition with *D. simulans* over the period 1956 to 1966 in the area around Alexandria, Egypt. It is known that *D. melanogaster* tends to be more common in northern United States of America than *D. simulans*, and in the south the reverse occurs (Wallace, 1968). In Brazil, Freire-Maia (1964) reported that *D. simulans* is commoner in the colder regions than *D. melanogaster*, which differs from Wallace's (1968) observation. However, as will be seen in further discussions on these two species in the section to follow on physiological traits, the explanation may lie in the degree of temperature fluctuations that the two species can tolerate in the wild.

Levins (1968) has attacked the problem of the distribution of different species by attempting to assess niche breadth. One source of data comes from habitat or food selectivity in multiple choice experiments. He reports on data collected by himself and coworkers where several kinds of bait (banana, tomato, potato, and oranges) were set out in *Drosophila* traps less than 10 feet apart (see Fig. 1). A species such as *D. cardini*, which is attracted to each bait with almost equal frequency, would be said to have a broad niche for food as compared to *D. repleta*. Levins has developed formulas to measure niche diversity based on setting up several traps in a small area classifying the species taken in each trap separately. Such an approach, if carried out systematically at different times and localities, might well provide detailed information of a microecological type on the distribution of various cosmopolitan species, and thus seems a useful technique.

Using a measure of community diversity Levins found for his own and other published data that there was a relatively small contribution due to geographic site, which he suggests is a result peculiar to small animals. Within a forest in Puerto Rico, for example, there is considerable variation in temperature and humidity over very short distances. Thus it is likely that almost all the range of environments that can be found among forests at all elevations in Puerto Rico can occur somewhere within a limited region in the same forest. Similarly, the diversity found at Austin, Texas, is only increased by one third when the whole south-west of the United States is included. This would not be true for species with very narrow food preferences.

In laboratory experiments in several species of *Drosophila* it has been found that when offered a series of apparently similar oviposition sites, flies do not distribute their eggs at random. The main factor which brings about the aggregation of the eggs is the preference of female flies for ovipositing near the

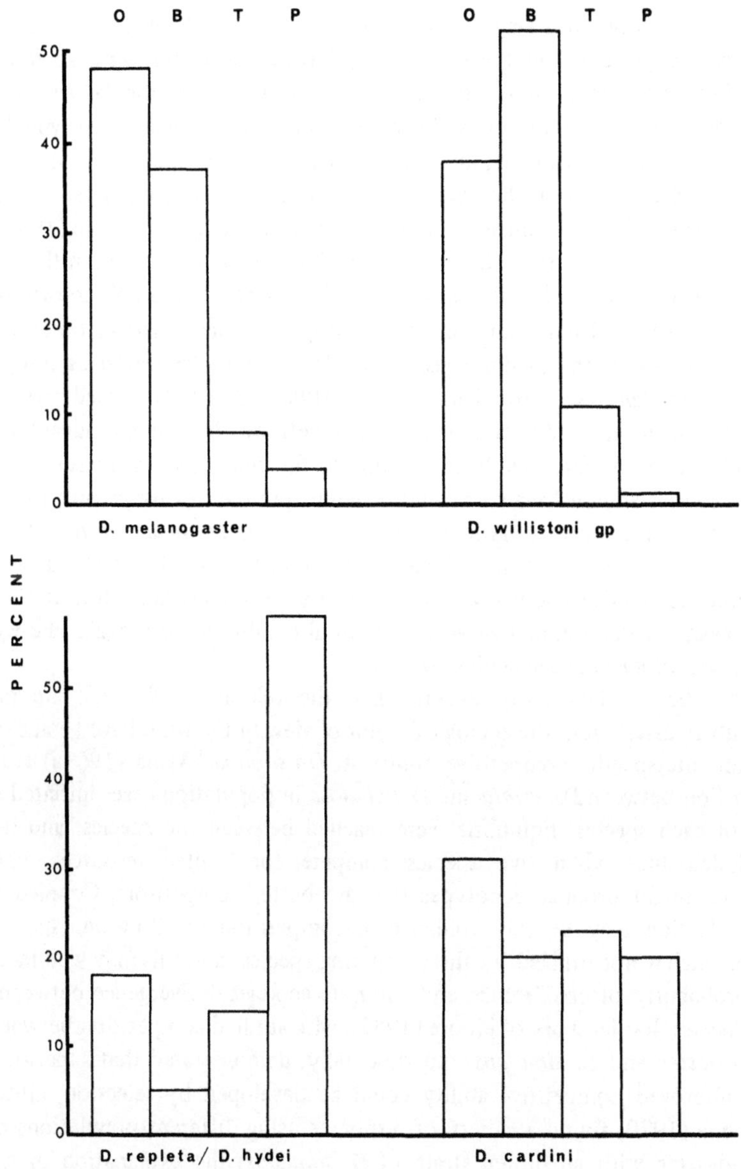

Fig. 1. Food preference histograms for four species of Drosophila. *O, orange; B, banana; T, tomato; P, potato (After Levins, 1968).*

sites previously used for oviposition by other females. It is further possible to direct the flies to a given site simply by placing some eggs on one of the several suitable sites offered to them (see del Solar and Palomino, 1966; del Solar, 1968). In *D. pseudoobscura*, the degree of aggregation can be changed by selection, and so is therefore genetically controlled. Probably there is a compromise between a single oviposition site, leaving all other potential food sources unexploited, and the situation of having many sites with so few eggs deposited that the food is not properly conditioned by the larvae. Natural selection strikes the compromise between some oviposition sites receiving larger numbers of eggs than others, and others remaining unoccupied. This mechanism may well also operate in the wild and be an important controlling factor, but at the moment this must remain as a pure supposition. Even so, Moore (1952) has presented evidence to show that in competition experiments between the sibling species *D. simulans* and *D. melanogaster, D. simulans* is better able to oviposit in the center of food cups, and on food having a surface crust; in other words desiccation makes the medium less favorable for *D. melanogaster* as compared with *D. simulans* (see also Sameoto and Miller, 1966). It turned out to be possible to reverse the outcome of competition by taking advantage of oviposition site preferences. Extrapolating to the wild, it seems reasonable that sibling species should have somewhat different preferences for oviposition sites. Other data have shown variations in pupation site in *D. melanogaster* (Wallace, 1968) for different strains, and also the proportion of drowned pupae in a medium that tends to liquify was found to vary between strains. Both traits may be relevant in determining survival in natural conditions, and again one would expect variations between sibling species.

A number of laboratory experiments, the relevance of which are rather difficult to assess from the ecological point of view in the wild, have been carried out on interspecific competitive ability in *Drosophila.* Ayala (1969a) studied competion between *D. serrata* and *D. nebulosa* in population cages initiated with 50% of each species. Equilibria were reached between the species, and it was concluded that when two species compete for limited resources, natural selection might produce genotypes that are better competitors. Coupled with this, selection may be for avoidance of competition to allow one species to exploit niches not utilized by the competing species, and this may well increase the probability of coexistence and promote ecological divergence between the two species. Earlier work of Moore (1952), who studied competition between *D. melanogaster* and *D. simulans* experimentally, demonstrated that superior flies with improved competitive ability could be developed by selection, although Futuyma (1970) found a variety of responses using different populations of *D. melanogaster* with an inbred strain of *D. simulans*, the explanation of which requires invoking qualitatively different changes in different populations. Futuyma concluded that the genetic variance for traits affecting the outcome of competition was mainly nonadditive, so preventing rapid adaptation to environmental change.

DISTRIBUTION OF GENOTYPES WITHIN SPECIES

Inversion Polymorphisms and Natural Selection

Many species of the Diptera have inversion polymorphisms. It seems likely that inversion polymorphisms can become more easily established in the Diptera than in many other groups because of the absence of crossing-over in the males, and the selective elimination of dicentric and acentric fragments during meiosis in females, such that the balanced noncrossover gametes pass selectively to the egg nucleus (Sturtevant and Beadle, 1936). Furthermore, crossing-over may be suppressed within the inversion. Inversion polymorphisms frequently occur in other insects, e.g., grasshoppers (White 1956, 1962), but have been very thoroughly studied in *Drosophila* (Dobzhansky, 1951, 1957a,b; Da Cunha, 1955). In *D. willistoni* as many as 50 different inversions have been found in nature, such that in central Brazil an individual is heterozygous for an average of nine inversions (see Dobzhansky, 1957b, 1959). Presumably the inversions are maintained in the population by natural selection.

Convincing evidence for the selective control of inversion frequencies comes from observations on the change in frequency of some inversions after their introduction into population cages. In *D. pseudoobscura*, two sequences, Standard (*ST*) and Chiricahua (*CH*), gave stable equilibria of about 70% *ST* and 30% *CH* irrespective of the initial inversion frequencies at 25°C (Wright and Dobzhansky, 1946). Other pairs of inversions also give characteristic equilibria. Stable equilibria under the relatively constant environment of the population cage implies advantage of the inversion heterozygotes over the homozygotes. At 16.5°C, however, little change occurred, showing the dependence of the equilibrium on the environment, in this case temperature. At the intermediate temperature of 22°C, it is of interest that in some cases polymorphism occurs, and in some cases monomorphism for Arrowhead (*AR*) and *CH*. (Van Valen et al., 1962).

Considering the 25°C cages, the F_2 between *ST* and *CH* and some other pairs of polymorphic inversions show an excess of heterozygotes (Dobzhansky, 1947a). Samples of eggs were deposited on food in the population cages and then the larvae were grown under optimal conditions. Agreement with Hardy-Weinberg proportions was good, although there was a very slight excess of heterozygotes. However, in the adult flies from the cages there was a very much greater excess of heterozygotes. Thus, selective elimination of the homozygotes must occur during the larval stages under the high level of competition in the cages (Dobzhansky, 1947a).

At 25°C Beardmore et al. (1960) compared population cages polymorphic for *AR* and *CH* chromosomes with cages monomorphic for these chromosomes. They found that (1) the polymorphic populations produced more individuals than the monomorphic ones, (2) the number of individuals hatching per egg-laying cup was less variable in the polymorphic populations, (3) the mean weights of individuals were similar, but the variances of weights of individuals were lower from the polymorphic populations, (4) there was a greater biomass from the polymorphic populations, and (5) wing lengths were similar, but wing length variances and asymmetry were lower from the polymorphic populations. Thus the polymorphic populations were found to be more productive. The likely interpretation of the lower variability and wing asymmetry of the polymorphic populations is that they have superior developmental homeostasis, i.e., they are better buffered against environmental variations and so vary less. In these experiments, larval crowding was high and the adults were relatively uncrowded. Further experiments (Dobzhansky and Pavlovsky, 1961) have shown that the polymorphic populations were superior to the monomorphic ones when the adults were crowded and the larvae relatively uncrowded.

Beardmore et al. (1960) carried out their experiments at 25°C, but they have been repeated at 16°C (Battaglia and Smith, 1961). At 16°C the heterozygotes *AR/CH* were not heterotic as they were at 25°C, as would be expected from the population cage experiments (Wright and Dobzhansky, 1946; Dobzhansky, 1947b, 1948a), where fitnesses of karyotypes were much more similar at 16.5°C as compared with 25°C. There was some slight evidence that the polymorphic population was superior in fitness to the monomorphic ones. This was based mainly on (1) more individuals being produced from the polymorphic population, and (2) asymmetry of wings being greater in the monomorphic populations. It is to be expected that different environments will give different results, since the fitnesses of genotypes must be a function of their precise environment.

The greater fitness of polymorphic *Drosophila* populations as compared with monomorphic is considered by Beardmore (1963) to be evidence for mutual facilitation between gentoypes, and that the polymorphic population enables more ready adaptation to a variety of ecological niches than the monomorphic populations. For example, populations of *D. pseudoobscura* polymorphic for *AR* and *CH* were superior to monomorphic populations *AR* or *CH*, when they competed for the available food resources with another species, *D. serrata* (Ayala, 1969b). Mutual facilitation has been shown experimentally between strains of *D. melanogaster* for larval viability, perhaps due to diffusible metabolic products produced by the larvae in the food medium from one genotype affecting another (Dawood and Strickberger, 1969). Modifications of fitness due to interactions between genotypes, the exact result not necessarily being predictable based on the fitnesses of the genotypes tested in isolation, have been shown by others, e.g., Lewontin (1955) in *D. melanogaster*, and Lewontin and Matsuo (1963) in *D. bucksii.* These results have analogies with the polymorphic situation in that the different genotypes may have slightly different ecological requirements, so that mutual facilitation (or interference) would result. Beardmore (1961) showed that *D. melanogaster* populations of common ancestry had a higher additive genetic variance in a diurnally varying temperature condition, than in a constant temperature. He interpreted this as being due to a greater variety of niches in the fluctuating environment. Additional evidence with a general discussion is given by Beardmore (1970). A recent experiment in *D. willistoni* (de Souza et al., 1970) has lent support to the idea that polymorphic populations are fitter than monomorphic, for a polymorphism involving a behavioral trait, namely whether larvae prefer to pupate on the food in a population cage, or on the bottom of the cage; the difference being controlled by a pair of major genes at a locus. Polymorphic populations with both types of flies reach larger population sizes and have a greater biomass than monomorphic populations. That the polymorphism depends on the multiniche environment is shown by the rapid elimination of genes of preference for pupation outside the food in populations kept in 250 cm^3 bottles. Such a system could also permit two distinct species to coexist as pointed out by Ayala (1969a–see previous section).

It is not surprising therefore that studies on natural populations of *D. pseudoobscura* (Dobzhansky, 1947b,c) have shown that cyclical changes in the frequencies of these inversions occur annually. Part of the explanation lies in the differential adaptation of the inversion sequences to regular changes in temperature and competition each year (Birch, 1955). The observed regular altitudinal changes in inversion sequence frequencies described by Dobzhansky (1948b) in *D. pseudoobscura* and *D. persimilis* are then not unexpected. Thus the polymorphism is an efficient way of accommodating regular environmental variations and other types of environmental heterogeneity, by giving the population greater powers to adapt to a multiplicity of environments. Similarly, geographical clines have been often reported (see for example Mayhew et al., 1966). In other species such as *D. flavopilosa* (Brncic, 1962), altitudinal clines have been reported, which are associated with fitness differences between karyotypes in two temperatures in which the larvae were permitted to develop in the laboratory (Brncic, 1968). On the other hand in *D. pavani*, Brncic (1969) has shown that there are no seasonal or geographical variations. Thus in some species, chromosomal polymorphism is flexible and leads to varying frequencies of inversions in different environments due to variations in selective values of homo- and heterokaryotypes, whereas in other species the chromosomal polymorphism is more rigid, presumably because of less variation. Examples of rigid species, apart from those already mentioned, include *D. subobscura* and *D. willistoni*, and of flexible species *D. robusta* (see Krimbas, 1967, and Brncic, 1969). Another factor which may be relevant in explaining the distribution of the karyotypes may be the presence or absence of different bacteria and yeasts found in the wild in the intestinal canal. In *D. pseudoobscura*, da Cunha (1951) found that the relative fitnesses of *ST/ST*, *ST/CH* and *CH/CH* varied according to the microorganism supplied to population cages. This has been explored little further but may be of some significance.

In *D. funebris*, a further flexible species, Dubinin and Tiniakov (1945, 1946a, b, c, 1947) found that inversion heterozygotes were more frequent in the populations of cities than in the country populations. The city of Moscow (see Table 1) showed a high level of polymorphism, and furthermore, larger cities and towns showed more polymorphism than smaller ones. The incidence of the inversions underwent cyclic seasonal changes; in Moscow the frequency being

Table 1. Frequencies of inversions in *D. funebris* in the vicinity of Moscow (data of Dubinin and Tiniakob, 1947 – summarized by Dobzhansky, 1965)

Populations	Inversions (%)
Central part of city of Moscow	88.1
Residential part of same city	55.5
Near suburbs	12.1
Villages 20 to 200 km distant	1.8
Villages 200 to 500 km to the north	0

lowest in winter and early spring (March) and apparently trebling by late summer and autumn (August-September). By exposing flies to low temperatures, +3°C and -2°C, Dubinin and Tiniakov seemed to find that during the warm period certain inversions are favored, and in the cold season this reversed; also, the cold-resistant genotypes are more fecund than the cold-susceptible ones. There is clearly a situation here of considerable interest from the ecological genetic point of view, which would repay much more investigation.

Dobzhansky (1950) compared pairs of arrangements from different ecogeographic populations under experimental conditions in *D. pseudoobscura.* On the whole, heterozygote superiority broke down completely. This indicates that the gene arrangements within the inversions are unique to specific populations and are mutually adjusted. Dobzhansky (1950) calls this mutual adjustment of gene complexes within the same population "coadaptation." Clearly the gene complexes from different populations will not have had the same opportunity for mutual adjustment of their genic contents, so that heterozygosity for inversion sequences from different localities will not be expected to lead to increased fitness.

The mutual adjustment of gene complexes occurring within populations must have evolved by natural selection in the environments of the populations in question. Dobzhansky and Levene (1951) have demonstrated the development of heterozygote superiority in population cages for pairs of inversion sequences derived from different localities which did not at first show any superiority. They took *ST* from Piñon Flats, California, and *CH* from Mexico. In the F_2 of the interracial cross, fitnesses such that $ST/ST > ST/CH > CH/CH$ were indicated. In a population cage it would be expected that *ST* would be fixed. However, after some 14 months an equlibrium of 69% *ST* and 31% *CH* was obtained with fitnesses becoming $ST/CH > ST/ST > CH/CH$, showing that the hybrid California/Mexico population had developed overdominance during the course of the experiment. Thus heterozygote fitness was selectively modified by natural selection operating on the increased genetic variance generated by segregation and recombination in the hybrid population (see also Dobzhansky and Pavlovsky, 1953).

Levene, Pavlovsky, and Dobzhansky (1954) studied experimental populations of *Drosophila pseudoobscura* containing all three of the arrangements *ST, AR*, and *CH*, and showed that their results could not be explained by a model assuming constant fitnesses. It was expected from the known fitnesses that *CH* would eventually be eliminated and that a balanced polymorphism of *ST* and *AR* would remain (see also Bodmer and Parsons, 1969). However, *CH* was not eliminated, and a reasonable fit to the observed equilibrium frequencies showed an increase of *ST/CH* fitness, with a minor change in *AR/AR* and *CH/CH* fitnesses. The increase in the fitness of the heterozygote *ST/CH* may have occurred through the selection of modifiers that interact with the genic contents of the inversions, or by selection of inversions carrying altered genic contents, which may have arisen by recombination. Thus we have another case of the possible modification of fitnesses by natural selection under experimental conditions, and analogous phenomena may be expected in the wild.

In a theoretical study Haldane (1957) examined the necessary conditions for the establishment of new inversion sequences. He concluded that there must be a general trend towards overdominance. Similar conclusions have been reached by Bodmer and Parsons (1960, 1962). A general trend towards overdominance has been found in the experimental situations discussed. The trend could occur by the incorporation of modifiers that must interact with the existing gene contents of the inversion sequences. In this way the overdominance of each particular inversion heterozygote will depend on the population in which it exists, since in different environments different modifiers will be incorporated. When the gene complexes are disrupted by recombination overdominance may disappear, as observed for example by Dobzhansky and Pavlovsky (1958), who found evidence for a lowering of heterozygote fitness in population cages begun with flies heterozygous for inversions from different localities in *D. paulistorum* and *D. willistoni.* The cages with flies from different localities soon showed a preponderance of homozygotes, while flies from the same locality maintained a high frequency of heterozygotes.

Of interest so far as natural environments are concerned, are the observations of Dobzhansky, Anderson, et al. (1964), who reported changes in the relative frequencies of the third chromosomes with different gene arrangements over the period 1940 to 1963 in California. In most localities the changes occurred in the same direction, i.e., *ST* increased and *CH* decreased, although in some localities they took place earlier or more rapidly than in others (see also Dobzhansky, 1947c). Similar changes seem to have occurred throughout the Pacific States of the U. S. (Dobzhansky, Anderson, and Pavlovsky, 1966). The causes are obscure, but Dobzhansky, Anderson, et al. (1964) considered that the massive use of pesticides in Californian agriculture was a contributing factor, or perhaps the rapid spread of new gene complexes may have been a cause. Oshima and Watanabe (1966) showed that *ST* seems fitter than *AR* when exposed to traces of the insecticides DDT and dieldrin, and this in fact parallels the changes in the American Southwest. On the other hand, Anderson et al. (1968) found no evidence that the genetic changes observed in natural populations might have been caused by the insecticides DDT and dieldrin.

While these studies have been primarily based on laboratory experimentation, attempts have been made to correlate variations in chromosomal arrangement frequency with environmental factors faced by the natural populations in the field. Work on laboratory populations alone, with subsequent extrapolation to the field, is not possible without the introduction of possible sources of error, as no laboratory "environment" as yet constructed can simulate macro- and perhaps more importantly microfluctuations encountered by a natural population. Furthermore, changes in the arrangement frequency can be an extremely rapid process, whether one considers seasonal (Strickberger and Wills, 1966; Dobzhansky, 1947b), altitudinal (Stalker and Carson, 1948), or geographical variation (Carson, 1958a), which would suggest that a given arrangement is best adapted to a given environmental situation. Accordingly, one can only argue that the results gained from a laboratory population are directly applicable to natural populations, if the physiological and adaptive functions of the organisms in question are susceptible to the environmental fluctuations ot stresses imposed.

Marginal and Central Populations

In certain species of *Drosophila*, notably *D. willistoni* and *D. robusta*, inversion heterozygosity decreases rapidly as the margin of the species range is approached; in other words central populations may be better adapted to their diverse environments by means of a high level of inversion heterozygosity. It has been suggested that in marginal regions there is a greater need for more flexibility in order to adapt to new environmental conditions. The low level of inversion heterozygosity permits a higher level of recombination, thus providing novel gene combinations more rapidly than in populations containing high frequencies of inversions (see for example Dobzhansky, Burla, and da Cunha, 1950; da Cunha and Dobzhansky, 1954; da Cunha et al., 1959). Marginal populations may thus be better able to respond to evolutionary opportunities, and can probably form new species more readily than populations from the center of the distribution of the species. In *D. robusta* central populations have seven to nine inversions, while peripheral populations have one to six inversions (Carson, 1958a, b). Low levels of polymorphism also occur in island populations of *D. willistoni*, and populations on small islands tend to be less polymorphic than on large islands (Dobzhansky, 1957b). The amount of polymorphism in a population therefore tends to be related to ecological diversity and the environmental opportunities that the population exploits. Prevosti (1966) has given similar evidence in *D. subobscura* in Western Mediterranean populations.

Cosmopolitan species such as *D. melanogaster* are relatively free from natural chromosomal polymorphisms (*D. robusta* and *D. willistoni* are restricted to parts of North and Tropical South America, respectively). The genetic system of *D. melanogaster* thus resembles the marginal populations of *D. robusta* and *D.*

willistoni, and so should be able to respond rapidly to new conditions. *D. pseudoobscura* seems to be intermediate in this respect, as inversion heterozygosity is a feature of the species, but it is restricted to the third chromosome. Restriction of inversions to one chromosome seems to occur more frequently than would be expected by chance, in species with small numbers of inversions. Perhaps this is because the organism compensates for the low rate of recombination in the chromosome with inversions, by a high rate in other chromosomes (Schultz and Redfield, 1951). Free recombination in chromosomes without inversions permits flexibility for adaptation to new environments, whereas the inversion sequences are probably adapted to a given series of predictable, but often widely varying environments. Inversion heterozygosity for one chromosome is therefore an example of the way in which natural selection may accommodate the opposing evolutionary requirements of fitness in one environment and flexibility to adapt to others. This argument may be an oversimplification because other methods of varying overall recombination exist, e.g., chiasma localization and chromosome number variations (see Darlington, 1937; White 1958, 1962), although within a genus such as *Drosophila*, where inversions assume so much importance, the argument may be of relevance.

In examples cited in this section, we are not discussing what environmental variables are involved, but are discussing the consequences of the whole environment on a species and the genotype so developed. Appropriate laboratory experiments may allow a clearer definition of marginal vs. central environments, but as already pointed out such experiments are liable to error, because of the impossibility of simulating natural environments, which have a level of unpredictability difficult to create under experimental conditions. A start has, however, been made, since Band (1963) compared the frequency of drastic variants (lethals and semilethals having a fitness up to 50% of normal) in *D. melanogaster* populations collected from South Amherst, Massachusetts in three environments, a constant temperature 25°C, a fluctuating environment with a wide range 14°C between the daily maximum and minimum (which approximates the wide-range temperatures observed in the natural environment during the precollecting periods), and a fluctuating environment with a narrower range of 8°C between daily maximum and minimum (which approximates the narrow-range temperatures observed in the natural environment). In all cases the average overall temperature was 25°C. It was found that the frequency of lethals and semilethals was higher for the narrow-range temperature than for the wide-range temperature or the constant environment. Thus, extreme environments lead to a reduction of genetic diversity, as does a constant environment, which implies less ecological heterogeneity. This accords with the evidence of Beardmore (1961) already cited, who found a higher additive genetic variance in *D. melanogaster* in a narrow-range, diurnally varying temperature condition 25 ± 5°C. Similarly, Oshima (1969) found that fitnesses of heterozygotes for lethal, semilethal, and quasinormal genes were greater in fluctuating environments of 25 ± 5°C with a mean of 25°C, as compared with a constant 25°C. Thus, greater genetic diversity may be associated with favorable environments, which may comprise fluctuations without their being too extreme. More work needs to be done in this area, especially since the interpretation of the data is not easy, since the amplitude of the fluctuations seems to be of importance, and could vary in its effect both within and between species. Further consideration will be given to this problem in the section on physiological traits.

Similarly, marginal populations of *D. willistoni* tend to contain fewer drastic variants than populations at the central part of the species distribution (Cordeiro et al., 1958; Townsend, 1952). In *D. robusta* (Nair, 1969) the genetic load, estimated by comparing the egg- to adult-lethality in inbred and outbred crosses of F_1 lines derived from wild flies collected during consecutive years, was found to be more or less uniform in the central populations from year to year, so perhaps reflecting the homeostatic properties characteristic of a large population. In marginal populations there was a considerable difference in the loads between years.

Dobzhansky, Hunter, et al. (1963) have made similar observations in *D. pseudoobscura*, where a marginal population in the Andean highlands around Bogota, separated from the main body of the species by 1,500 miles, was found to have the lowest proportion of lethals and semilethals of any population studied. Conversely the heaviest loads are found in the most prosperous (and central) populations. United States populations that are comparable with the Colombian populations come from the Death Valley and Mount San Jacinto regions of California. In these we are probably dealing with ecological rather than geographical marginality. The Death Valley populations have to cope with extreme summer aridity, and at high elevations in the mountains with long and severe winters. These populations are presumably reduced to very few survivors in unfavorable seasons so allowing periodic inbreeding, drift, and hence a general reduction in the genetic load. To a lesser extent this is true of Mt. San Jacinto and the Colombian highlands. Using 24 assayable and hopefully randomly chosen loci for enzymes and larval proteins, Prakash et al. (1969) have shown that the isolated Bogota population has only about one third the variability shown by the main body of the species, and shows the greatest amount of genic divergence of all populations.

In *D. melanogaster*, Band and Ives (1968) found that lethal and semilethal frequencies were significantly positively correlated with total rainfall in summer. In the period 1947-1964, the annual rainfall per summer was almost two inches (5 cm) less than in the 1938-1946 period, and in the latter period the frequencies of lethals and semilethals were lower than in the former. Thus, the drier environment, which is probably, but cannot be proven to be, more marginal is associated with lower frequencies of lethals and semilethals, although as will be shown in the section on physiological traits, desiccation is a stress to which *Drosophila* is susceptible.

In conclusion, greater genetic diversity is associated with central environments, implying diversity of ecological niches, favorable temperatures, or other climatic conditions, while in marginal environments, implying limited ecological niches, unfavorable temperatures, or other climatic conditions, there is less genetic diversity–or put succinctly, the amount of genetic variability in a population is a function of the diversity of the environment in which the population lives. Beardmore (1970) has also argued from his experimental work on *D. melanogaster* that the amount of heterogeneity of the habitat is generally correlated with the amount of variability in gene pools. However, genetic variability as a function of the environment does *not* always imply a causal relationship, since reduced variability can be a function of reduced population size, as has been discussed already. It is therefore clear that the interpretation of much of the experimental data on marginal and central populations poses problems of peculiar difficulty, and the story is far from complete.

The Integrated Genotype Within and Between Species

The frequencies of lethal and semilethal chromosomes are probably higher in some species than in others, such as the more widespread and versatile species such as *D. pseudoobscura* and *D. willistoni* (see for example Krimbas, 1959). Another manifestation of the ecological versatility of these two species, compared with *D. persimilis* and *D. prosaltans* for example, is that the former two have been shown to release more variability by recombination than the latter two. The differences in the amount of variability released by recombination presumably reflects a difference in the amount of concealed genetic variability in these species; the higher genetic variability occurring in the ecologically most versatile species (Spassky et al., 1958; Spiess, 1958, 1959; Dobzhansky, Levene, et al., 1959; Levene, 1959; Krimbas, 1961). This result stresses the highly integrated nature of the genotype, and its dependence on the species and its ecology.

Within species, studies have made the concept of a single wild-type genotype untenable. Instead, the details of the integrated genotype must be regarded as a characteristic of local populations, rather than of the whole species, so that there is a multiplicity of wild-type genotypes each adapted to local conditions. Some evidence for this has already been cited. In addition Vetukhiv (1954) and Wallace and Vetukhiv (1955) found that the F_2's between population hybrids in several species were generally inferior in survival and fecundity as compared with hybrids within populations. This presumably occurs because in the F_2's the highly integrated genotypes evolved within populations are broken down by recombination in between-population crosses, but in within-population crosses, no such breakdown occurs. Vetukhiv (1957) has made similar observations for longevity in *D. pseudoobscura*. These and other results (Brncic, 1954, 1961; Anderson, 1968; and see Bodmer and Parsons, 1962) make the concept of a wild type genotype of a species untenable. A recent study of Anderson (1969) comfirms this for the selection coefficients for lethal heterozygotes in data of Dobzhansky and Spassky (1968). His analysis showed that in a genetic background from their own populations, 5 out of 45 lethal chromosomes were statistically significantly heterotic, and none were significantly deleterious, whereas in foreign genetic backgrounds about 10 of 45 lethal chromosomes were significantly deleterious and none were significantly heterotic. In other words it seems that heterosis has evolved for lethals in their own populations, but is lost when the lethals are put into a foreign genetic environment. These results show that it may be of importance in the future to carry out detailed studies of genes (and polygenes) differentiating local populations. On the other hand, in *D. subobscura*, McFarquhar and Robertson (1963) did not find any evidence for differentiation into local races as assessed by F_2 data. While this may be so and may be a feature of the genetic system of *D. subobscura*, the evidence advanced above for other species does seem convincing. Thus it seems that the gene pools of some species are more prone to evolving coadapted gene complexes, such that the gene pool of *D. pseudoobscura* can change readily evolving geographic races, whereas that of *D. subobscura* is relatively rigid.

Wigan (1944) studied the effect, on sternopleural chaeta number, of chromosomes extracted from wild populations. It was found that individuals carrying pairs of homologues derived from individuals from the same locality were less variable for chaeta number than were individuals in which the members of a homologous pair were derived from different localities. Thus within each locality the genotype has evolved relational balance, or coadaptation, for the sternopleural chaeta number adapted to that locality, whereas between localities more variability is apparent, as the relationally balanced systems differ between localities.

These studies show the genotype to be integrated within species at one level, and within populations at another, evolving in response to the environments prevailing upon species and populations. Although the environment has hardly been discussed specifically, if we are to assume that it is a relevant factor in the evolution of the integrated genotype, studies of environment in relation to genotype, or more generally local populations, are of crucial importance.

MORPHOLOGICAL, PHYSIOLOGICAL, AND BEHAVIORAL TRAITS

Morphological Traits

Some studies have been carried out on morphological variation in *Drosophila* in different environments. For example, in Europe in *D. subobscura*, Misra and Reeve (1964) found that five dimensions, namely wing length, wing width, thorax length, head width, and tibia length show strong positive correlations with latitude and slightly lower negative correlations with mean annual temperature. The flies had been maintained in the laboratory for 8 to 11 generations before testing. There is therefore a uniform cline in the five dimensions going northwards (see also Prevosti, 1955 for earlier data). This differs from the pattern found for essentially the same dimensions in the American species *D. robusta*, where all correlations with latitude were lower and those for head and thorax size of opposite sign in the two species (see Stalker and Carson, 1948, 1949). Even so, flies from high altitudes in northern regions tend to have greater wing sizes than those further south. Seasonal shifts in morphology also have been found such that there was a regular highly significant shift towards a southern morphology during the summer months June to August. This shift is in the expected direction, assuming the southern morphology to have a high-temperature adaptive value. In *D. melanogaster* and *D. simulans*, Tantawy and Mallah (1961) found that wing and thorax lengths were greater at lower temperatures. It is therefore reasonable that there should be a seasonal change in morphological differences; e.g., wing length declines during summer months in wild Egyptian populations (Tantawy, 1964).

In *D. melanogaster* and *D. simulans*, the number of flies with more than four scutellar chaetae was highest in Australia in a collection site with a Mediterranean climate, and lowest in a site with a subtropical environment (Fraser, 1963). Since the flies on which the observations were made were cultured under standard conditions in the laboratory, it is reasonable to suggest that the variations found were not fortuitous, but have some adaptive value. Perhaps the changes may be correlated with fly size, since it is known that when flies of a given strain are grown under high temperatures, fly size, and sternopleural and scutellar chaeta numbers are lower than at low temperatures (Parsons, 1961; Fraser et al., 1968). Therefore in natural populations, variations in fly size that can be induced experimentally in the laboratory are fixed in different regions presumably because of somewhat differing genetic architectures.

Levins (1969), using flies collected in the wild, found a body weight cline for *D. melanogaster* in Puerto Rico such that flies caught at cooler, higher localities were larger than those taken in the hot lowlands. However, when raised under identical conditions in the laboratory, the coastal flies were larger, i.e., the general expectation that phenotypic differences in nature are in the same direction as in the laboratory is contradicted.

Sokoloff (1966) studied *D. pseudoobscura* captured in nature for body weight and/or wing length, wing width, and tibia length, and found that within a given year, despite the wide range of ecological conditions prevalent in their areas of capture, flies were fairly uniform. There was, however, a distinct difference in size of flies captured in two successive years, and this has been correlated with climatic conditions prevailing in the two years. The uniformity within years allows one to conclude that this species has adapted to a wide variety of environments by individuals belonging to the various races of this species evolving favorable genotypes through natural selection, which enable them to develop an optimal phenotype in the face of a wide range of environments. One can also argue that clines might normally be expected as described for other species in the face of climatic variations, and it seems that different species adapt in different ways—especially since in *D. pseudoobscura* there are known and definite clines for chromosomal variants. In *D. pseudoobscura* therefore, it seems that morphological variations may be independent of the types of inversions that characterize the various races of this species, since the morphological phenotypes do not seem to vary in the same way as the inversions, which are relatively environment-dependent in their frequencies. Even so, Anderson (1968) has reported that natural populations from Canada to Mexico vary in body size, such that populations from the Pacific coast are generally smaller than those from the interior provinces.

Another problem is that some of the observations are based on flies collected in the wild, and others are based on cultures set up from flies collected in the wild, often reared under relatively known and standard conditions. For most of the year, flies collected in the wild are usually intermediate between flies reared in the laboratory under optimal and minimal conditions, although during the summer the size of wild adults is lower, and this is attributed to high temperatures (Tantawy and Mallah, 1961). This may be a partial explanation of Levins' (1969) data above, since under less stressed conditions in the laboratory, coastal flies may be expected to be larger (but see also Levins, 1969 for further discussions).

Work on morphological traits per se thus allows certain inferences about populations in relation to their environments, but it is difficult to be precise, and we therefore turn to physiological traits.

Physiological Traits

Indirect effects of temperature were considered above, but here direct effects of physiological stresses as assessed by mortality will be considered. Genetic differences between strains, each derived from a single inseminated female collected in the wild, have been reported for morphological, behavioral, and physiological traits. The presence of genetic variability between strains presumably arose from genetic differences between founder females, and hence the base population must be polymorphic for some of the genes that determine these traits (Parsons et al., 1967). Thus Hosgood and Parsons (1968) showed differences between strains for the ability of adult flies to withstand high temperature shock (33.5°C for 24 hours for 7-day old flies), and Parsons (1969) showed the same for the ability of newly hatched larvae to emerge as adults at 30.5°C, so that the strains differ for genes controlling resistance to high temperatures. There is other evidence in the literature for temperature sensitive strains (Ogaki and Nakashima-Tanaka, 1966) in *Drosophila*. Tantawy and Mallah (1961) found superiority at high temperatures for populations of *D. melanogaster* and *D. simulans* from Uganda, an area of high temperature. The temperatures used by Hosgood and Parsons (1968) are fairly extreme, but they would occur in the locality from which the flies were derived (Leslie Manor,

Victoria), which is characterized by relatively hot summers, so it is reasonable to find evidence for some genetic control of the trait (assuming that the flies do not completely avoid high temperatures by behavioral or other means). It is unlikely that such genes would be fixed in the population, because natural selection would not favor them except at limited times during the year. Such polymorphisms would allow a rapid adjustment to environmental changes as they occur.

The data of Tantawy and Mallah (1961) show that the percentage of emergence of flies from eggs was higher for *D. simulans* in the range 18° to 25°C, whereas over the *entire* range studied, 10° to 31°C, *D. melanogaster* had higher average survival (see also Wallace, 1968). Thus, *D. simulans* might be expected to outnumber *D. melanogaster* in regions where temperature extremes are not encountered, but in regions where there are rather wide temperature fluctuations, one might expect *D. melanogaster* to displace *D. simulans.* Wallace (1968) points out that data in the U.S.A. fit this expectation fairly well, since *D. simulans* is the more common species of the two in the south where the climate is equitable throughout the year, and *D. melanogaster* is more common in the north where winters are severe and wide daily fluctuations in temperature occur. Hosgood and Parsons (1966) set up four strains of *D. melanogaster* and three of *D. simulans* each derived from single inseminated females collected in the wild, at 29.5°, 27.5°, 25°, 20°, and 15°C in the laboratory. After five generations, all strains of *D. melanogaster* were living at all temperatures, whereas for *D. simulans* the three strains at 20° were living and only one at 25°C (which in fact survived for 24 generations). At 29.5° and 15°C all of the *D. simulans* strains had died out by the second generation, and at 27.5°C by the third generation. Thus *D. simulans* is much more restricted in its tolerance to diverse temperatures than *D. melanogaster*, and this agrees with the United States distribution data reported by Wallace (1968). One problem not answered by these experiments is the effect of fluctuating temperatures, either diurnally or on a longer term basis, since the length of exposure to extreme temperatures, and the amphitude of the temperature variation may be of importance.

Because of the genetic heterogeneity for resistance to temperature, the early observations of Timofeeff-Ressovsky (1940) in *D. funebris* of "temperature-races" is of considerable interest. The relative viability of *D. funebris* was studied in flies from regions covering the major climatic zones of Europe, northern Africa and into Asiatic Russia at 15°, 22°, and 29°C. It was found that the northern populations were more resistant to cold temperatures, and southern more resistant to high temperatures. The eastern populations showed resistance to both high and low temperatures, which can be accounted for in terms of the more "continental" climate of these regions with very low mean winter temperatures and very high mean summer temperatures. Presumably the temperature races differ by the accumulation of appropriate temperature genes into their genome. In *D. flavopilosa*, Brncic (1969) has found certain heterokaryotypes A to be favored when larvae develop at 25° in the laboratory and others B favored at 16°C. The data agree with what is observed in the corresponding natural populations, where A heterokaryotypes are favored in the summer and B in the winter (Brncic, 1966). As a result of the increase of A concomitant to the decrease in B at 25°C, and the reverse at 16°, the heterozygosity for inversions is maintained at a more or less constant level. In their studies of the Hawaiian Drosophilidae, Carson, Hardy, et al. (1970) believe temperature to be an important limiting factor. Thus, most species are found between 1,000 feet (330 m) to up to 5,000 feet in elevation, and also significantly some of the plant species that serve as larval and adult food sources, are abundant in these regions.

Another environmental stress of some ecological importance is desiccation. Little work has been done on this trait, although Kalmus (1945) and Waddington et al. (1954) described differences between strains of *D. melanogaster* for preferences for environments with different humidities. Pittendrigh (1958) found that *D. persimilis* lost water by cuticular transpiration more rapidly than *D. pseudoobscura*, and that males lost water more quickly than females. In *D. pseudoobscura*, Heuts (1948) found that the relative humidity during the pupal stage at 25°C differentially affected the carriers of different gene arrangements: pupae homozygous for *AR* had higher hatching percentages at low humidities, *CH* homozygotes at high humidities, and *ST* homozygotes were intermediate. Subsequently Spassky (1951) and Levine (1952) found certain karyotypes to be differentially affected by certain combinations of temperature and humidity.

Probably the best work in *D. pseudoobscura* is that presented in a Ph.D. thesis by Thomson (1971) for desiccation of *ST/ST, ST/CH,* and *CH/CH* showing distinct heterokaryotype advantage (see Fig. 2) for survival 24 to 42 hours after the commencement of desiccation. In fact, the only survivors at 42 hours were heterokaryotypes. The heterokaryotypes were heavier than the homokaryotypes, suggesting a relationship between ability to survive and body weight (Table 2). Parsons (1970) found that strains set up from single inseminated females of *D. melanogaster* derived from two populations differed in ability to withstand desiccation, as measured by mortalities after 16 hours in a dry environment, thus indicating genetic variability in wild populations for ability to withstand desiccation. Those strains with high wet and dry weights lost water relatively less rapidly and had lower mortalities than strains with lower wet and dry weights. Quite high heritabilities were found for all traits. Levels of resistance to desiccation may have some influence in determining the distribution of the species in the wild. As with temperature stress, it is reasonable that there should be genes segregating in the population for resistance to desiccation, because during parts of the year populations would be exposed to desiccation stresses, during which time genetic variability to adapt may be desirable. One might often expect a correlation between the two stresses, since in the region of

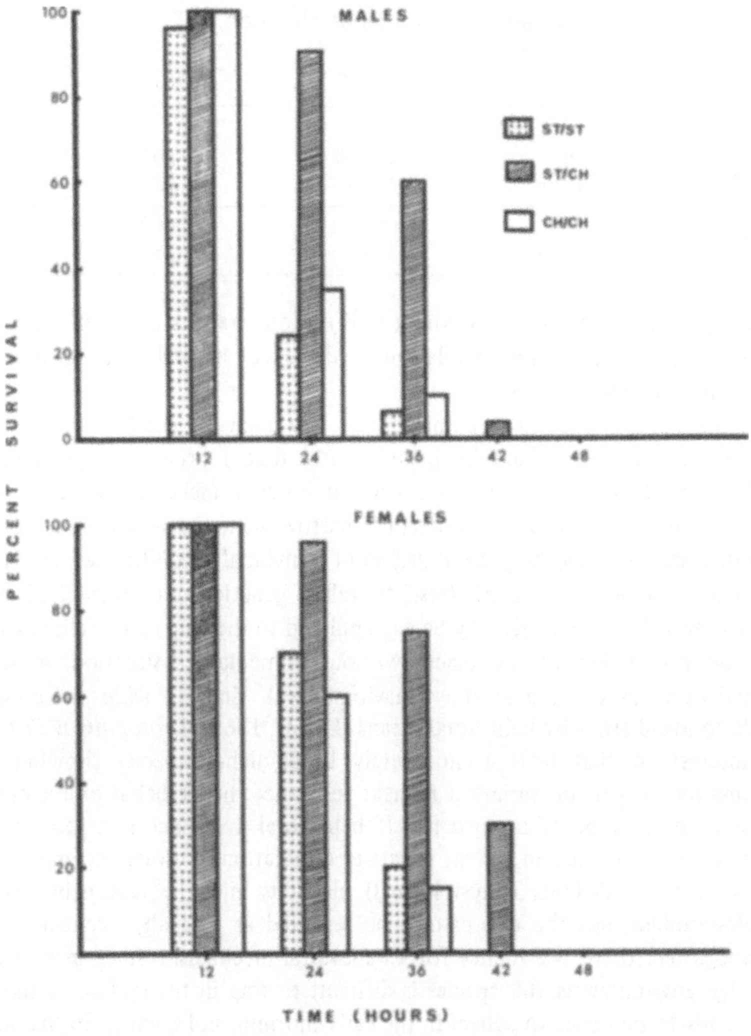

Fig. 2. Comparison of percentage survivals of ST/ST, ST/CH, *and* CH/CH *genotypes when desiccated based on pooled results from 4 replicate samples of 25 flies of each sex and genotype (after Thomson, 1964).*

Australia where these strains were collected, a correlation between high temperature and low humidity would be common, although in some parts of the world, especially in the tropics, this might not be expected. However, in the Australian strains tested, a significant correlation was found between the two traits. Clearly more work would be rewarding, especially if, following Levins (1969) and the work cited above, body weight can be regarded as an adaption to

Table 2. Means and variances of live weights expressed in micrometer units of individual flies of *D. pseudoobscura* (after Thomson, 1971)

		ST/ST	*ST/CH*	*CH/CH*
Females	mean	80.410	89.505	85.505
	variance	247.86	238.24	524.29
Males	mean	66.075	73.880	72.425
	variance	175.84	183.26	339.43

desiccation stress. However, Levins (1969) quotes results indicating the likely complexity of these traits in relation to the exact natural environments from which the flies were derived.

Levins (1969) studied acclimation to dry heat (a combination of high temperature and low humidity), and found that a process of physiological acclimation tended to occur especially in broader niche species such as *D. melanogaster*, and tended not to occur in narrow niche species e.g. *D. prosaltans.* In other words, there may be a degree of individual flexibility, which enables immediate adaptation to heat stress, as well as genetic variation as cited above. Levins feels that this work may be extrapolated to the wild, and acclimatization may permit feeding in dry places without immediate desiccation. A further complication may come at the behavioral level, since *D. willistoni* probably tends to avoid stress by behavioral means (Levins, 1969). *Drosophila willistoni* is of interest in that it is a moderately broad-niched species showing little acclimation or genetic variation in heat resistance; hence behavioral avoidance mechanisms may be of importance. If behavioral avoidance is a general trait, then Levins' conclusion of a negative correlation between strong habitat selection and individual (physiological) plasticity may be reasonable for the species studied, but the two cannot be regarded as mutually exclusive. In *D. melanogaster*, there is evidence for a behavioral mechanism, since in extremely hot dry environments this species is difficult to trap in the middle of the day, and tends to be easier to collect in the early morning and evening. In the middle of the day when temperatures may exceed 40°C in parts of Victoria, flies might tend to avoid this temperature by seeking cooler niches, emerging when the environment becomes more favorable. Levins (1969) found that *D. melanogaster* taken at midday at exposed traps in Puerto Rico were 5% larger than those collected in the morning or late afternoon, and in protected sites there was no difference. In other words only those flies which are probably genetically most resistant to desiccation appear at midday, by virtue of their large size (Parsons, 1970), and those which are genetically less resistant do not. Thus *D. melanogaster* shows genetic heterogeneity, physiological plasticity, and behavioral mechanisms in coping with hot dry environments, and it might be expected that in the various species all three may play a role, but that between species the relative importance of each may vary in different ways.

Turning to cold tolerance, laboratory work has indicated that varying degrees of tolerance to cold stress occur at different stages of the life cycle of *D. pseudoobscura.* Crumpacker and Marinkovic (1967) found in a region with a cold winter and considerable diurnal fluctuation that adults were more resistant to cold stress than larvae or pupae, which in turn were more tolerant than eggs. It was also noted that adults collected immediately after the overwintering period tended to be old individuals of the previous autumn. In *D. flavopilosa* (Brncic, 1968) and *D. robusta* (Carson, 1958a), the adults tend to hibernate, and it has been observed that the females of *D. robusta* caught in September have extremely small ovaries and very large quantities of body fat. It is inferred that these changes represent a biological adjustment of the species to overwintering as adult flies (Carson, 1958a). In *D. persimilis* (Mohn and Spiess, 1963) both the pupal and adult stages may overwinter best, and in some other species it is likely that overwintering may occur in nonadult stages (See Crumpacker and Marinkovic, 1967). In all of these cases habitat selection may be of considerable importance. In *D. pseudoobscura*, for example, no supercooling system, which requires glycerol in the metabolic system of the insect (Salt, 1961), and which would allow some release from a dependence on environmental sources, has been found. In *D. nigrospiracula*, which has as its host plant the saguaro (*Cereus giganteus*), supercooling occurs such that the LD_{50} for 1 hour of cold treatment was $-7.73°$, the LD_0 between $-5°$ and $-6°C$, and the LD_{100} between $-9°$ and $-10°$ (Lowe et al., 1967). In nature the flies have been found in a predawn minimum air temperature of $-5.1°C$, and begin flying shortly after the sunlight reached them at the cactus. This enables the flies to constitute a more or less continuously breeding species during the winter in a relatively cold winter part of the Sonoran Desert.

Marinkovic et al., (1969) found substantial heterozygote advantage for cold temperature resistance in *D. pseudoobscura*. For chromosome III the survival of homozygotes was 50% less than that of the heterozygotes, and a much smaller differential was uncovered for chromosome II although it was in the same direction. The pronounced heterozygote advantage associated with genes on chromosome III is another example of heterosis at suboptimal environments (high or low temperatures and desiccation in ectotherms). An early observation of Heuts (1948) is in agreement in showing that longevity of adults in the presence of food at low temperatures (0° to 4°C) is greater for *ST/CH* heterokaryotypes than the corresponding homokaryotypes.

Behavioral Traits

Genetic heterogeneity has been found within certain populations for some behavioral traits, especially those associated with sexual behavior. Thus, strains set up from single inseminated females from the same wild population of *D. melanogaster* and *D. pseudoobscura* (for a given karyotype) are genetically heterogeneous for mating speed and duration of copulation (Parsons and Kaul, 1967; Hosgood and Parsons, 1967). Considerable differences in mating speed have been found between homokaryotypes in *D. pseudoobscura* and *D. persimilis* (Spiess and Langer, 1964a,b; Parsons, 1967). Male heterokaryotype advantage has been reported for mating activity in *D. pavani* (Brncic and Koref-Santibañez, 1964), and mating speed in *D. pseudoobscura* (Spiess et al., 1966; Parsons, 1967). Apart from showing heterokaryotype advantage under some circumstances, the evidence for coadapted complexes adapted to the environments in which the populations live is limited for behavioral traits (see Parsons, 1967). One manifestation of a heterozygous coadapted gene complex is the ability to tolerate a wide range of environments, which is reduced on inbreeding. Similar observations occur for heterokaryotypes, as compared with homokaryotypes in *D. pseudoobscura*, for many fitness traits including mating speed and duration of copulation, for which heterokaryotype advantage has been found at 25° and not 20°C (Parsons and Kaul, 1966; Parsons, 1967). A good example of higher fitness of a behavioral trait when polymorphic, compared with when it is monomorphic, is the work of de Souza, da Cunha, and dos Santos (1970) who reported a behavioral polymorphism for pupation site (de Souza et al., 1970).

Two South American sibling species, *D. gaucha* and *D. pavani* (Koref-Santibañez, 1964), show little discrimination in courtship behavior, and crosses yield abundant but sterile progenies. Thus the main isolating mechanism is hybrid sterility. In nature the lack of discrimination in courtship behavior is unimportant, since except for a small part of Argentina where the two species are found together, they are allopatric. It would be expected that greater divergence in courtship behavior and isolation would occur in sympatric rather than allopatric species, as found in flies from the small region of Argentina where the two species are sympatric, so as to avoid the production of large numbers of relatively unfit hybrids (see also Koref-Santibañez and del Solar, 1961). Thus interaction between species is relevant in determining the degree of isolation. This type of interaction has also been found in *D. paulistorum*, a taxon that contains a complex of geographic races or incipient species, which are mainly but not always allopatric. Joint isolation coefficients have been computed (Ehrman, 1965) for pairs of races that have been found to occur both sympatrically and allopatrically. In allopatric crosses the average isolation coefficient was 0.67, and in sympatric crosses 0.85. Thus, pairs of races occurring sympatrically exhibit more sexual isolation than the same pairs occurring allopatrically, or races coexisting geographically tend to be reproductively more isolated than those that do not, which is reasonable since the production of large numbers of rather unfit hybrids would be inefficient from the population point of view. Thus, behavioral isolation helps to maintain the integrity of gene pools of geographical races and species, and the degree of behavioral isolation may be determined by the ecological situation and so be affected by natural selection.

The pair of sibling species *D. pseudoobscura* and *D. persimilis* are sympatric in some regions. In regions where the species are sympatric, isolation is maintained by several factors, some of which are:

(1). Sexual isolation. Thus it has been shown that the males of the two species differ in the frequency of underlying vibrations, the time between vibration pulses and the duration of the pulse (Waldron, 1964). Clearly such differences may be relevant in favoring isolation at the behavioral level.

(2). Somewhat different habitat preferences such that *D. persimilis* favors cooler moister niches, and *D. pseudoobscura* warmer drier niches.

(3). Different diurnal periods of maximum activity.

All of these factors have behavioral components, and taken together apparently lead to isolation in nature, even though the two species can mate and produce hybrids in the laboratory (for further details see Dobzhansky, 1951; Parsons, 1967). The sibling species *D. melanogaster* and *D. simulans* also show courtship behavior differences, and some ecological differences as pointed out already in this review (see Parsons, 1967 for a more detailed discussion of various *Drosophila* species). Thus, behavioral variations associated with ecological variations are of importance in reproductive isolation between sibling species.

Environmental Stresses Generally

In this review the effects of certain environmental stresses have been considered, beginning with the population cage experiments of Wright and Dobzhansky (1946) and Dobzhansky (1947b, 1948a) in *D. pseudoobscura*, who reported large fitness differentials between hetero- and homokaryotypes leading to heterokaryotype advantage and a stable polymorphism at 25° and not 16.5°C, with an intermediate situation at 22° (Van Valen et al., 1962). For *D. pseudoobscura*, 25°C represents a more extreme environment than 16.5°C. Other examples of heterokaryotype advantage in extreme environments already cited in *D. pseudoobscura* include desiccation (Thomson, 1971), cold tolerance (Heuts, 1948), and mating speed and duration of copulation at high temperatures (Parsons and Kaul, 1966). Thus it could be argued that in more optimal environments heterokaryotype advantage is minimal, and is expressed only in extreme environments. If this is a general phenomenon it would provide a further explanation, in addition to those already advanced in the literature (see for example Sved et al., 1967; Wallace, 1968), for the high level of genetic polymorphism in natural populations being associated with a not excessive genetic load. Under more optimal environments, the species might have a high level of polymorphism without a concomitant genetic load, but in times of environmental stress, heterozygote advantage would occur for some polymorphisms according to the stress. Essentially one can imagine a situation where a population is subjected to a number of environmental stresses during its existance, during which time heterozygote advantage for certain loci or inversions might be favored according to the stress, so enabling more ready adaptation to the stress. Since a high level of heterozygote advantage would be restricted more or less to stress environments, which would probably make up a minor part of the population, the overall genetic load would not be high assuming a multiplicity of extreme ecological niches in which different heterozygotes would be advantageous.

Analogous to this are the several observations of heterosis for hybrids between inbred lines under extreme environments in this case high temperatures (e.g., larval survival [Parsons, 1959] and longevity [Parsons, 1966] in *D. melanogaster*, and growth rates in *Arabidopsis thaliana* [Langridge, 1962] and maize [McWilliam and Griffing, 1965]) and low temperatures (viability in *D. melanogaster* [Fontdevila, 1970] and growth rates in maize [McWilliam and Griffing, 1965]). Furthermore, a similar phenomenon has been found for the effects of homozygotes of chromosome II as compared with the corresponding heterozygotes of *D. pseudoobscura* with respect to viability at 25°C (Dobzhansky et al., 1955), for cold temperature resistance (Marinkovic et al., 1969), and for preadult survival for chromosome II in *D. melanogaster* at 29°C (Tobari, 1966). These are not natural population situations, but are analogous in that for a normally outbred species, homozygotes and heterozygotes are being compared and show that the genetic load, in this case the inbreeding load, is higher in adverse environments than in favorable ones. Therefore the more severe the physical environment, the greater the level of heterosis (and genetic load) developed.

So far as temperature-dependent heterosis is concerned, Langridge (1968) has proposed an explanation at the molecular level, namely that heat sensitive enzymes are the most common consequences of mutations that do not inactivate the enzyme, and that some of these mutations are expressed in the organism only at high temperatures and that complete dominance of the normal phenotype is expected in a heterozygote. Langridge has found some support for this hypothesis for mutations affecting the structural gene for β-galactosidase *Escherichia coli* K_{12}. In agreement, too, is a recent report of Gibson and Miklovich (1971) on alcohol dehydrogenase in *D. melanogaster*, which shows in some cases that enzyme activity is significantly heterotic after incubation of homogenates for 10 minutes at 40°C, as compared with prior incubation at 25°C. This mechanism could account for desiccation-dependent heterosis, if we can assume a reasonably high correlation between the two stresses as found in *D. melanogaster* (Parsons, 1970). So far as cold temperatures are concerned, low-temperature-sensitive enzymes seem relatively rare (Langridge, 1968); thus perhaps heterosis might be expected to be less extreme at low temperatures, but it does exist as discussed above.

An alternative hypothesis to explain heterosis in extreme-environment comes from evidence for the coadaptation or the relational balance of gene complexes between chromosomes in heterozygotes, which is broken down in homozygotes.

Because homozygotes are less commonly subjected to natural selection, their fitnesses will normally be lower, and probably this can be expected to be particularly marked in extreme environments. Because heterozygotes are the genotypes most commonly exposed to natural selection in outbred species, the general observation that they are less variable in a multiplicity of environments is then not unexpected–and a concomitant of this is the increased level of heterosis found in stress environments. Needless to say, neither of these hypotheses is mutually exclusive, since the latter could be put into molecular terms. Even if the molecular hypothesis is too restrictive, it is likely that the major environmental stresses that a wild population of *Drosophila* is likely to be subjected to, will be tied up with variations in temperature and humidity. The term "extreme-environment heterosis" is adopted since it encompasses all stresses, but still admits the importance of temperature (and humidity).

CORRELATIONS WITH ENZYME VARIANTS

In recent years there has been a spectacular increase in analyzing individuals of many species (see Stone, Kojima, and Johnson, 1969) for enzyme and protein variants by gel electrophoresis, which enables a quick survey of size and electric charge differences in the protein molecules in different individuals, the differences reflecting altered molecular structures. Such studies have revealed that a large fraction of loci controlling enzyme and protein type specificity are polymorphic (Lewontin and Hubby, 1966). Based on such studies, the proportion of polymorphic to total loci in the genome has been conservatively estimated as 30% to 70%, although this is subject to some assumptions such as the randomness of sampled enzymes and proteins. Considerable discussion has appeared in the literature as to how the population can tolerate the "genetic load" implied by this level of polymorphism, especially if it is assumed that the polymorphisms are maintained by the classical overdominance fitness model. To some extent this can be explained (see Sved et al., 1967; Wallace, 1968; Crow and Kimura, 1970), and it is not the place of this review to discuss this generally, although a specific suggestion was made above on environment-dependent heterosis. There is little evidence for overdominance for enzyme polymorphisms except for a brief report by Powell and Richmond (1970) for a sex-linked enzyme locus, tetrazolium oxidase in *D. paulistorum.* However, assuming the existence of environment-dependent heterosis, overdominance may not be expected for most experiments, which would normally be carried out in nonextreme environments. On the other hand, frequency-dependent selection has been shown to be responsible for the maintenance of biochemical polymorphism in at least one system, Est-6 in *D. melanogaster* (Kojima and Yarbrough, 1967), and this could well turn out to be increasingly important in the maintenance of polymorphisms (see Wallace, 1968 for a further discussion).

It is likely that the next few years will provide a great deal of information for selective differentials in different environments for enzyme and protein variants. In *D. ananassae*, there is probably a cline for the major allele in the Est-C system in the Pacific, such that its frequency increases from north to south towards the equator, and then decreases again south of the equator (Stone, Kojima, and Johnson, 1969). In *D. melanogaster,* Beardmore (1970) has presented experimental data to show that the frequency of alleles at the esterase-6-locus depends on temperature, such that the esterase-6^F allele tends to be at a higher frequency in experimental populations at high temperatures. A good example of a possible association with environment comes from the harvester ant, *Pogonomyrmex barbatus*, from Texas (Johnson et al., 1969) for two esterases and a malic dehydrogenase. Principle component analysis was used to compare the patterns of variability of the allelic frequencies with environmental factors (plant life areas, elevation, rainfall, January and July temperatures, and growing season). Significant correlation was particularly evident with respect to weather and the pattern of variability in both esterases, from which it may be inferred that natural selection is important in determining the allele frequency patterns. Overdominance was not found, and it was concluded that all the enzyme systems studied were affected by the environmental factors. It will be of interest to extrapolate such studies to *Drosophila* spp.; the species *Pogonomyrmex barbatus* was chosen because of its abundance over much of Texas, and because sampling could be done so as to avoid most migration difficulties. Also compared with *Drosophila*, populations are stable and are maintained at relatively high densities throughout the year, and individuals live considerably longer than most *Drosophila*. Another example of a dependence on temperature comes from the freshwater fish *Catostomus clarkii*, where the frequency of alleles for polymorphic serum esterases varies with latitude (Koehn and Rasmussen, 1967; Koehn, 1969). The activity of the allele more frequent in southern populations increases as temperatures increase from 0° to 37°C, whereas the activity of the allele more frequent in northern populations increases as temperature decreases. In *D. melanogaster*, Gibson (1970) has shown that ethanol supplementation in the diet affects gene frequencies at a locus for alcohol dehydrogenase. This may be of some ecological importance because there is probably considerable variation in the amounts of ethanol produced by yeast fermentation in *Drosophila* habitats. Finally, Beardmore (1970) has shown that the frequency of the Esterase-6^F allele is correlated with population sizes in the laboratory, but it is difficult to relate this to natural populations at this stage.

Much evidence shows that various inversions in some *Drosophila* species are overdominant within populations. Prakash and Lewontin (1968) have shown considerable genetic divergence between inversions of *D. pseudoobscura* for enzyme and protein variants, and the available evidence suggests that the genic differences date from the origin of the inversions, predating the present distribution of the species. Even though no *direct* studies of environmental variables were carried out, there is the possibility that loci controlling enzyme and protein variants associated with different inversions could be of importance in explaining the associations of inversions with season, altitude, and environmental factors. The likelihood of this is enhanced if a molecular basis, either partly or wholly, for environment-dependent heterosis is accepted. Information on this is likely in the near future, and some of the definitive work may involve comparisons of enzyme and protein variants under extreme and more optimal environments.

MIGRATION AND DISPERSAL

If the physiological state of an animal is such that it can survive certain deprivations and its vagility is reasonable, then it is possible that the restrictions of an unsuitable habitat may be overcome by migration to a new habitat void of such restrictions. Migration should therefore be considered, although Wallace (1966a) does not consider the dispersal abilities of *Drosophila* to be very great; this limits areas that can be used as dispersal routes in times of macroenvironmental crisis. Wallace examined dispersal data of Dobzhansky and Wright (1943) in *D. pseudoobscura* and he concluded that as many as one quarter of all flies in a single highly localized collection may have emerged from pupae within 25 yards (23 m) of the collection site, and available data suggest that *D. melanogaster, D. willistoni,* and *D. funebris* are much more restricted, since 60% to 80% of individuals of these species may have a point of origin within the 25 yards (Fig. 3). This is based on a model where the logarithm of the number of flies recaptured at various distances from the point of release decreases linearly with the square root of the distance. There was also some evidence that crowding led to an initial exceptionally rapid dispersal rate. It should, however, be noted that Wright (1968) has suggested that variations in the results of the separate experiments are responsible for the linearity. A similar model has been shown to apply to *D. aldrichi* by Richardson (1969) using a technique of chemically labelling flies. This seems a good general technique not yet much used, provided that it can be shown that the label has no effect on flies. Wallace (1966b) studied the allelism of lethals in *D. melanogaster* males captured near Bogota, Colombia at short distances (0, 30, 60, and 90 m) from one another within a 2 hour period. Eighty percent of the 119 genomes tested proved to be lethal, and he found that the model, which assumes that the logarithm of the frequency of allelism due to common descent declines linearly with the square root of distance, fitted his data reasonably. This type of model has been found to be applicable to migration data in other species (Bateman, 1950). Cavalli-Sforza (1962) has discussed similar models for the probability of marriage in man in relation to the distance between marriage partners.

Migration permits the movement of genes from one locality or population to another, and so the migrated genotype interacts with the new environment. If

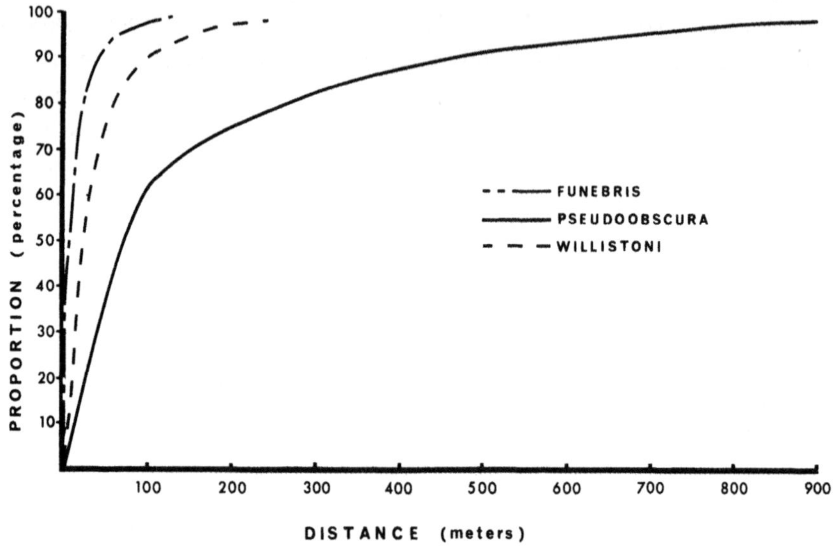

Fig. 3. Proportion of individuals of several Drosophila *species whose calculated point of origin is expected to be less than a given distance from the point of capture (after Wallace, 1966a).*

differential migration of genotypes occurs such that certain genotypes tend to migrate more than others, we have, following Fisher (1930), migrational selection, and this may occur as a result of environmental heterogeneity (see also Parsons, 1963). Fisher developed his arguments in terms of geographical distance. A likely conclusion, if there is a gradient of ecological conditions over a geographical transect, will be a gradient of fitnesses so leading to clines in gene frequencies, and the integrity of the cline may be maintained by migration. If active migration of genes continues between sections of a cline, there will be little tendency for parts to become isolated. *D. pseudoobscura* shows such geographical clines, and it must perhaps be concluded that this is aided by a degree of migration between adjacent populations.

In natural conditions, migration cannot therefore be overlooked, and so it is of interest to consider the performance of different genotypes in the laboratory where migration is possible, and this may be achieved using a migration tube system (Fig. 4). This allows individuals to move from one food chamber to another, and different strains show different powers of dispersal (Sakai et al., 1958). Sakai and coworkers divided migrations in *D. melanogaster* into two kinds; namely, random migration, which occurs as a result of random movement of individual flies; and mass migration as a result of pressure of population density. Interstrain differences for both activities were very pronounced, and it is assumed that the two types of activities are under genetic control. Furthermore, in a highly inbred and domesticated strain, the intensity of

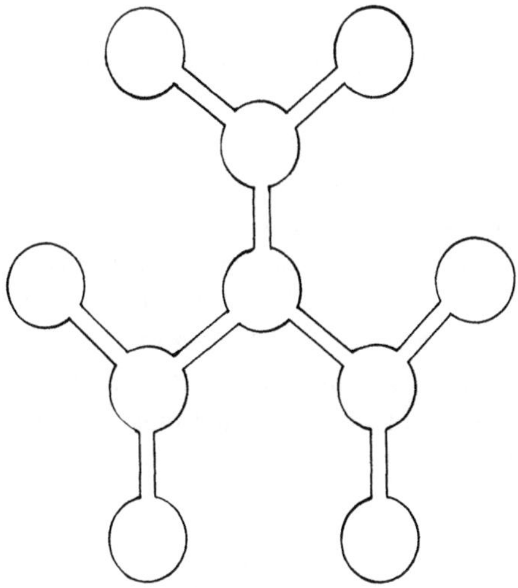

Fig. 4. Arrangement of 10 migration-tubes from above, showing the connections between them (after Narise, 1968. Courtesy the Editor of Evolution).

random migration was as low as 2.25% of the total population per day, while in wild strains the figure was 14% to 20%. The low random migration rate in the domesticated strain may well have been developed in response to a closed environment, where migratory ability would not be favored. Narise (1968, 1969) found that while *vg* flies were eliminated in a population cage, an equilibrium value was obtained when migration was possible, so indicating that migration can be an extremely powerful factor in the continuance of a species, or of different genotypes of the species. Since different strains of the species *D. melanogaster* show differences in vagility (Sakai et al., 1958), it is not unlikely that different species will differ in dispersal capabilities.

Population dispersal of *Drosophila* in the wild is therefore far from being understood, and considerable controversy exists regarding the data available. Most work is based on the release and recapture of mutant eye varieties (Dobzhansky and Wright, 1943). Traps are set in patterns allowing a mathematical analysis of dispersal rates and distances from the number of flies caught. To what extent this can reflect what actually happens in wild *Drosophila* is debatable, since the individual genotype affects vagility (Narise, 1968), and also the mutants are released from a crowded situation, which would tend to yield an overestimate of dispersal, since as discussed above, population density is related to dispersal rate. Thus the biological interpretation of dispersal rates, even if they can be described mathematically as has been done by Wallace (1966a) is a matter of argument. Perhaps, by the use of labeled isotopes some of these problems could be minimized, so long as labeled individuals could be shown to be equally vagile as unlabeled individuals, but it may be difficult to get complete biological reality from any capture-recapture system since the releasing of labeled flies adds a parameter to the ecosystem that does not exist in the wild. Richardson (1969) used flies labeled in the wild by attracting them to a trap, so partially avoiding this problem.

In conclusion, migration is affected by factors such as population density, genotype, competition with other genotypes, the nature of the surrounding environment, and the living space of the species, but much more remains to be done. One factor which might increase the relative effectiveness of migration is the observation of Petit and Ehrman (1968) and Ehrman (1969) and others in *D. pseudoobscura* and other species of *Drosophila* that the mating success of certain genotypes of males increases as the males become rare relative to other males. If common, this density-dependent effect could substantially increase the effectiveness of migration, as migrant males would leave more than their proportionate share of offspring.

POPULATION SIZE

One measure used when discussing population sizes is the innate capacity for increase r_m, which is the rate of increase of a population having a constant age schedule of births and deaths, and increasing in numbers in unlimited space. In *D. birchii*, r_m tended to increase in experimental populations at both 20° and 25°C for flies collected in the wild going from north to south from Rabaul in New Britain through Port Moresby (New Guinea) to Cairns (Australia); while for *D. serrata*, r_m tended to decrease from north (Brisbane) to south (Sydney) (Birch et al., 1963). Not enough is known about the ecology of the two species to speculate on the possible adaptive significance of this pattern, but variations in r_m in different climates and laboratory environments must have general ecological significance, and are worth detailed study.

Carson (1961a) detected differences in productivity and size in experimental populations of *D. robusta* derived from different localities, such that the performance of a strain collected at the center of the geographic distribution performed better than a marginal strain as expected from considerations presented earlier. The same seems to apply to *D. serrata.* In *D. serrata* and *D. birchii*, hybrid populations between strains were generally larger and with higher productivity than the parents (Ayala, 1966), and in *D. pseudoobscura*, size and productivity were higher in populations polymorphic for two chromosomal rearrangements compared with monomorphic populations (Dobzhansky and Pavlovsky, 1961; Dobzhansky, Lewontin and Pavlovsky, 1964). The behaviorally polymorphic population of *D. willistoni* for pupation site showed a greater biomass than monomorphic ones (de Souza et al., 1970). In *D. melanogaster*, the average population size of a strain carrying several mutant genes was about one third that of a hybrid population of the mutant gene stock and an Oregon-R strain (Carson, 1961b). Ayala (1965) studied populations of *D. serrata* in the laboratory and found a slow overall increase in the number of

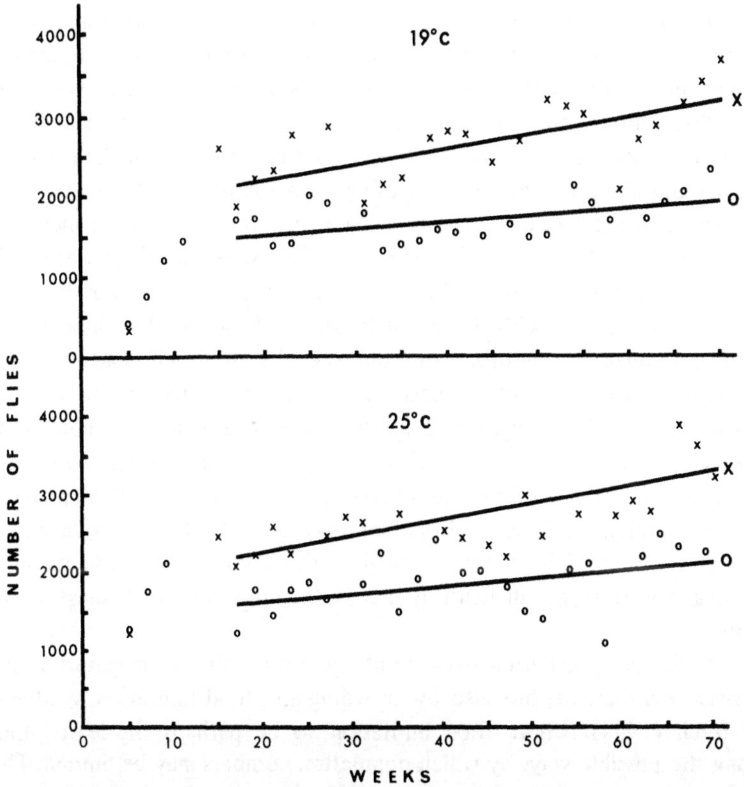

Fig. 5. Population size of 4 experimental populations of D. serrata *at two temperatures: O, Popondetta (New Guinea) population; X, Sydney X Popondetta hybrid population. (After Ayala, 1968. Copyright © 1968 by the American Association for the Advancement of Science.)*

flies over a 70 week period, which was greater in a hybrid population between flies collected from two localities compared with one locality (Fig. 5), perhaps because of the greater genetic variability of the former. The hybrid population therefore shows superior genetic adaptation to the experimental environment, and the results above generally indicate hybrid superiority.

Animals can survive and multiply within certain temperature ranges, as has been shown. The innate capacity for increase in numbers of various geographic strains of *D. serrata* and *D. birchii* was nearly double at 25° as compared with 20°C (Birch et al., 1963), and *D. pseudoobscura* of various genetic constitutions had an average capacity for increase more than double at 25° than at 16° (Dobzhansky, Lewontin, and Pavlovsky, 1964). These experiments are difficult to extrapolate directly to the wild, especially in view of evidence for extreme environment heterosis, and the known complexity of the environment of *D. pseudoobscura* in the wild, which has been found to live in slime fluxes in association with a number of organisms that may restrict its population size.

When a population is growing with a limited amount of food and space, the number of animals per unit of food and space often increases. Eventually an equilibrium occurs when the number of births per unit time equals the number of deaths, although the actual population size for a given number of births per unit time varies because of differences in overall longevity. In the limited space of a population cage, 400 cm^3, increasing the amount of food from two to three units of food per week in *D. serrata* led to the total population increasing by only 17%, even though nearly 50% more flies were added per week (Ayala, 1968). The increased amount of food caused an increase in the number of births which was associated with decreased longevity. It seems that population sizes under experimental conditions are controlled by food and space, and that as food is progressively increased space becomes a limiting factor, the consequence of which is to reduce longevity. Experiments with several geographic strains of *D. serrata*, and with *D. birchii, D. melanogaster*, and *D. pseudoobscura* have shown the importance of food and space (Ayala, 1968). In nature, predators or other components of the environment may take the limiting role played by living space in the laboratory, although without further experiments and field investigations it seems difficult to assess the importance of these results in nature.

In conclusion, population sizes are affected not only by the genotype and the physical environment, but also by crowding and food limitation. Andrewartha and Birch (1954) believe food limitation to be perhaps the least important among the possible ways by which population numbers may be limited. The real problem that emerges is the extreme difficulty of extrapolating these experiments to the wild, since a trait like population size, while having a meaning in the laboratory, may have little meaning in the wild. This is because population size must depend on *all* the variables discussed in this review and others as well, and is one manifestation of their interaction.

CONCLUSIONS

From a diffuse literature the following conclusions emerge:

(1). Little is known about the factors controlling the distribution of different *Drosophila* spp. in nature, although variations due to environmental and seasonal changes can be detected. A problem is that samples detected by trapping could be nonrandom. Care in trapping procedures should help to minimize this, but the differential attraction of various species to different baits is a problem. The simplest situations concern those species of *Drosophila* restricted to specific plant or animal species as hosts, while more cosmopolitan species are more difficult to study because they occupy complex ecological niches, but on the other hand, the cosmopolitan species are often easier to breed in the laboratory. Thus knowledge of the ecology of well-studied laboratory species is poor.

(2). In many species frequencies of inversions can be correlated with environmental variables. The frequent greater fitness of populations polymorphic for inversions as compared with monomorphic can be regarded as evidence of mutual facilitation between genotypes, so that the polymorphic population may more readily be able to adapt to a wide variety of ecological niches than the monomorphic population. The mutual adjustment of gene complexes occurring within populations must have evolved by natural selection in the environments of the populations in question, as shown by the lowered fitness of hybrid populations where chromosomes were taken from different localities.

(3). In some species the level of inversion heterozygosity decreases as the margin of the species range is approached. It is often inferred that central populations are better adapted to their diverse environments by a high level of inversion heterozygosity, while in marginal populations there is a premium on flexibility to adapt to new environments that would occur in populations with a lower level of inversion heterozygosity and a consequent higher level of recombination. Similarly, marginal populations tend to have fewer lethal and semilethal genes than central populations. More generally the amount of genetic variability in a population may be a function of the diversity of the environment in which the population lives.

(4). Within many *Drosophila* species populations possess integrated genotypes adapted to the various environments prevailing within the populations, although the environments can often not be defined.

(5). Some morphological traits vary between environments in the wild, but it is difficult to interpret such variations except as correlated with larger fly sizes, which are usually found in cooler, and smaller fly sizes in warmer, environments. Clines in morphological traits have been reported in relation to altitude and latitude.

(6). Genetic heterogeneity is known in natural populations for physiological traits such as resistance to extreme temperatures and desiccation, and behavioral traits. In hot dry environments *D. melanogaster* probably adapts by a complex of genetic heterogeneity, physiological plasticity, and behavioral avoidance mechanisms. The relative importance of these factors probably varies between species.

(7). In *D. pseudoobscura* heterokaryotype advantage is more pronounced in extreme than optimal environemnts. Similarly, hybrids between inbred lines and other homozygotes tend to show heterosis in extreme environments in several species. Extreme environment heterosis could be associated with temperature-sensitive enzymes (since extreme environments often seem to involve, directly or indirectly, extreme temperatures), or may be due to the general poorer fitness of homozygotes as compared with the corresponding heterozygotes, although the two hypotheses are not mutually exclusive. Environment dependent heterosis provides an explanation of the high level of polymorphism in natural populations additional to those already advanced in the literature, since heterosis is strongly manifested only in extreme environments.

(8). Some attempts have been made to relate the enzyme and protein variants in natural populations of *Drosophila* to environment, but results are sparse, although some direct associations with the environment are emerging especially for temperature.

(9). In the species of *Drosophila* studied, migration is limited. Laboratory experiments in *D. melanogaster* show that migration can be divided into that which occurs by the random movement of individual flies, and the mass migration of flies as a result of population density. Interstrain differences in both activities were observed to be probably under genetic control.

(10). The innate capacity for increase of a population and its ultimate size depends on factors in the laboratory such as temperature, genotype, food, and space. While being difficult to assess the importance of these experiments in nature, it is considered that further work will be rewarding.

Some links between the artificial environments of the laboratory and the wild have been found often by inference. The linking of the two will slowly come with time with the development of more sophisticated laboratory experiments aimed at simulating what may happen in the wild, and by more precise studies of flies in nature. The genetic sophistication possible in the laboratory in many species of *Drosophila* makes it worth aiming at such a unification. We conclude by quoting G. G. Simpson (1953) in his book *The Major Features of Evolution*: "Adaptation is a fitting together of organisms and, in the very broadest sense, environments. It is false to ascribe this action either to the organisms or to the environments. Both organisms and environments have their characteristic actions, in other words functions, in the process, and adaptation is a characteristic interaction of the two." It is our view that the genus *Drosophila*, in spite of some disadvantages, may permit the elaboration and verification of this statement.

Acknowledgements

We are grateful for helpful and detailed comments on the whole manuscript from Dr. I. T. MacBean of La Trobe University, and Dr. W. R. Scowcroft of the C.S.I.R.O., Division of Plant Industry, Canberra. Other helpful comments have come from Dr. J. Bundgaard, Dr. R. S. Holmes, and Mr. B. A. Conroy of La Trobe University. We would like to thank Miss Glenda Wilson for secretarial help and Mrs. G. Greer for drawing the diagrams. Financial support has been received from the Australian Institute of Nuclear Science and Engineering, and the Australian Research Grants Committee.

REFERENCES

Anderson, W. W. 1968. Further evidence for coadaptation in crosses between geographic populations of *Drosophila pseudoobscura*. Genet. Res., 12:317-330.

——1969. Genetics of natural populations. XLI. The selection coefficients of heterozygotes for lethal chromosomes in *Drosophila* on different genetic backgrounds. Genetics, 62:827-836.

____C. Oshima, T. Watanabe, Th. Dobzhansky, and O. Pavlovsky. 1968. Genetics of natural populations. XXXIX. A test of the possible influence of two insecticides on the chromosomal polymorphism in *Drosophila pseudoobscura.* Genetics, 58:423-434.

Andrewartha, H. G., and L. C. Birch. 1954. The Distribution and Abundance of Animals. Chicago, University of Chicago Press.

Ayala, F. J. 1965. Evolution of fitness in experimental populations of *Drosophila serrata.* Science, 150:903-905.

____1966. Dynamics of populations I. Factors controlling population growth and population size in *Drosophila serrata.* Amer. Natural., 100:333-344.

____1968. Genotype, environment, and population numbers. Science, 162:1453-1459.

____1969a. Evolution of fitness. IV. Genetic evolution of interspecific competitive ability in *Drosophila.* Genetics, 61:737-747.

____1969b. Genetic polymorphism and interspecific competitive ability in *Drosophila.* Genet. Res., 14:95-102.

Band, H. T. 1963. Genetic structure of populations II. Viabilities and variances of heterozygotes in constant and fluctuating environments. Evolution, 17:307-319.

____and P. T. Ives. 1968. Genetic structure of populations IV. Summer environmental variables and lethal and semilethal frequencies in a natural population of *Drosophila melanogaster.* Evolution, 22:633-641.

Bateman, A. J. 1950. Is gene dispersion normal? Heredity (London), 4:353-363.

Battaglia, B., and H. Smith. 1961. The Darwinian fitness of polymorphic and monomorphic populations of *Drosophila pseudoobscura* at 16°C. Heredity (London), 16:475-484.

Beardmore, J. A. 1961. Diurnal temperature fluctuation and genetic variance in *Drosophila* populations. Nature (London), 189:162-163.

____1963. Mutual facilitation and the fitness of polymorphic populations. Amer. Natural., 97:69-74.

____1970. Ecological factors and the variability of gene-pools in *Drosophila. In* Hecht, M. K., and W. C., Steere, eds., Essays in Evolution and Genetics in Honor of Theodosius Dobzhansky, pp. 299-314. New York, Appleton-Century-Crofts.

____Th. Dobzhansky, and O. A. Pavlovsky. 1960. An attempt to compare the fitness of polymorphic and monomorphic experimental populations of *Drosophila pseudoobscura.* Heredity (London), 14:19-33.

Birch, L. C. 1955. Selection in *Drosophila pseudoobscura* in relation to crowding. Evolution, 9:389-399.

____Th. Dobzhansky, P. O. Ellott, and R. C. Lewontin. 1963. Relative fitness of geographic races of *Drosophila serrata.* Evolution, 17:72-83.

Blight, W. C., and A. Romano. 1953. Notes on a breeding site of *Drosophila americana* near St. Louis, Missouri. Amer. Natural., 87:111-112.

Bodmer, W. F., and P. A. Parsons. 1960. The initial progress of new genes with various genetic systems. Heredity (London), 15:283-299.

____and P. A. Parsons. 1962. Linkage and recombination in evolution. Advances Genet., 11:1-100.

Brncic, D. 1954. Heterosis and the integration of the genotype in geographic populations of *Drosophila pseudoobscura.* Genetics, 39:77-88.

____1961. Integration of the genotype in geographic populations of *Drosophila pavani.* Evolution, 15:92-97.

____1962. Chromosomal structure of populations of *Drosophila flavopilosa* studied in larvae collected in their natural breeding sites. Chromosoma, 13:183-195.

____1966. Ecological and cytogenetic studies of *Drosophila flavopilosa*, a neotropical species living in *Cestrum* flowers. Evolution, 20:16-29.

____1968. The effects of temperature on chromosomal polymorphism of *Drosophila flavopilosa* larvae. Genetics, 59:427-432.

____1969. Long-term changes in chromosomally polymorphic laboratory stocks of *Drosophila pavani.* Evolution, 23:502-508.

____1970. Studies on the evolutionary biology of Chilean species of *Drosophila. In* Hecht, M. K., and W. C. Steere, eds., Essays in Evolution and Genetics in Honor of Theodosius Dobzhansky, pp. 401-436. New York, Appleton-Century-Crofts.

____and S. Koref-Santibañez. 1964. Mating activity of homo- and heterokaryotypes in

Drosophila pavani. Genetics, 49:585-591.

Carson, H. L. 1951. Breeding sites of *Drosophila pseudoobscura* and *Drosophila persimilis* in the transition zone of the Sierra Nevada. Evolution, 5:91-96.

____1958a. The population genetics of *Drosophila robusta.* Advances Genet., 9:1-40.

____1958b. Response to selection under different conditions of recombination in *Drosophila.* Sympos. Quant. Biol., 23:291-306.

____1961a. Relative fitness of genetically open and closed experimental populations of *Drosophila robusta.* Genetics, 46:553-567.

____1961b. Heterosis and fitness in experimental populations of *Drosophila melanogaster.* Evolution, 15:496-509.

____1967. The association between *Drosophila carcinophila* Wheeler and its host, the land crab *Gecarcinus ruricola* (L). Amer. Midl. Natural., 78:324-343.

____D. E. Hardy, H. T. Speith and W. S. Stone. 1970. The evolutionary biology of the Hawaiian Drosophilidae. *In* Hecht, M. K. and W. C. Steere, eds., Essays in Evolution and Genetics in Honor of Theodosius Dobzhansky, pp. 437-543. New York, Appleton-Century-Crofts.

____and M. Wasserman. 1965. A widespread chromosomal polymorphism in a widespread species, *Drosophila buzzatii.* Amer. Natural., 99:111-115.

____and M. R. Wheeler. 1967. *Drosophila endobranchia*, a new Drosophila associated with land crabs in the West Indies. Ann. Entom. Soc. Amer., 61:675-678.

Cavalli-Sforza, L. 1962. The distribution of migration distances: models, and applications to genetics. *In* Sutter, J., Les Déplacements Humains–Aspects Methodologiques de Leur Mesure, pp. 139-158. Paris, Hachette.

Cordeiro, A. R., J. I. Townsend, J. A. Petersen, and E. C. Jaeger. 1958. Genetics of Southern marginal populations of *Drosophila willistoni.* Proc. X. Int. Congr. Genet., 2:58-59.

Crow, J. F., and M. Kimura. 1970. An Introduction to Population Genetics Theory. New York, Harper & Row.

Crumpacker, D. W., and D. Marinkovic. 1967. Preliminary evidence of cold temperature resistance in *Drosophila pseudoobscura.* Amer. Natural., 101:505-514.

Da Cunha, A. B. 1951. Modification of the adaptive values of chromosomal types in *Drosophila pseudoobscura* by nutritional variables. Evolution, 5:395-404.

____1955. Chromosomal polymorphism in the Diptera. Advances Genet., 7:93-138.

____and Th. Dobzhansky. 1954. A further study of chromosomal polymorphism in *Drosophila willistoni* in its relation to the environment. Evolution, 8:119-134.

____Th. Dobzhansky, O. Pavlovsky, and B. Spassky. 1959. Genetics of natural populations. XXVIII. Supplementary data on the chromosomal polymorphism in *Drosophila willistoni* in its relation to the environment. Evolution, 13:389-404.

____Th. Dobzhansky, and A. Sokoloff. 1951. On food preferences of sympatric species of *Drosophila.* Evolution, 5:97-101.

Darlington, C. D. 1937. Recent Advances in Cytology. 2nd ed. London, Churchill.

Dawood, M. M., and M. W. Strickberger. 1969. The effect of larval interaction on viability in *Drosophila melanogaster* III. Effects of biotic residues. Genetics, 63:213-220.

del Solar, E. 1968. Selection for and against gregariousness in the choice of oviposition sites by *Drosophila pseudoobscura.* Genetics, 58:275-282.

____ and H. Palomino. 1966. Choice of oviposition sites in *Drosophila melanogaster.* Amer. Natural., 100:127-133.

de Souza, H. M. L., A. B. da Cunha, and E. P. dos Santos. 1970. Adaptive polymorphism of behavior evolved in laboratory populations of *Drosophila willistoni.* Amer. Natural., 104:175-189.

Dobzhansky, Th. 1947a. Genetics of natural populations. XIV. A response of certain gene arrangements in the third chromosome of *Drosophila pseudoobscura* to natural selection. Genetics, 32:142-160.

____1947b. Adaptive changes induced by natural selection in wild populations of *Drosophila.* Evolution, 1:1-16.

____1947c. A directional change in the genetic constitution of a natural population of *Drosophila pseudoobscura.* Heredity (London), 1:53-64.

____1948a. Genetics of natural populations. XVIII. Experiments on chromosomes of

Drosophila pseudoobscura from different geographic regions. Genetics, 33:588-602.
____1948b. Genetics of natural populations. XVI. Altitudinal and seasonal changes produced by natural selection in certain populations of *Drosophila pseudoobscura* and *Drosophila persimilis.* Genetics, 33:158-176.
____1950. Genetics of natural populations. XIX. Origin of heterosis through natural selection in populations of *Drosophila pseudoobscura.* Genetics, 35:288-302.
____1951. Genetics and the Origin of Species, 3rd ed. New York, Columbia University Press.
____1957a. Mendelian populations as genetic systems. Sympos. Quant. Biol., 22:385-393.
____1957b. Genetics of natural populations. XXVI. Chromosomal variability in island and continental populations of *Drosophila willistoni* from Central America and the West Indies. Evolution, 11:280-293.
____1959. Evolution of genes and genes in evolution. Sympos. Quant. Biol., 24:15-30.
____1965. "Wild" and "domestic" species of *Drosophila. In* Baker, H. G. and G. L. Stebbins, eds., The Genetics of Colonizing Species, pp. 533-546. New York, Academic Press.
____1968. On some fundamental concepts of Darwinian biology. Evol. Biol, 2:1-34.
W. W. Anderson, and O. Pavlovsky. 1966. Genetics of natural populations. XXXVIII. Continuity and change in populations of *Drosophila pseudoobscura* in Western United States. Evolution, 20:418-427.
____W. W. Anderson, O. Pavlovsky, B. Spassky, and C. J. Wills. 1964. Genetics of natural populations. XXXV. A progress report on genetic changes in populations of *Drosophila pseudoobscura* in the American Southwest. Evolution, 18:164-176.
____H. Burla, and A. B. daCunha. 1950. A comparative study of chromosomal polymorphism in sibling species of the *willistoni* group of *Drosophila.* Amer. Natural., 84:229-246.
____A. S. Hunter, O. Pavlovsky, B. Spassky, and B. Wallace. 1963. Genetics of natural populations. XXXI. Genetics of an isolated marginal population of *Drosophila pseudoobscura.* Genetics, 48:91-103.
____and H. Levene. 1951. Development of heterosis through natural selection in experimental populations of *Drosophila pseudoobscura.* Amer. Natural., 85:247-264.
____H. Levene, B. Spassky, and N. Spassky. 1959. Release of genetic variability through recombination. III. *Drosophila prosaltans.* Genetics, 44:75-92.
____R. C. Lewontin, and O. Pavlovsky. 1964. The capacity for increase in chromosomally polymorphic and monomorphic populations of *Drosophila pseudoobscura.* Heredity (London), 19:597-614.
____and C. Pavan. 1950. Local and seasonal variations in relative frequencies of species of *Drosophila* in Brazil. J. Anim. Ecol., 19:1-14.
____and O. Pavlovsky. 1953. The indeterminate outcome of certain experiments on *Drosophila* populations. Evolution, 7:198-210.
____and O. Pavlovsky. 1958. Interracial hybridization and breakdown of coadapted gene complexes in *Drosophila paulistorum* and *Drosophila willistoni.* Proc. Nat. Acad. Sci. U.S.A., 44:622-629.
____and O. Pavlovsky. 1961. A further study of fitness of chromosomally polymorphic and monomorphic populations of *Drosophila pseudoobscura.* Heredity (London), 16:169-179.
____O. Pavlovsky, B. Spassky, and N. Spassky. 1955. Genetics of natural populations. XXIII. Biological role of deleterious recessives in populations of *Drosophila pseudoobscura.* Genetics, 40:781-796.
____and B. Spassky. 1968. Genetics of natural populations. XL. Heterotic and deleterious effects of recessive lethals in populations of *Drosophila pseudoobscura.* Genetics, 59:411-425.
____and S. Wright. 1943. Genetics of natural populations. X. Dispersion rates in *Drosophila pseudoobscura.* Genetics, 28:304-340.
Dubinin, N. P., and G. G. Tiniakov. 1945. Seasonal cycles and the concentration of inversions in populations of *Drosophila funebris.* Amer. Natural., 79:570-572.
____and G. G. Tiniakov. 1946a. Structural chromosome variability in urban and rural populations of *Drosophila funebris.* Amer. Natural., 80:393-396.
____and G. G. Tiniakov. 1946b. Natural selection and chromosomal variability in populations of *Drosophila funebris.* J. Hered., 37:39-44.

——and G. G. Tiniakov. 1946c. Inversion gradients and natural selection in ecological races of *Drosophila funebris*. Genetics, 31:537-545.

——and G. G. Tiniakov. 1947. Inversion gradients and selection in ecological races of *Drosophila funebris.* Amer. Natural., 81:148-153.

Ehrman, L. 1965. Direct observation of sexual isolation between allopatric and between sympatric strains of the different *Drosophila paulistorum* races. Evolution, 19:459-464.

——1969. The sensory basis of mate selection in *Drosophila.* Evolution, 23:59-64.

Fisher, R. A. 1930. The Genetical Theory of Natural Selection. Oxford, Clarendon Press.

Fontdevila, A. 1970. Genotype-temperature interaction in *Drosophila melanogaster.* I. Viability. Genetica, 41:257-264.

Ford, E. B. 1964. Ecological Genetics. London, Methuen.

Fraser, A. 1963. Variation of scutellar bristles in *Drosophila.* I. Genetic leakage. Genetics, 48:497-514.

——L. Erway, and W. Brenton. 1968. Variation of scutellar bristles in *Drosophila.* XIV. Effects of temperature and crowding. Aust. J. Biol. Sci., 21:75-87.

Freire-Maia, N. 1964. Chromosomal monomorphism in Brazilian and Argentine populations of *Drosophila simulans* and *Drosophila repleta.* Genetics, 50:1447-1448.

Futuyma, D. J. 1970. Variation in genetic response to interspecific competition in laboratory populations of *Drosophila.* Amer. Natural., 104:239-252.

Gibson, J. 1970. Enzyme flexibility in *Drosophila melanogaster.* Nature (London), 227:959-960.

——and Miklovich, R. 1971. Modes of variation in alcohol dehydrogenase in *Drosophila melanogaster.* Experientia, 27:99-100.

Haldane, J. B. S. 1957. The conditions for coadaptation in polymorphism for inversions. J. Genet., 55:218-225.

Heuts, M. J. 1948. Adaptive properties of carriers of certain gene arrangements in *Drosophila pseudoobscura.* Heredity (London), 2:63-75.

Hoenigsberg, H. F. 1968. An ecological situation which produced a change in the proportion of *Drosophila melanogaster* to *Drosophila simulans.* Amer. Natural., 102:389-390.

——L. A. Granobles, and L. E. Castro. 1969. Population genetics in the American tropics. IV. Temporal changes effected in natural populations of *Drosophila melanogaster* from Colombia. Genetica, 40:210-215.

Hosgood, S. M. W., and P. A. Parsons. 1966. Differences between *D. simulans* and *D. melanogaster* in tolerances to laboratory temperatures. Drosophila Information Service, 41:176.

——and P. A. Parsons. 1967. Genetic heterogeneity among the founders of laboratory populations of *Drosophila melanogaster.* II. Mating Behavior. Aust. J. Biol. Sci., 20:1193-1203.

——and P. A. Parsons. 1968. Polymorphism in natural populations of *Drosophila* for the ability to withstand temperature shocks. Experientia, 24:727-728.

Johnson, F. M., H. E. Schaffer, J. E. Gillaspy, and E. S. Rockwood. 1969. Isozyme genotype-environment relationships in natural populations of the harvester ant, *Pogonomyrmex barbatus*, from Texas. Biochem. Genet., 3:429-450.

Kalmus, H. 1945. Adaptive and selective responses of a population of *Drosophila melanogaster* containing *e* and *e+* to differences in temperature, humidity and to selection for developmental speed. J. Genet., 47:58-63.

Koehn, R. K. 1969. Esterase heterogeneity: dynamics of a polymorphism. Science, 163:943-944.

——and D. I. Rasmussen. 1967. Polymorphic and monomorphic serum esterase heterogeneity in catostomid fish populations. Biochem. Genet., 1:131-144.

Kojima, K. I. and K. M. Yarbrough. 1967. Frequency-dependent selection at the esterase 6 locus in *Drosophila melanogaster.* Proc. Nat. Acad. Sci. U.S.A., 57:645-649.

Koref-Santibañez, S. 1964. Reproductive isolation between the sibling species *Drosophila pavani* and *Drosophila gaucha*. Evolution, 18:245-251.

——and E. del Solar. 1961. Courtship and sexual isolation in *Drosophila pavani* Brncic and *Drosophila gaucha* Jaeger and Salzano. Evolution, 15:401-406.

Krimbas, C. B. 1959. Comparison of the concealed variability in *Drosophila willistoni* with that in *Drosophila prosaltans.* Genetics, 44:1359-1369.

____1961. Release of genetic variability through recombination. VI. *Drosophila willistoni.* Genetics, 46:1323-1334.

____1967. The genetics of *Drosophila subobscura* populations. III. Inversion polymorphism and climatic factors. Molec. Gen. Genet., 99:133-150.

Langridge, J. 1962. A genetic and molecular basis for heterosis in *Arabidopsis* and *Drosophila.* Amer. Natural., 96:5-27.

____1968. Thermal responses of mutant enzymes and temperature limits to growth. Molec. Gen. Genet., 103:116-126.

Levene, H. 1959. Release of genetic variability through recombination. IV. Statistical theory. Genetics, 44:93-104.

____O. Pavlovsky, and Th. Dobzhansky. 1954. Interaction of the adaptive values in polymorphic experimental populations of *Drosophila pseudoobscura.* Evolution, 8:335-349.

Levine, R. P. 1952. Adaptive responses of some third chromosome types of *Drosophila pseudoobscura.* Evolution, 6:216-233.

Levins, R. 1968. Evolution in Changing Environments. Some Theoretical Explorations. Princeton, Princeton University Press.

____1969. Thermal acclimation and heat resistance in *Drosophila* species. Amer. Natural., 103:483-499.

Lewontin, R. C. 1955. The effects of population density and composition on viability in *Drosophila melanogaster.* Evolution, 9:27-41.

____and J. L. Hubby. 1966. A molecular approach to the study of genic heterozygosity in natural populations. II. Amount of variation and degree of heterozygosity in natural populations of *Drosophila pseudoobscura.* Genetics, 54:595-609.

____and Y. Matsuo. 1963. Interaction of genotypes determining viability in *Drosophila bucksii.* Proc. Nat. Acad. Sci. U.S.A., 49:270-278.

Lowe, C. H., W. B. Heed, and E. A. Halpern. 1967. Supercooling of the saguaro species *Drosophila nigrospiracula* in the Sonoran Desert. Ecology, 48:984-985.

McFarquhar, A. M., and F. W. Robertson, 1963. The lack of evidence for co-adaptation in crosses between geographical races of *Drosophila subobscura.* Coll. Genet. Res., 4:104-131.

McWilliam, J. R. and B. Griffing. 1965. Temperature-dependent heterosis in maize. Aust. J. Biol. Sci., 18:569-583.

Marinkovic, D., D. W. Crumpacker, and V. M. Salceda. 1969. Genetic loads and cold temperature resistance in *Drosophila pseudoobscura.* Amer. Natural., 103:235-246.

Mayhew, S. H., S. K. Kato, F. M. Ball, and C. Epling. 1966. Comparative studies of arrangements within and between populations of *Drosophila pseudoobscura.* Evolution, 20:646-662.

Misra, R. K., and E. C. R. Reeve. 1964. Clines in body dimensions in populations of *Drosophila subobscura.* Coll. Genet. Res., 5:240-256.

Mohn, N. and E. B. Spiess. 1963. Cold resistance of karyotypes in *Drosophila persimilis* from Timberline of California. Evolution, 17:548-563.

Moore, J. A. 1952. Competition between *Drosophila melanogaster* and *Drosophila simulans.* I. Population cage experiments. Evolution, 6:407-420.

Nair, P. S. 1969. Genetic load in *Drosophila robusta.* Genetics, 63:221-228.

Narise, T. 1968. Migration and competition in *Drosophila.* I. Competition between wild and vestigial strains of *Drosophila melanogaster* in a cage and migration-tube population. Evolution, 22:301-306.

____1969. Migration and competition in *Drosophila.* III. Effect of genetic background on migratory behavior of *D. melanogaster.* Jap. J. Genet., 44:297-302.

Ogaki, M., and E. Nakashima-Tanaka. 1966. Inheritance of radioresistance in *Drosophila.* I. Mutat. Res., 3:438-443.

Oshima, C. 1969. Persistence of some recessive lethal genes in natural populations of *Drosophila melanogaster.* Jap. J. Genet., 44, Suppl. 1:209-216.

____and T. Watanabe. 1966. The effect of insecticide selection on experimental populations of *Drosophila pseudoobscura.* Drosophila Information Service, 41:140-141.

Parsons, P. A. 1959. Genotypic-environmental interactions for various temperatures in *Drosophila melanogaster.* Genetics, 44:1325-1333.

——1961. Fly size, emergence time and sternopleural chaeta number in *Drosophila.* Heredity (London), 16:455-473.

——1963. Migration as a factor in natural selection. Genetica, 33:184-206.

——1966. The genotypic control of longevity in *Drosophila melanogaster* under two environmental regimes. Aust. J. Biol. Sci., 19:587-591.

——1967. The Genetic Analysis of Behavior. London, Methuen.

——1969. A correlation between the ability to withstand high temperatures and radioresistance in *Drosophila melanogaster.* Experientia, 25:1000.

——1970. Genetic heterogeneity in natural populations of *Drosophila melanogaster* for ability to withstand desiccation. Theoret. Appl. Genet., 40:261-266.

——S. M. W. Hosgood, and B. T. O. Lee. 1967. Polygenes and polymorphism. Molec. Gen. Genetics, 99:165-170.

——and D. Kaul. 1966. Mating speed and duration of copulation in *Drosophila pseudoobscura.* Heredity (London), 21:219-225.

——and D. Kaul. 1967. Variability within and between strains for mating behaviour parameters in *Drosophila pseudoobscura.* Experientia, 23:131-132.

Petit, C., and L. Ehrman. 1968. Le rôle de la selection sexuelle dans l'évolution des populations: L'avantage des types rares dans le groupe *willistoni* (Genre *Drosophila*, sous-genre *Sophophora*). Bull. Biologique, 102:433-446.

Pipkin, S. B. 1953. Fluctuations in *Drosophila* populations in a tropical area. Amer. Natural., 87:317-322.

——1965. The influence of adult and larval food habits on population size of neotropical ground-feeding *Drosophila.* Amer. Midl. Natural., 74:1-27.

——R. L. Rodríguez, and J. León. 1966. Plant host specificity among flower-feeding neotropical *Drosophila* (Diptera: Drosophilidae). Amer. Natural., 100:135-156.

Pittendrigh, C. S. 1958. Adaptation, natural selection, and behavior. *In* Roe, A., and G. G. Simpson, eds., Behavior and Evolution, pp. 390-416. New Haven, Yale University Press.

Powell, J. R., and R. C. Richmond. 1970. Evidence of heterosis associated with an enzyme locus in a natural population of *Drosophila.* Genetics, 64:s51.

Prakash, S., and R. C. Lewontin. 1968. A molecular approach to the study of genic heterozygosity in natural populations. III. Direct evidence of coadaptation in gene arrangements of *Drosophila.* Proc. Nat. Acad. Sci. U.S.A., 59:398-405.

——R. C. Lewontin, and J. L. Hubby. 1969. A molecular approach to the study of genic heterozygosity in natural populations. IV. Patterns of genic variation in central, marginal and isolated populations of *Drosophila pseudoobscura.* Genetics, 61:841-858.

Prevosti, A. 1955. Geographical variability in quantitative traits in populations of *Drosophila subobscura.* Sympos. Quant. Biol., 20:294-299.

——1966. Chromosomal polymorphism in western Mediterranean populations of *Drosophila subobscura.* Genet. Res., 7:149-158.

Richardson, R. H. 1969. Migration, and enzyme polymorphisms in natural populations of *Drosophila.* Jap. J. Genet., 44, Suppl. 1: 172-179.

Sakai, K. I., T. Narise, Y. Hiraizumi, and S. Y. Iyama. 1958. Studies on competition in plants and animals. IX. Experimental studies on migration in *Drosophila melanogaster.* Evolution, 12:93-101.

Salt, R. W. 1961. Principles of insect cold-hardiness. Ann. Rev. Entomol., 6:55-74.

Sameoto, D. D., and R. S. Miller. 1966. Factors controlling the productivity of *Drosophila melanogaster* and *D. simulans.* Ecology, 47:695-704.

Schultz, J., and H. Redfield. 1951. Interchromosomal effects on crossing over in *Drosophila.* Sympos. Quant. Biol., 16:175-197.

Simpson, G. G. 1944. Tempo and Mode in Evolution. New York, Columbia University Press.

——1953. The Major Features of Evolution. New York, Columbia University Press.

Sokoloff, A. 1964. Studies on the ecology of *Drosophila* in the Yosemite region of California. V. A preliminary survey of species associated with *D. pseudoobscura* and *D. persimilis* at slime fluxes and banana traps. Pan-Pacific Entomol., 40:203-218.

——1966. Morphological variation in natural and experimental populations of *Drosophila pseudoobscura* and *Drosophila persimilis.* Evolution, 20:49-71.

Spassky, B. 1951. Effect of temperature and moisture content of the nutrient medium on the viability of chromosomal types in *Drosophila pseudoobscura.* Amer. Natural., 85:177-180.

____N. Spassky, H. Levene, and Th. Dobzhansky. 1958. Release of genetic variability through recombination. I. *Drosophila pseudoobscura.* Genetics, 43:844-867.

Spiess, E. B. 1958. Effects of recombination on viability in *Drosophila*. Sympos. Quant. Biol., 23:239-250.

____1959. Release of genetic variability through recombination. II. *Drosophila persimilis.* Genetics, 44:43-58.

____and B. Langer. 1964a. Mating speed control by gene arrangements in *Drosophila pseudoobscura* homokaryotypes. Proc. Nat. Acad. Sci. U.S.A., 51:1015-1019.

____and B. Langer. 1964b. Mating speed control by gene arrangement carriers in *Drosophila persimilis.* Evolution, 18:430-444.

____B. Langer and L. D. Spiess. 1966. Mating control by gene arrangements in *Drosophila pseudoobscura.* Genetics, 54:1139-1149

Spieth, H. T. 1951. The breeding site of *Drosophila lacicola* Patterson. Science, 113:232.

Stalker, H. D., and H. L. Carson. 1948. An altitudinal transect of *Drosophila robusta* Sturtevant. Evolution, 2:295-305.

____and H. L. Carson. 1949. Seasonal variation in the morphology of *Drosophila robusta* Sturtevant. Evolution, 3:330-343.

Stone, W. S., W. C. Guest, and F. D. Wilson. 1960. The evolutionary implications of the cytological polymorphism and phylogeny of the *virilis* group of *Drosophila*. Proc. Nat. Acad. Sci. U.S.A., 46:350-361.

____K. I. Kojima, and F. M. Johnson. 1969. Enzyme polymorphisms in animal populations. Jap. J. Genet., 44, Suppl. 1:166-171.

Strickberger, M. W., and C. J. Wills. 1966. Monthly frequency changes of *Drosophila pseudoobscura* third chromosome gene arrangements in a California locality. Evolution, 20:592-602.

Sturtevant, A. H., and G. W. Beadle. 1936. The relations of inversions in the X chromosome of *Drosophila melanogaster* to crossing over and disjunction. Genetics, 21:554-604.

Sved, J. A., T. E. Reed, and W. F. Bodmer. 1967. The number of balanced polymorphisms that can be maintained in a natural population. Genetics, 55:469-481.

Tantawy, A. O. 1964. Studies on natural populations of *Drosophila.* II. Morphological and genetic differences of wing length in *Drosophila melanogaster* and *D. simulans* in relation to season. Evolution, 18:560-570.

____and G. S. Mallah. 1961. Studies on natural populations of *Drosophila*. I. Heat resistance and geographical variation in *Drosophila melanogaster* and *D. simulans.* Evolution, 15:1-14.

____A. M. Mourad, and A. M. Masri. 1970. Studies on natural populations of *Drosophila.* VIII. A note on the directional changes over a long period of time in the structure of *Drosophila* near Alexandria, Egypt. Amer. Natural., 104:105-109.

Thomson, J. A. 1971. Association of karyotype with body weight and resistance to desiccation in *Drosophila pseudoobscura.* Canad. J. Genet. Cytol., 13:63-69.

Timofeeff-Ressovsky, N. W. 1940. Mutations and geographical variation. *In* Huxley, J., The New Systematics, pp. 73-136. Oxford, Oxford University Press.

Tobari, I. 1966. Effects of temperature on the viabilities of homozygotes and heterozygotes for second chromosomes of *Drosophila melanogaster.* Genetics, 54:783-791.

Townsend, J. I. 1952. Genetics of marginal populations of *Drosophila willistoni.* Evolution, 6:428-442.

van Valen, L., L. Levine, and J. A. Beardmore. 1962. Temperature sensitivity of Chromosomal polymorphism in *Drosophila pseudoobscura.* Genetica, 33:113-127.

Vetukhiv, M. 1954. Integration of the genotype in local populations of three species of *Drosophila.* Evolution, 8:241-251.

____1957. Longevity of hybrids between geographic populations of *Drosophila pseudoobscura.* Evolution, 11:348-360.

Waddington, C. H., B. Woolf, and M. M. Perry. 1954. Environment selection by *Drosophila* mutants. Evolution, 8:89-96.

Waldron, I. 1964. Courtship sound production in two sympatric sibling *Drosophila* species. Science, 144:191-192.

Wallace, B. 1966a. On the dispersal of *Drosophila.* Amer. Natural., 100:551-563.

____1966b. Distance and the allelism of lethals in a tropical population of *Drosophila melanogaster.* Amer. Natural., 100:565-578.

——1968. Topics in Population Genetics. New York, W. W. Norton and Company.
——and M. Vetukhiv. 1955. Adaptive organization of the gene pools of *Drosophila* populations Sympos. Quant. Biol., 20:303-310.
White, M. J. D. 1956. Adaptive chromosomal polymorphism in an Australian grasshopper. Evolution, 10:298-313.
——1958. Restrictions on recombination in grasshopper populations and species. Sympos. Quant. Biol., 23: 307-317.
——1962. Genetic adaptation. Aust. J. Sci., 25:179-186.
Wigan, L. G. 1944. Balance and potence in natural populations. J. Genet., 46:150-160.
Wright, S. 1968. Dispersion of *Drosophila pseudoobscura.* Amer. Natural., 102:81-84.
——and Th. Dobzhansky. 1946. Genetics of natural populations. XII. Experimental reproduction of some of the changes caused by natural selection in certain populations of *Drosophila pseudoobscura.* Genetics, 31:125-156.

5

Population Biology and the *Tribolium* Model

CHARLES E. KING
Department of Biology, Yale University
New Haven, Connecticut

PETER S. DAWSON
Department of Zoology, Oregon State University
Corvallis, Oregon

PROLOGUE

The three most fundamental processes involving populations are natural selection, population growth, and competition. Classically, the first of these has been assigned to the province of the population geneticist and the latter two to that of the population ecologist. In part because of this division, the quantitative theory of both fields has developed rapidly. Simplifying assumptions have been made and the resultant models have served to identify the various sources perturbing either gene frequencies or population numbers.

Consider, for instance, the Malthusian growth model of population ecology: for N individuals each having R offspring and discrete generations, the change in population size in a single generation is

$$\Delta N = N(R - 1) \tag{1}$$

or, over T generations,

$$\Delta N = N(R^T - 1) \tag{2}$$

This model is applicable to the geometric increase of organisms reproducing by fission or with some other form of nonoverlapping generations, as exemplified by annual plants. The analogous model for continuous population growth is

$$\frac{dN}{dt} = Nr \tag{3}$$

in which r is the intrinsic rate of natural increase.

Both Equations (2) and (3) define populations growing in unlimited environments, and for this reason both have been set aside by ecologists in favor of models that account for density-dependent restraints on population growth. The most important of the growth models that does not assume an unlimited environment is the "logistic" equation of Verhulst (1839) and Pearl and Reed (1920).

The logistic model may be derived by defining s as the reduction of the effective population growth rate caused by the addition of a single individual to N. That is,

$$\frac{dN}{dt} = N(r - sN) \tag{4}$$

so that the *effective growth rate* of the population, as indicated by the expression in parentheses, is under density control. Equation (3) may be rearranged to

$$\frac{dN}{dt} = rN\left(1 - \frac{sN}{r}\right)$$

Obviously, when $sN = r$, dN/dt will be zero. We may therefore define K, the carrying capacity of the environment, as that value of N at which $sN = r$ and state the logistic in its usual form

$$\frac{dN}{dt} = rN\left(\frac{K-N}{K}\right) \tag{5}$$

There are many criticisms that may be directed at the logistic equation (see, for instance, Andrewartha and Birch, 1954; Slobodkin, 1962). However, a serious flaw seldom explicitly mentioned by the ecologist is the implicit assumption of genetic uniformity among individuals. Thus r and its two components, the instantaneous rates of death and birth, are assumed to be fixed quantities not subject to selection during population growth and equilibrium. However, examples of the influence of genetic factors on both survivorship and fecundity are legion. Moreover, there is ample evidence for a genetic component in the determination of K itself. Southwick (1955), for instance, initiated three "replicate" populations of house mice in each of two types of population cages. Food and water were always available in excess. Each population was started with four pairs of mice, and population size was followed for a two-year period. At the end of that time, the numbers had more or less stabilized in each cage, but varied from less than 25 individuals in the cage with the lowest density to more than 140 individuals in the most densely populated cage. The type of cage did not appear to influence population density and, in fact, the only consistent pattern that Southwick could find was the level of aggression among cages. In each cage, the number of fights per mouse per hour increased with population size and growth continued until this measure of aggression reached approximately one fight/mouse/hour. When that level was attained, population growth ceased. This result may be taken to indicate that the constant s in Equation 3 was higher for cages with a low equilibrium size than for cages with a high equilibrium size. (Since $s = r/K$, we expect low K to give high s if Equation 3 is valid.) If the variance in s was genetic, the variance in K was determined by the differing genetic structure of the six founding populations.

Clearly, it is not possible to assay genetic variation at all of the appropriate loci in many studies of population growth. There is, therefore, a practical or operational justification for trying to analyze population growth in the absence of a knowledge of the underlying genetic variation in the population. However, it is more difficult to justify the observation that much of the *theory* of population ecology is based on a model that ignores the universality of Mendelian inheritance.

There is an equally untenable assumption in the basic theory of population genetics. For the simple model of natural selection in a population with discrete breeding intervals, the basic argument takes the form: Let there be two alleles, A and B, with relative frequencies p and q at a single autosomal locus. The per-individual reproductive outputs (fitness values) of the three possible genotypes are symbolized as W_{AA}, W_{AB}, and W_{BB}. Commonly the model is developed by using relative frequencies of individuals and relative (Darwinian) fitnesses or selective values. In fact, relative frequencies are often preferable to absolute frequencies because the desired output is generally the change in allele frequencies over T generations. The mathematics of this calculation are simpler and the results are the same if relative frequencies are used in lieu of absolute frequencies. However, the "relative" aspect of the model is a derivative simplification; the most general statement is in the form:

AA	AB	BB	Genotypes
p^2N	$2pqN$	q^2N	absolute frequencies before selection
W_{AA}	W_{AB}	W_{BB}	fitness values
$(p^2 W_{AA})N$	$(2pqW_{AB})N$	$(q^2 W_{BB})N$	absolute frequencies after selection

In this model, the mean fitness, $\bar{W}$, is the sum of the frequencies after one generation of selection divided by N. The change in population size in a single generation under this model is given by

$$\Delta N = N(\bar{W} - 1)$$

which is precisely equivalent to Equation (1) because $\bar{W} = R$. Under this model, fitness values are constant even though they are known to change with genotype frequency, population density, and alterations of the physical environment. Thus, since the basic selection model of population genetics is one of exponential growth, it is as unrealistic as the logistic model, which assumes that selection does not occur during population growth.

This discussion is not meant to imply that ecologists feel genetic variation to be unimportant or that geneticists do not believe in environmental variation. Many members of both groups have clearly stated that variation within a population arises from both genetic and environmental sources. However, until recently, these verbal affirmations have not appeared in the mathematical theory on which the two fields are based.

The field of population biology had its genesis when the complementary fallacies of population genetics and ecology were set aside in favor of a broader perspective. Undeniably, the fusion has proceeded at the expense of precise, numerically refined predictions on the outcome of events subject to both genetic and environmental control. Various starting points can be assigned to this union, but the one we prefer to choose is a lucid series of articles by Richard Levins culminating in his book *Evolution in Changing Environments* (1968). In this series, a highly intuitive and abstract framework has been sketched for the simultaneous manipulation of natural selection and population growth under both spatial and temporal environmental variation. At approximately the same time, MacArthur (1962), and MacArthur and Wilson (1967) developed the contrasting notions of r and K selections (which act in populations free of and subject to density control respectively). More recently, the interactions between r and K have been investigated in simulated populations undergoing age-specific selection by Anderson and King (1970) and King and Anderson (1971); and Anderson (1971) has formulated a maximization principle relating genetic equilibrium and population size under r and K selection. Another view of K selection has been advanced by Wallace (1968), who used the interplay between the theories of density-dependent and density-independent selections to demonstrate that the genetic load controversy may be an artificial consequence of the failure to consider the ecology of natural selection.

These recent developments are bounded by a common feature: all seem reasonable, none has been tested. A major point of the present paper is to show that these theories are testable and that flour beetles of the genus *Tribolium* probably constitute the most nearly ideal organisms to use in theoretically based research that seeks to combine genetic and ecological approaches to an understanding of the biology of populations.

TRIBOLIUM – THE ORGANISM

In a gentler age, when physicists were more concerned with the nature of reality than the reality of nature, it was quite fashionable for biologists to frame their concepts as analogs of physical processes. For instance, Chapman (1928) developed his concept of population size equilibrium as a process analogous to the thermodynamic and electrical conductance equilibria of Fourier and Ohm. In Chapman's model, the equilibrium population size was determined by the ratio of the biotic potential (maximum population growth rate under a given set of conditions) to the environmental resistance. His model in essence states that every population has some (genetically determined) maximum capacity for increase in a particular environment. Opposing this potential is a set of physical and biotic restraints which form an environmental resistance. As the population grows the resistance increases until it matches the biotic potential and prevents further growth. At this point, equilibrium is attained.

Chapman's model has been related to the logistic equation (Equation 5) and has been criticized by several authors (e.g., Nicholson, 1933; Smith, 1935; Andrewartha and Birch, 1954). This aspect of his study is not particularly germane to the present paper. What is relevant is Chapman's test of his hypothesis using the flour beetle *Tribolium confusum.* Chapman established a geometric series of environments containing 4, 8, 16, 32, 64, and 128 g of whole-wheat flour at a constant depth of 2 cm. The initial density of organisms was in each case one pair of adult beetles per 4 g flour; the humidity was kept constant and the temperature held at 27°C. The purpose of this experiment was to determine whether an equilibrium population size was achieved and, if so, how it varied with the size of the environment.

Although later workers were to have reservations on this study, an apparent equilibrium was attained (Table 1). Because all life history stages contribute to

Table 1. Numbers of *Tribolium confusum* per gram flour at day 156 in Chapman's (1928) experiment

Quantity of Flour	Eggs	Larvae	Pupae	Adults	Total
4	23.3	0.0	0.0	16.3	38.6
8	29.5	0.3	0.0	19.6	49.4
16	34.0	0.5	0.4	10.8	45.7
32	26.2	0.2	0.1	13.9	40.4
64	29.9	0.7	0.1	14.1	44.8
128	32.0	0.5	0.1	14.7	47.3

this equilibrium, Chapman based his subsequent analysis on the summed abundances of eggs, larvae, pupae, and adults. He found that a high potential for increase was always present, as indicated by the large number of eggs in each culture. Moreover, when deaths reduced the population below its carrying capacity, it was inferred that the environmental resistance had also been lowered, thereby permitting the preadult stages to develop and replace the dead adults. Chapman also asserted that the environmental resistance in his *Tribolium* populations was determined by the level of adult cannibalism on the preadult stages. At low densities, the egg predation rates were also low. As the population grew larger, it was inferred that the amount of cannibalism increased in a density-dependent fashion. Thus the equilibrium population size in these experiments was attributed to an interaction between adult fecundity and egg, larval, and pupal mortality.

The advantages of *Tribolium* are obvious for work of this type. Flour beetles have long been associated with man; Andres (1931) recorded the occurrence of a species of *Tribolium* in flour urns from a pharaonic tomb of the VI Dynasty (about 2,500 B.C.). These individuals are almost certainly *T. confusum* (Hinton, 1948). Because of this long selective history, it can be argued that flour constitutes one of the "natural" habitats of the genus. It is, of course, easy to vary a flour beetle's environment by altering the size and shape of the culture vessel, by substituting different types of flour or additives to the flour, or by changing the climatic conditions in which the cultures are placed. Furthermore, there are a large number of different genetic strains and mutants of flour beetles, so that the gene pool of the founding population can also be varied. In addition, *Tribolium* has a rather short generation time (approximately one month under optimal conditions), and all life history stages can readily be recovered and enumerated by passing the culture medium through sieves of appropriate pore sizes. Finally, the techniques used in flour beetle culture are simple enough so that they can be learned in a short period of time and applied without a great deal of elaborate equipment.

Flour beetles of the tenebrionid genus *Tribolium* have a world-wide distribution. There are approximately 26 species in the genus (Hinton, 1948), but most of the research to be reviewed in this paper has been done with *T. castaneum* Herbst and/or *T. confusum* Duval. In the interest of brevity, throughout the remainder of this paper we shall adopt the convention of using "CF" as an abbreviation for *Tribolium confusum* and "CS" for *Tribolium castaneum.*

The geographic distribution of both CF and CS is cosmopolitan, since they are easily transported from one place to another in stored food products (Good, 1936; Hinton, 1948). Temperature, however, does seem to have quite an effect on distribution; CS is essentially a subtropical species, whereas CF is more frequently found in temperate regions. Of course, both species are found throughout this range of conditions because they can survive and reproduce in heated granaries, food stores, etc. In general, however, natural populations of CF probably have to withstand more severe cold conditions and, in particular, greater fluctuations in temperature and humidity.

This difference is reflected by the results of laboratory studies on survival, productivity and developmental rate measured under different temperature and humidity regimens (Gray, 1948; Howe, 1956, 1960; Dawson, 1965b). CF is found to be more tolerant of extremes, particularly of cold, than CS. On the other hand, CS is generally found to survive better and develop faster at higher temperatures and humidities. As we shall see later, differences in interspecific competitive ability may also be interpreted as reflecting the differences in distribution of the two species in nature. The two species are closely related and very similar morphologically, but are completely reproductively isolated. This attribute, of course, is essential for their extensive use in studies on the nature of the competition process.

Smith (1952), on the basis of a comparative cytological study, suggested that CF, which has nine pairs of chromosomes, was derived from an ancestral form similar cytologically to CS (10 pairs) by a translocation involving a pair of autosomes and the sex chromosomes. The sex chromosomes of CF are considerably larger than those of CS, and the neo-Y in CF is heterochromatic. Evidence supporting Smith's hypothesis was provided by the discovery of an autosomal homeotic mutation in CS that closely resembles a sex-linked gene in CF (Dawson, 1968c).

The life cycles of both CF and CS have been described in detail in several publications (e.g., Good, 1933; Park, 1934a) and will only be briefly presented in this paper. Both species have very similar life histories, the stages of which are depicted in Fig. 1. The eggs have a very sticky external surface and adhere to the medium in or on which they are laid. Under standard culture techniques this means that they become coated with flour. The developmental time of the eggs (measured as the mean number of days between egg deposition and hatching) is approximately 4 to 5 days at 34°C, 5 to 6 days at 30°C, 6.5 to 7.5 days at 26°C, and 13 to 15 days at 22°C. CS eggs develop faster than CF eggs and thus are skewed towards the lower end of the indicated ranges. The number of larval instars varies from six to eleven or more, but is normally either seven or eight. This variation is due to such external environmental factors as food, temperature, and humidity, as well as intrinsic individual variation (Good, 1933). Young (1970) has recently published the schedule presented in Table 2 for one genetic strain of each of the two species under one particular set of environmental conditions.

At 29°C, 70% relative humidity, CS females are capable of being fertilized within three hours after eclosion and begin to lay fertile eggs at an age of 96 to 108 hours (Dawson, 1964a). The corresponding times for CF are 17 to 20 hours for fertilization and 114 to 126 hours for egg deposition. Erdman (1962, 1964) reported that CF males were fertile within 24 hours of eclosion and females produced progeny on the sixth day of adult life; comparable figures for CS were two and four days. Since CS females matured sexually two days earlier than CF females, Erdman suggested that this might confer an advantage on CS in competitive populations. This, of course, will depend also on the total developmental period; for the two strains listed in Table 2, CF reaches the adult stage three full days earlier than CS, thereby negating a two-day difference in sexual maturation. It should be emphasized that the two genetic strains shown in

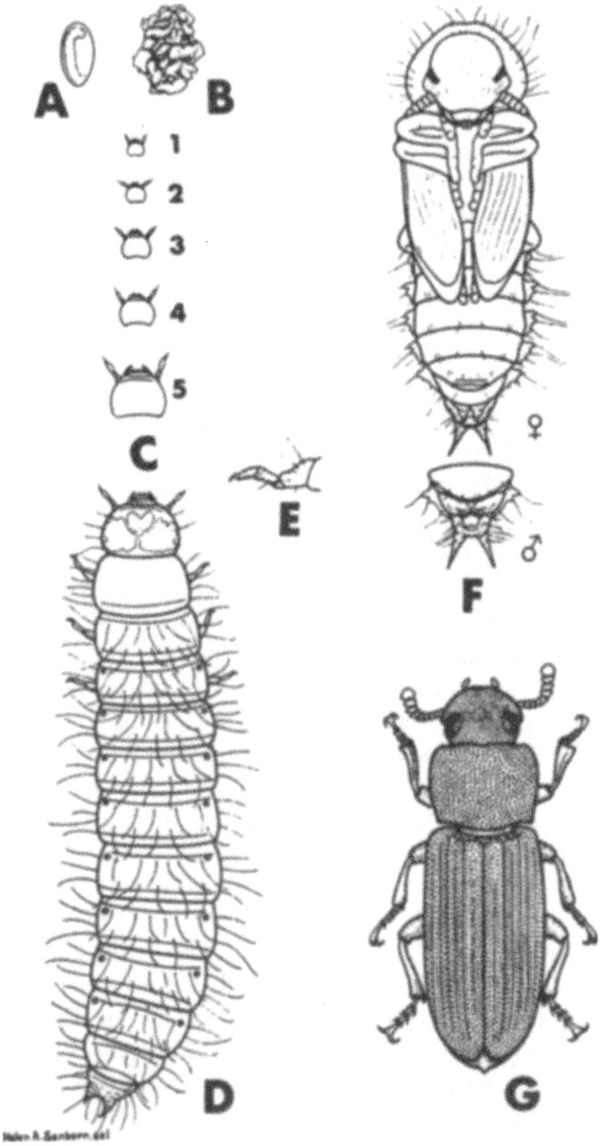

Fig. 1. *Tribolium confusum.* A, naked egg; B, egg coated with flour; C, larval instars; D, mature larva; E, larval leg; F, female pupa (top) and terminal segments of male pupa (bottom); G, adult. (Chapman, 1918)

Table 2. The duration of development for one strain each of *T. confusum* (CF) and *T. castaneum* (CS). (Data of Young, 1970 and personal communication)

Stage	CF Days	CF %	CS Days	CS %
Egg	5.9	19.7	4.7	14.4
Larva	16.3	54.3	20.3	62.1
pupa[a]	5.9	19.7	6.7	20.5
SUBTOTAL	28.1	93.7	31.8	97.0
Callow[a]	1.9	6.3	1.0	3.0
TOTAL	30.0	100.0	32.8	100.0

[a] significant difference between species.

Table 2 are not representative in this regard; for outbred populations under these conditions, the total developmental period of CF is usually longer than that of CS (Park and Frank, 1948; Dawson, 1965a, and unpublished).

Data relevant to this point, presented in Table 3, were taken from an experiment on the effects of parental age on developmental time, using the

Table 3. Mean developmental time (days from egg to pupa) and mean productivity (number of pupae recovered from 40 pairs of adults allowed to oviposit for one day) for the Berkeley outbred stocks of CS and CF cultured at 29°C, 70% relative humidity

		n	Developmental Time $\bar{x} \pm$ S.E.	Productivity $\bar{x} \pm$ S.E.
CS	Males	2614	23.18 ± 0.05[a]	
	Females	2624	23.44 ± 0.05	
	Total	5238	23.31 ± 0.03[b]	327.4 ± 11.4[b]
CF	Males	2116	24.84 ± 0.03[a]	
	Females	2039	24.48 ± 0.04	
	Total	4155	24.66 ± 0.02	259.7 ± 5.9

[a] Significantly different ($p < 0.001$) from females by Mann-Whitney Rank Test.
[b] Significantly different from CF ($p < 0.001$) by Mann-Whitney Rank Test.

Berkeley outbred strains (see Lerner and Ho, 1961). Forty pairs of beetles of four different ages (20, 30, 40, and 50 days) were transferred daily to fresh media for several days, and four replicate cultures for each age were assayed for developmental time by daily removal of pupae. Mean productivity is higher, and mean developmental time is about one day faster in CS than in CF. CF males develop faster than females, whereas the reverse is the case for CS. These data include only the egg and larval stages, but when the pupal stage is included the differences remain virtually the same.

Survival and reproduction schedules of *Tribolium* will be considered in the next section. At this point it need only be mentioned that under favorable conditions the mean lifespan of the two species is generally about 5 to 10 months (with CS at the lower end and CF at the upper end of this range) and, depending on age and a number of other factors, a female will usually produce 5 to 15 eggs per day.

Adults of the two species are readily distinguished under a dissecting microscope on the basis of antennal morphology, the degree of ridging of the anterior-dorsal margin of the head, and the relative extent of separation between the eyes. The wild type body color of adults of both species is a reddish-bronze; in some studies in which the two are reared together, an autosomal recessive marker gene (*sooty*, which produces a black body color) has been incorporated into CS. With this technique the adults can be separated without the aid of a microscope.

It is more difficult to distinguish between the pupae of the two species. Mertz (1961) has described one technique based on the patterns of arrangement and relative lengths of the lateral bristles. Ho (1960) has identified species from the pupae by differences in the shape of the genae and by eye characteristics. Larvae can be distinguished also (Ho, 1967), but the technique is cumbersome. No method has been published which is of use in distinguishing betwen the eggs of CS and CF.

Males and females are readily separated in either the pupal (Fig. 1) or adult stages. The sexes of the adults are generally distinguished on etherized individuals either by gently expressing the genitalia or by checking the ventral surface of the femur of a front leg for a shallow pit which is present in males and absent in females (Hinton, 1942).

As is the case for most bisexual animals, the sex ratio in *Tribolium* populations is usually unity. Some exceptions to this general phenomenon, involving the Chicago strains of CS and CF, are summarized by Sokoloff, Slatis, and Stanley (1960), who suggested that sex-ratio genes or sex-linked lethals may have been involved in some of their crosses. Dawson (1962) reported some peculiar sex ratios in crosses involving a sex-linked lethal gene in CS, but has not analyzed the phenomenon any further. Holdaway and Smith (1933) found that the sex ratio of CF could be altered by starvation of larvae and suggested that this resulted from sex reversal. However, Englert and Bell (1962) were unable to repeat this result with CS.

Tribolium can be reared on a variety of different flours. Both species are also able to survive and reproduce on whole, uncracked grains of wheat, although at much lower rates than in pulverized flour (Birch, 1947). Generally, stone ground flour is used in the laboratory because the fine-ground, "all purpose" flour available in grocery stores will not support the tunnels that *Tribolium* makes as it moves through the flour. When given a choice between fine and coarse ground flours, CF selects the coarser grind even when the type of flour is not as nutritionally suitable (Chapman, 1918). Although most laboratory studies with *Tribolium* have used wheat flour, the fecundity of both CS and CF is higher in oat flour (Sinha, 1969).

In most of the experimental work with *Tribolium*, wheat flour has been supplemented with 5% (by weight) of dried brewer's yeast. This nutritional supplementation increases the productivity of cultures, although neither species will actively select a flour plus yeast habitat over an unsupplemented medium when given a choice. However, the distribution of the two species, and more notably the distributions of males and females within each species, do differ when the animals are placed in a four-celled choice chamber containing 0%, 10%, 20%, and 30% yeast-to-wheat flour mixtures (unpublished observations of the authors). These results have not been extensively repeated and may be quite specific for the strains employed. They nevertheless indicate that both CS and CF adults are capable of distinguishing between different degrees of supplementation. (One point should be mentioned as an aside. Most ecological studies have been performed with 5% yeast and 95% flour. Most genetically oriented studies of *Tribolium* are made with flour that is supplemented with 5% yeast so that the actual concentration is 95.3% flour, 4.7% yeast).

Tribolium can live in almost any type of pulverized grain (Chittenden, 1896), and a variety of foods has been employed in different studies. The list of foods assembled from the literature for CF by Park (1934a) includes whole-wheat flour, bran, rice flour, rye flour, corn meal, barley flour, oat meal, chocolate, spices (red pepper), various nuts, dried specimens in insect collections, snuff, oris root, baking powder, ginger, slippery elm, peas, and beans. In addition, Park and Burrows (1942) devised a semisynthetic medium which was described as adequate but not optimal for the culture of CF. This medium consisted of a base of nonnutritive wood dust and a mixture of inorganic salts, dextrose, casein, and dry yeast.

Sinha (1966, 1968) has tested both CS and CF for their ability to survive and reproduce on pure cultures of each of 23 species of fungi, selected because of their frequent occurrence in stored grains. CS was found to complete its life cycle on eight, and CF on seven of the species; only five of the fungal species were suitable for both. These results are of interest because *Tribolium* occurs in woods under the bark of trees and rotting logs (Good, 1933). Fungi exist abundantly in both of these habitats and it is likely that they serve, along with decaying organic matter, as an important source of nutrients for *Tribolium* in nature.

Experiments with *Tribolium* have been performed under a wide range of physical and biotic conditions. The "typical" procedure is to start an experiment with males and females that have been identified in the pupal stage. Commonly either two or ten pairs of beetles are introduced into an 8-dram shell vial containing 8 g of a flour and yeast mixture. The various life history stages in the replicate vials are enumerated at monthly intervals, at which time the medium is replaced with fresh flour. If the purpose of the experiment is to study the effects of selection, adults may be discarded at the monthly census so that long life spans do not greatly slow down the rates of gene substitution. Adults are usually retained in ecological studies.

The most serious drawback to the use of *Tribolium* in laboratory experiments is the incidence of parasitism by a variety of sporozoans (Park, 1948; Park and Frank, 1950; Dissanaike, 1955; Laird, 1959; Ashford, 1968; Lipa, 1968). The parasites are spread both by reproduction (the microsporidia adhere to the sticky surface of the eggs) and by subsequent cannibalism of eggs or of larvae that have been killed by the parasite. CF tends to be much more resistant than CS to the effects of the parasites on survivorship and natality (Park, 1948). Stanley (1961) has described methods of checking for parasites and for eliminating the sporozoans from cultures. The latter technique involves cleaning the outer egg surface by prolonged shaking in a detergent germicide and then rearing the emergent larvae in isolation from infected stocks. This procedure is quite effective but very laborious when large numbers of culture need to be treated.

A second liability to research with flour beetles is the rather high risk of developing an allergy to flour that has been conditioned by the beetles.

THE ECOLOGY OF TRIBOLIUM

A population is usually defined by the geneticist as a group of organisms that share a common gene pool. This definition is without ambiguity as long as the individuals in the population (1) are bisexual, (2) mate randomly throughout the population or are geographically restricted to a single homogeneous environment, and (3) are free from external gene flow. Of course these criteria are seldom met in nature. In contrast, the ecologist has tended to view a population as a group of coexisting organisms that share a common set of environmental restraints on their survivorship and reproduction. This second definition also has difficulties when applied to populations in nature because there are many factors that can affect both natality and mortality. Moreover, some of these factors are time or age specific, and others may exert different levels of effectiveness over different parts of the spatial unit occupied by the population. However, the requirements of both definitions are fully met under the standard culture protocol used in most work with flour beetles.

The simplest possible environment would seem to be a small jar of flour set in a dark incubator at constant temperature and humidity. This may well be true, but it has taken more than 40 years and the cumulative efforts of a large number of investigators to develop a working knowledge of the population ecology of

Tribolium under these conditions. However, as a result of these efforts–more specifically as a result of the professional lifetime of one individual, Thomas Park of the University of Chicago–we know more about the regulation of fecundity and survivorship of CF than of any other organism. Our knowledge of the population ecology of CS in simple environments is nearly as complete, again because of the efforts of Park and his students.

Regulation of Fecundity

Chapman's (1928) study indicated that the initial growth rate of CF differs in environments of different size, even when the initial density of beetles is held constant at 0.5/g flour. There were two major variables in Chapman's experiments: the total volume of flour and the total number of beetles. Although the same equilibrium point was eventually reached in all of the environments, the number of eggs and larvae increased more rapidly under intermediate initial densities than under extreme densities. This effect is seen more clearly in another of Chapman's experiments presented in the same paper. In the second experiment, he kept the amount of flour constant at 32 g and varied the initial density of animals. Estimates of the reproductive rate were then taken over an 11-day period and over a 25-day period (Fig. 2). Chapman's investigation stimulated Gause (1931) to examine the data presented in our Table 1 with the

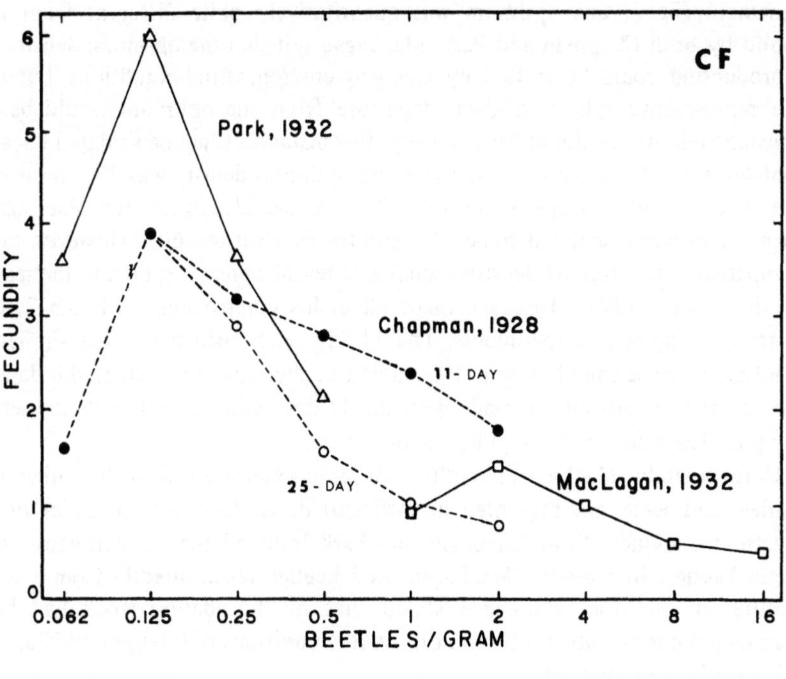

Fig. 2. The response of fecundity to changes in CF adult density recorded by Chapman (1928), MacLagan (1932), and Park (1932).

logistic model (Equation 5) and Stanley (1932) to initiate a long series of papers, which model the population growth of *Tribolium* in response to interactions among individuals and between the population and its environment.

More importantly, Chapman's study stimulated both MacLagan (1932) and Park (1932, 1933) to investigate the relationship between initial density and initial population growth rate. Both used a similar approach in their research and obtained qualitatively similar results. However, their interpretations of the results were very different.

Park varied the initial density of beetles in 32-g containers and showed that the production of eggs and larvae in successive 11-day periods was greater for populations started with two pairs of beetles (0.125 beetles/g) than for populations initiated with one, four, or eight pairs (Fig. 2). Park's explanation for this observation was that at a lower than optimum density (i.e., one pair/32 g), the amount of copulation was insufficient to simulate the fecundity and fertility of the females. At higher than optimum densities, the rate of survival of eggs was decreased because of egg cannibalism. Thus, in Park's study the observed reproductive rate at different densities was interpreted as being determined by an interaction between the density dependent rates of cannibalism and copulation.

MacLagan's experiments led to the same general results in that they also indicated an intermediate optimum density for early reproduction. But, as shown in Fig. 2, this optimum was quantitatively quite different from that found by both Chapman and Park. MacLagan felt that the optimum density for reproduction could be shifted by changing environmental conditions, but that the reproductive rate at a given departure from the optimum would bear a constant relation to the optimal density. For instance, Chapman's data indicated that the rate of reproduction at twice the optimum density was 79% as great as the rate at the optimum density. The comparable figure for MacLagan's experiment was calculated to be 73%, and for Park's it was 60%. However, other comparisons at different density increments reveal a greater spread in the results of the three studies. MacLagan made all of his comparisons with the 25-day series of Chapman's experiments. The 11-day series, which is more similar to MacLagan's experiment, was not evaluated in the same context. If the density increment comparisons are made with the 11-day rather than the 25-day series, the postulated uniformity rapidly disappears.

Unfortunately, MacLagan's methods differed extensively from the other two studies and were not presented in sufficient detail to inform a reader of his precise techniques. Both Chapman and Park initiated their experiments with virgin beetles. In contrast, MacLagan used beetles taken directly from a stock culture. In addition, MacLagan stated that his foundation stock had been maintained continually for six months under conditions of intense crowding. On this basis he asserted that

> the changed optimum density was not due so much to a difference in physical

> environmental factors (although perhaps partly so) as to *the evolution of a strain of beetles . . . better adapted to withstand crowded conditions* . . . as evidenced by a slightly higher rate of reproduction in the highest densities as well as a shifting of the optimum to a more intense degree of crowding than obtaining in Chapman's experiments [italics added].

The italicized portion of this quotation constitutes the first postulation of *K* selection known to us, as well as the first introduction of genetic factors to the analysis of *Tribolium* population research. It is not clear, however, that MacLagan's analysis was correct. For instance, the magnitude of differences between the few comparable points may simply reflect variation in experimental conditions. The possibility also exists that MacLagan's data represent a high density tail of Chapman's and Park's curves.

Park (1933) further investigated the counterbalancing effects he had postulated to explain the *existence* (but not position) of an optimum density. In this study he tried to determine the relative egg cannibalism rates of males, virgin females, and fecundated females and, although later improvements in technique have shown that some of his conclusions were erroneous, the effects of cannibalism were again demonstrated. He also found that successive recopulations stimulated the rate of egg laying.

MacLagan and Dunn (1935) used the egg cannibalism rates presented by Park (1933) to show that the reduction in fecundity at high population densities which Park observed was too great to be accounted for by egg mortality alone. They therefore suggested that the actual rate of fecundity of *Tribolium* was decreased at high population densities. As Chapman and Whang (1934) pointed out, Park's technique confounded the separate effects of egg laying and egg eating, since the total number of eggs observed at any time reflects both activities.

These experiments were repeated by Crombie (1943), who clearly identified the factors affecting the number of eggs recovered at different adult densities. His experiments differed from Park's in two ways. First, Crombie used eight individuals in each experiment and varied the amount of flour in order to alter population density. Second, instead of following Park's technique of initiating the experiments with virgins, Crombie let the females associate with males for a period of two weeks before the experiment was begun. Under these conditions, there was no intermediate optimum density but, instead, the number of eggs recovered per female per day was highest at the lowest density. Although no control was run starting with virgin females, Crombie's observation provided additional support for Park's hypothesis based on the effects of insufficient copulation. Crombie suggested that two factors were involved in the decrease in apparent fecundity at higher densities. As density increases, the rate of cannibalism increases. However, at still higher densities the apparent fecundity was low because of the simultaneous effects of egg cannibalism and a reduction in the actual rate of oviposition.

Park and Frank (1948) obtained comparative data on the fecundity of CF and CS at three different temperatures, 24°, 29°, and 34°C, under a relative humidity of about 70% to 75%. They set up 20 vials for each species and temperature with each vial containing one pair of beetles and 8 g of a mixture of 95% wheat flour, 5% yeast. The animals were transferred to fresh vials at 3-day intervals for a total of 30 days. After each transfer, the number of eggs laid in the previous interval was counted. Since density and egg cannibalism were low and the medium was frequently renewed, the results, presented in Fig. 3,

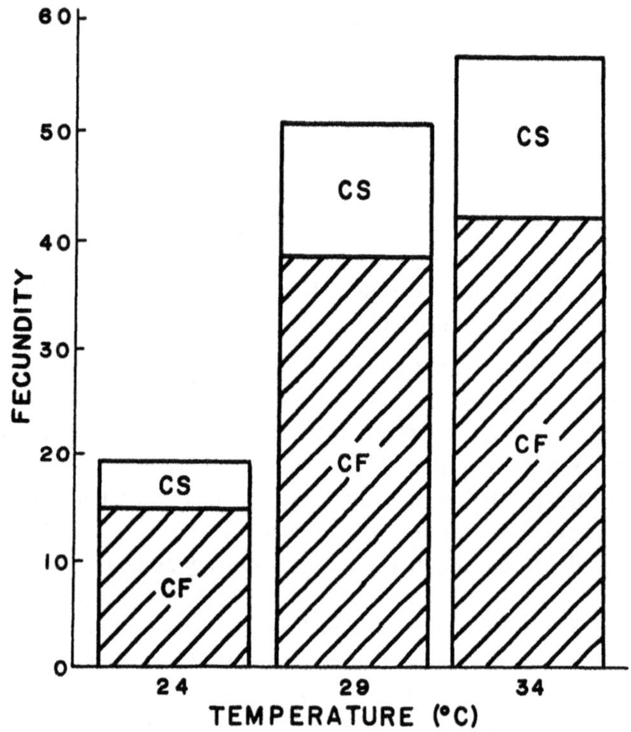

Fig. 3. The fecundity of CF and CS at three different temperatures. (data of Park and Frank, 1948)

approximate an optimum oviposition rate. CS had a consistently higher mean rate of oviposition than CF, and the absolute difference between the two increased with increasing temperature. Of more interest, however, is the observation that the relative difference between the two species was almost constant over the 10-degree temperature span; at each of the three temperatures CF produced approximately 75% as many eggs as CS. A similar conclusion on the relative fecundities of young beetles of the two species had been reached by Park and Davis (1945) in studies conducted at 29°.

These results were extended at 29°C by Birch, Park, and Frank (1951), and later by Park, DeBruyn, and Bond (1958), who found that the rate of oviposition of CS was greater than that of CF for the first 45 days of life. However, the two species were found to oviposit at approximately the same rate in the interval between 45 and 60 days; thereafter CF had a higher fecundity rate than CS. Over the entire period of reproduction, which was longer for CF than CS, the average net reproduction (eggs/female lifetime) for CF was 1,424 eggs, whereas the comparable measure for CS was 1,016 eggs.

Park, Mertz, and Petrusewicz (1961) published the results of a study on the fecundity, fertility, developmental rate, and survivorship of four strains of both CS and CF. Of the 16 possible statistical comparisons between the fecundities of the CS strains versus the fecundities of the CF strains over the first 90 days of adult life, 11 of the comparisons were significant. In 8 of these 11, CS had higher rates of fecundity than did CF.

In addition to temperature and density, the effects on fecundity of a number of other factors have been studied. In experiments on crowding of CF and CS by both species, Birch et al. (1951) showed that in general CF has a greater inhibitory effect on the fecundity of CS than on itself, or than CS has on CF. Furthermore, the reduction in fecundity with increased density in single-species populations was much greater when the sex ratio was equal than it was when density was increased by the addition of males only. In the latter case, real fecundity may be unchanged, since the reduction in net fecundity could be completely accounted for by cannibalism. However, the major effect of crowding by females was postulated to be a reduction in real fecundity.

Sonleitner (1961) analyzed the effects of density, age structure, conditioning of the medium, surface numbers, and egg cannibalism on fecundity of CS. The results of this study will be discussed later.

In other studies, McFadden (1963) used data obtained in a laboratory study of the effects of temperature, density, and "time" on fecundity of CF to generate artificial "field" data on egg production that could be compared to the observed data. Park, DeBruyn, and Bond (1958) and Erdman (1963) have reported effects of X-irradiation on fecundity, and Erdman has also studied the joint effects of X-irradiation and temperature (1965), and X-irradiation and DDT (1966).

Regulation of Survivorship

The "life table" approach to the study of population dynamics is readily applied to a variety of grain-inhabiting beetles. Measures of survivorship and/or fecundity have been used by Crombie (1942, 1943, 1947) to study populations of *Oryzaephilus surinamensis* and by Birch (1948, 1953a,b,c) in his investigations of *Calandra oryzae* and *Rhizopertha dominica.* One study has been made on the survivorship of the black flour beetle, *Tribolium madens* (Park, 1945). However, considerably more information is available for both CF and CS than for all other species combined, and it is on these two beetles that our attention will be focused.

The first complete survivorship curve of CF adults was published by Pearl, Park, and Miner in 1941. Although the overall shape of the survivorship curve was intermediate between the standard diagonal and negative skew forms, a high early mortality rate was observed. Ten percent of the population died in the first month of adult life (which had a mean duration of about 88 days and a maximum observed duration of 540 days). It was hypothesized that the early deaths were the result of innate or genetic factors derived from a "defective biological organization" and therefore represented the effects of natural selection. The initial part of the survivorship curve received a great deal of attention in this study because of the purported resemblance of the early death rates of CF and man. However, later studies of CF (e.g., Lloyd and Park, 1962; Mertz et al., 1965; Young, 1970) have not confirmed the observations of Pearl and coworkers. Instead, the mortality rate in the first month or two is generally found to be about 1% to 3%.

Males and females were separated as pupae and reared in different vials in the experiments of Pearl and coworkers. It was found that males had a lower death rate than females in the first 85 days of adult life, but that the female death rates were lower in the interval between 85 and 290 days. Female CF were also observed to have a longer mean life span than males. In the final part of the survivorship curve, from 290 to 540 days, the females again had the highest mortality rates.

The survivorship patterns of the four strains of CF and of CS described by Park, Mertz, and Petrusewicz (1961) were investigated by Mertz, Park, and Youden (1965). Again, males and females were treated separately in the experimental design. The male mortality rate was initially higher than that of the females for two of the four strains of CF and for three of the four strains of CS. In the latter part of the survivorship curve, the females in all eight of the strains tended to have higher mortality rates than the males. Thus, in general, it was found that the median age at death was lower than the mean age at death for the males, whereas the reverse was true for the females. Although the authors were unable to explain the physiological basis of these sexual differences in mortality, their data attest to the reality of the patterns and a general consistency from strain to strain.

A recent study by Young (1970) presents data on both adult and preadult mortality schedules for two of these strains. The survival rate to the adult stage was found to be about 95% for the CF strain and about 93% for the CS strain. Over the entire life cycle, the average life span was 303 days for CF males and 318 days for CF females. Both sexes of CS had a much shorter mean lifespan: 119 days for males and 191 days for females.

Lloyd and Park (1962) used both CF and CS to test the effects on survivorship of pairing males with males, females with females, and males with females in each of the 16 possible combinations of sexes and species. It was found that the presence of CF males reduced the survivorship of either sex of both species. To account for this effect, the authors conjectured that the heightened mortality was due to physical injury from the aedeagus (the sclerotized male intromittent organ) during copulation or attempted homosexual copulation. (Both CF and CS males attempt to copulate with females of the other species or with other males of both species.) No external evidence of injury could be found and it was therefore hypothesized that the possible copulatory injury was to the soft structures underlying the elytra. Such an injury might lead to a gradual loss of body fluids and death. Lloyd and Park stated that although the evidence for this hypothesis was entirely indirect, it was the most reasonable mechanism they could advance to explain their interesting results.

Regulation of Developmental Time

In their 1948 paper, Park and Frank reported that the developmental rate of CS was faster than that of CF at each of the three studied temperatures. Moreover, the difference between the two species tended to increase with increasing temperature. This observation was related to the distribution of natural populations of the two species in North America, as recorded by Good (1936). CS appears to be limited to warmer climates than CF. In the United States, CS is usually collected below 40° N latitude whereas CF occurs more frequently above this latitude. Between 37° and 40° N, both species are common, but CF declines in abundance to the south of this band. (Further evidence for greater adaptation of CF to cooler climates comes from studies of the effect of temperature on survivorship of the two species [Solomon and Adamson, 1955; Howe, 1956, 1960; Nowosielski-Slepowrow et al., 1968]. The latter authors, along with Edwards (1958), have also shown strain differences in ability to tolerate sub-zero temperatures.)

Howe (1956, 1960) reviewed the previous work on the effects of temperature and humidity on developmental rate in CS and CF, and added extensive data of his own. In general, the duration of egg and pupal periods decreases with increasing temperature and is independent of humidity. Both of these environmental factors are important in determining the length of the larval stage. Optimum temperatures for development are somewhat higher for CS than for CF.

Chapman (1918), Good (1936), Miller (1944) and others have reported that development is retarded on patent flour. Miller found that CS is much more affected by this poor diet than is CF. We have found a similar effect of corn flour.

Park, Mertz, and Petrusewicz (1961) assayed the rates of development of eggs, larvae, and pupae of their four strains of CS and CF at 29°C, 70% relative humidity. Numerous strain differences were found. Duration of both egg and pupal stages was longer for all CF strains than for any of the CS strains; this could be an important factor in mixed species populations since these two stages are the most subject to predation. No consistent differences were found for larval development time.

Developmental rates may also be inhibited by increasing densities of adult beetles. For instance, Park, Miller, and Lutherman (1939) set up a number of replicate cultures containing 10 larvae and a varying number of adults ranging from a density of 0.125 to 8 individuals per g. Under these conditions, the duration of the larval period increased with adult density. The adult density, however, had no effect on the duration of the pupal period.

Regulation of Population Size

The intrinsic rate of population growth (*r*) has been calculated for both CF and CS in a number of studies (e.g., Leslie and Park, 1949; Park, Leslie, and Mertz, 1964; Mertz, 1969; Young, 1970). These studies have demonstrated the large capacity of both species for rapid population growth. However, the actual duration of increase in numbers is found to be relatively short in most experiments. This observation indicates the active and powerful effects of mechanisms regulating population size.

Changes in the habitat. As the larvae and adults in a population move through their habitat, they change the physical and chemical structure of the medium in several ways. These changes include the formation of tunnels, addition of frass and excretory wastes, and removal of nutrients from the flour during feeding. It has also been demonstrated that adult beetles release gaseous methyl- and ethylquinone, which interact with the flour to give old *Tribolium* cultures their characteristic odor (Alexander and Barton, 1943; Loconti and Roth, 1953). Flour that has been altered by these activities is referred to as "conditioned flour."

If the flour in a laboratory culture of *Tribolium* is not changed periodically, the population slowly declines in numbers to extinction. Park (1934b) was the first investigator to explore the physiological basis of this decline. The problem as he saw it was: "Does there exist any relation between self-contaminated flour and physiological response on the part of the beetles?" The answer to this question turned out to be affirmative, but detailing the nature of the relation and the response required several years of intense investigation. Park (1934b) showed that the fecundity of CF living in conditioned flour was significantly reduced in comparison to control populations living in fresh, unconditioned flour. Further, it was demonstrated that the inhibitory effect on fecundity could be reversed by transferring the beetles from conditioned to fresh flour.

These conclusions were amplified by Park (1935). Both virgin and nonvirgin females produced fewer eggs in conditioned than in fresh flour. Moreover, the fecundity of females in fresh flour was significantly higher when mated with males reared in fresh flour than when mated with males that had been reared in heavily conditioned flour. This experiment demonstrated that the effect of conditioning acted on both sexes to reduce fecundity. In another experiment, it was found that the fecundity of females reared in conditioned flour was no higher when mated to males from fresh flour than when mated to males from conditioned flour. It therefore appears that the type of regimen under which the male is reared is irrelevant if the female is in a heavily conditioned medium. There was no significant difference between the fertility of eggs laid in conditioned as opposed to fresh flour.

Three other important effects were clearly demonstrated by Park's 1935 study. First, the rate of cannibalism of adult beetles on eggs was shown to be only half as great in conditioned flour as in fresh flour. Second, developmental rate was much slower; and third, preadult mortality rate was much higher in conditioned than in fresh flour.

The conclusions reported above apply only to the activities of CF in heavily conditioned flour. However, by mixing fresh and conditioned flour in varying proportions, Park (1936b) demonstrated that fecundity was also affected at lower concentrations of conditioned flour. Thus the fecundity of CF was significantly lower in 25% conditioned flour than in fresh flour. No difference in fecundity was demonstrated between 25% and 50%, but fecundity was significantly lower in 75% conditioned flour than in either of the two lower concentrations.

These results were briefly summarized by Park (1937) and then detailed at greater length by Park and Woollcott (1937). One study was also made in an effort to identify a direct effect of conditioning on the physiology of the beetles. Park (1936a) found that females tended to weigh more and have higher oxygen consumption rates than males, but neither of these physiological parameters was affected by the presence of conditioned flour.

In a later experiment, Park (1938a) showed that the population size of CF after one year of undisturbed growth in an unrenewed medium was inversely related to the degree of conditioning at the start of the experiment. Furthermore, the observed differences in population size were determined more by differences in the numbers of larvae and adults than by the numbers of eggs and pupae.

The effects of varying densities of larvae have also been investigated by Park (1938b) using CF in both renewed and unrenewed media. It was found that larval and pupal mortality rates as well as length of the developmental period increased with increasing larval density in unrenewed media. However, if the flour was changed at two-day intervals, all of these density responses disappeared. Therefore it seems that the direct effects of crowding are relatively unimportant in comparison to the effects of density on the rate and extent of environmental conditioning by the larvae.

Thus, in summary, as an unrenewed culture of CF ages, the flour becomes conditioned by the actions of the beetles. This environmental conditioning (1) sharply reduces the rate of fecundity, but not egg fertility, (2) lowers the rate of cannibalism on eggs, (3) increases mortality rates at metamorphosis, (4) lengthens the developmental period and, (5) lowers the rate of population growth. More extensive reviews of different aspects of this phase of the research with CF may be found in Park (1939, 1941, 1942).

Cannibalism. It has been recognized since Chapman's study in 1928 that a considerable amount of cannibalistic predation occurs with *Tribolium* populations. However, until relatively recently there have been few studies of this interaction and its quantitative significance in the regulation of population size. Both larvae and adults prey on eggs, pupae, and callows (very young adults with soft exoskeletons). The complexity of these biological interactions raises a number of important questions pertaining to the regulation of population size. For instance, does the growth rate of the population reflect cannibalism on the various life history stages? Is cannibalism an effective regulator of population size and, if so, which of the six possible prey-predator interactions is most important?

The first of these questions was answered indirectly by Watt (1955), who experimented with CF populations by acting as an "artificial predator." Watt's basic technique was to remove various fractions of larvae, pupae, and adults at each monthly census. He then measured the rates of entry to the different life history stages and found that these rates were directly related to the predation intensity. Thus population productivity was a simple function of rate of exploitation and, as an extrapolation, it is reasonable to suppose that CF is at least partially able to buffer the effects of cannibalism.

In the early studies of egg cannibalism by adult flour beetles, it was assumed (Chapman and Whang, 1934) or asserted (Park, 1933) that males and females ate eggs at the same rates. This assumption made it possible to measure the effects of cannibalism by placing predetermined numbers of eggs and male beetles in a culture vial and following the loss of eggs through time. The comparable experiment with females would, of course, be confounded by egg production during the experiment. However, Stanley (1942) suggested that females were more voracious than males, and Boyce (1946) estimated that rates of egg cannibalism for females were twice those of males.

The study of sex differences in cannibalism remained at this point until a method was devised by Rich (1956) to measure and separate the effects of egg laying and egg eating. Rich set up a simple differential equation describing the change in number of eggs (E) through time (t) as a function of population size (N) and the rates of egg laying (B) and egg cannibalism (C):

$$\frac{dE}{dt} = NB - NEC \tag{6}$$

which can be integrated and rearranged to

$$E_t = E_0{}^{-NCt} + \frac{B}{C}(1 - e^{-NCt}) \tag{7}$$

(The assumptions and restraints embodied in these equations are discussed and examined by Rich [1956].) Color-marked eggs were used to initiate Rich's experiments so that the survival of eggs from time 0 to t could be directly measured. At the termination of an experiment, two "types" of eggs were recovered: marked (red) eggs that had been added to the vial before introduction of the beetles, and unmarked (white) eggs that had been laid during the experiment. Knowing the sex ratio of the experimental animals, it is possible to calculate both B and C, and thus Rich was able to show that CF females had rates of egg cannibalism that were approximately 6.8 times those of CF males. In the same paper it was also reported that both the rates of egg laying and egg cannibalism decreased as the adults in a population became crowded. This result reinforces the conclusions reported earlier for CF by investigators studying density effects in fresh and conditioned flour.

Rich's methods were applied to CS by Sonleitner (1961). Again it was found that females were more voracious cannibals than males, but for CS the difference between the sexes was even greater than that reported by Rich for CF. The egg cannibalism rates of females were found to be about 17 times those of males. It was also demonstrated that crowding reduced fecundity in conditioned media. However, in fresh media the results were more complex. Young adults (~15 days old) showed a clear reduction in fecundity with increased density (measured at 0.5, 1, 2, 4, and 8 pairs/g). This effect diminished as the beetles aged and, by an age of 63 days, only the highest density gave a reduction in fecundity. At an age of 99 days, all of the densities had approximately the same rates of egg production per female. In contrast to the results obtained by various workers with CF, Sonleitner showed that the rates of egg cannibalism of CS adults increased in conditioned flour.

The marked egg technique was also used by Park, Mertz, et al., (1965) in an extensive study of the four strains of CF and CS cited earlier. The adults of three of the CF strains had higher rates of egg cannibalism than any of the CS strains. However, a considerable amount of variation was observed among the strains of both species, indicating the existence of genetic variation for cannibalistic activity.

The cannibalism of eggs by larvae has not been as extensively studied as the cannibalism of eggs by adult flour beetles. In the study of Park, Mertz, et al. (1965) it was shown that the larvae of CS were relatively more cannibalistic toward pupae than toward eggs, whereas the opposite was found for CF. These results parallel the findings on the relative rates of adult cannibalistic activity. Again, a great deal of variation in cannibalism rates was present among strains, but in this case all four strains of CS had higher rates of egg cannibalism than any of the CF strains. Thus the adults of CF are more voracious egg eaters than CS, but the opposite relation holds for predation by larvae on eggs. Similar results were obtained by Ho and Dawson (1966) using other strains. The difference in rates between the two species was greater for young larvae than for older larvae. However, the larger active larvae of both species ate more eggs than did individuals in the smaller, early larval instars. Toward the end of the larval period, the rate of egg cannibalism decreased as the larvae became quiescent prior to pupation.

Investigations have also been performed on the cannibalism of pupae by both larvae and adults. Park, Mertz, et al. (1965) presented data showing that CS larvae were active predators both on their own and on CF pupae. However, the results for CF larvae were quite different. Larvae from each of the four CF strains were tested with both CF and CS pupae. In all cases, the pupal mortality was quite low. Two of the strains were shown to increase the mortality of CS pupae, but not CF pupae; one of the strains increased the mortality of both CF and CS pupae, and larvae of the fourth strain had no significant effect on the mortality of pupae of either species.

Both CF and CS adults prey on pupae and callows of both species. This interaction appears to have a sizable genetic component because the different strains of each species had quite different effects on pupal mortality (Park, Mertz, et al., 1965). The study of adult predation on pupae was expanded by Park, Mertz, and Nathanson (1968), who assessed the effects of variation in prey species, spatial distribution, and density. Two of their findings repeated results of the 1965 study: CS is in general a more voracious predator than CF, and adults of both species tend to prey selectively on pupae of the other species.

Park, Mertz, et al. (1965) distributed the pupae evenly from top to bottom of the column of flour in the experimental vial. The prey distribution was varied in the 1968 study by setting up one series of replicates in which the pupae were introduced to the bottom half only, another series in which the top half received the prey, and a third series in which the pupae were distributed throughout the flour column. The adult predators were then added to the flour and allowed to feed for 10.5 days. At the end of that period sufficient time had elapsed to insure that all surviving pupae and callows would have developed into fully-hardened adults immune from further predation. The vials were then censused, and survival through the pupal and callow stages determined by counting the number of adults in each vial (100% predator survival was assumed). It was clearly demonstrated in this experiment that the survival of pupae of both species was lowest in the bottom zone when CS was the predator. In contrast, pupal survival was lowest in the top zone when CF was the predator. Thus the two species have very different patterns of predatory behavior.

The rate of cannibalism of adults on pupae is also dependent on the density of prey available to the predator. At low prey densities, Park, Mertz, and Nathanson (1968) found that a rather low proportion of pupae survived to become fully hardened adults. This proportion increased with increasing prey density. Thus the rate of adult cannibalism is inversely related to prey density and it appears that a successful predatory attack results in some degree of predator satiation. As Mertz and Davies (1968) pointed out, "A satiated adult may be thought of as an adult who is occupied with a prey—either in attacking, handling, digesting, or in some other time-consuming way." The pupal prey density experiment of Park, Mertz, and Nathanson (1968) was repeated by Mertz and Davies (1968), who also developed a stochastic predation model with the novel attribute of predator satiation. It was shown that the new model gave a better fit to experimental observations than either the deterministic model of Rich (1956) or a second stochastic model that did not consider predator satiation.

Two recent papers have considered separate aspects of the predatory interaction of *Tribolium* adults and pupae. Park, Nathanson, et al., (1970) assessed the cannibalism rates of CS and CF adults under two conditions. In the first, adults of the two species were separately tested in order to determine control rates of cannibalism on pupae of both species, singly and together. Then, in the second series, adults were combined so that competitive interactions might take place if such occur. In both series the pupae were distributed throughout the flour column. The results of this experiment indicated that in general the cannibalism rates determined from the control series underestimated the rates of pupal mortality when adults of both species jointly exploited the pupae. However, when the differences between the cannibalism rates of CF and CS in different regions of the flour column (Park, Mertz, and Nathanson, 1968) were taken into account, the agreement between control and experimental results was much improved for the predation on CF pupae. The same correction for CS failed to account for the increased mortality rate of pupae when both species were present. No explanation was offered for this observation, but, in any event, it is clear that the two species together exploit the environment more effectively than either does by itself.

In the second recent paper on cannibalism, Young (1970) devised a method to separately estimate the cannibalism on pupae and callows. In all of the work reported previously, the two were treated jointly. Young found that over 90% of all deaths occurred in the pupal stage when CF was the predator, but for CS the results were quite different; less than 30% of the total deaths were attributable to the pupal stage. In both cases, the pupae were of the same species as the predator. These results therefore indicate that there is little, if any, predation on CF callows by CF adults, whereas the mortality from predation is higher for CS callows than for CS pupae when the predators are adult CS.

Young also found that both species tended to prey more heavily near the surface than near the bottom of the experimental vials. He postulated that the reason for the divergence between this result and that of Park, Mertz, and Nathanson (1968) was that in his experiments complete populations (containing all life history stages) were used, whereas Park and colleagues had studied artificial populations composed of only pupae and adults.

The last major question to be posed in this section is "To what extent does cannibalism control population size at equilibrium?" Chapman (1928) and Chapman and Whang (1934) felt that cannibalism was essentially *the* regulator of population size. Yet, as we have seen, there is also a reduction of fecundity under a variety of environmental conditions (e.g., Park and Frank, 1950; Birch et al., 1951). Boyce (1946) argued that most of the regulation could be ascribed to decreased fecundity at high densities, and Park and Woollcott (1937) summarized experiments showing that both fecundity and cannibalism rates decreased in old CF cultures containing conditioned media. The same conclusion has been reached with the marked-egg technique by Sonleitner (1961) for CS. Several other studies bear directly on this question.

Brereton (1962a) presented evidence that the larval, pupal, and adult mortality of CF was independent of density over the first 190 days of population growth. By that time population size was approximately stable, and it was concluded that the major density-dependent control operated on the egg stage. Two types of control were implicated (but the first was not explicitly stated) by Brereton's study: increased cannibalism on eggs, and reduced fecundity at high densities. In a second paper, Brereton (1962b) argued that the self-regulation resulted from group (interdeme) selection. However, as Williams (1966) and Lewontin (1970) have pointed out, this evolutionary mechanism is not the only, or necessarily the best, explanation of such regulatory phenomena.

Taylor (1965) constructed a deterministic population growth model, incorporating egg cannibalism by adults, and, using data from Brereton's (1962a) experiments, found a reasonably good agreement between theoretical and experimental curves for CF.

Two investigations have recently provided a more precise understanding of the regulation of population sizes in CS. Lloyd (1968) studied two strains of CS by determining their basic reproductive characteristics and following population growth at weekly intervals for a period of 36 weeks. In the initial assay, one of his strains had higher fecundity, faster development, and higher survival rate in the larval stage than the other strain. Yet, in spite of these differences, it was found that both maintained the same equilibrium numbers of adults. The more productive strain in the initial assay was also shown to produce the greatest number of pupae in the population growth experiments. Lloyd interpreted this result to indicate that the risk of cannibalistic death was greater for pupae in the highly productive strain than it was for the strain with lower productivity. As Chapman and Whang (1934) had earlier demonstrated, the preadult life history stages are produced in waves if adults are used to initiate a population. Lloyd found that these waves were largely avoided when populations were initiated with a mixture of larvae of different sizes, callows, and mature adults. Under these conditions the rates of cannibalism and production of pupae were reasonably independent and the population approached equilibrium smoothly. However, populations initiated only with adults exceeded ("overshot") the equilibrium level, presumably because of the predator satiation discussed earlier.

These patterns were documented more fully by Mertz (1969). In this study the effects of varying the initial age distribution and density of adults were clearly shown to influence the growth curve of the population. The age structure was found to be the most significant of these two factors; populations initiated with young (15 days old) adults almost always overshot the equilibrium point, whereas such overshoots were unlikely if at least half of the founding population consisted of old individuals (90 days old). A second type of overshoot was caused by fluctuations in larval abundance and adult predatory satiation. Waves in the numbers of eggs were most closely related to the presence of large larvae which are highly cannibalistic on eggs. The other effect–waves in pupal numbers–was caused by fluctuations in the intensity of cannibalism on pupae. Mertz also found that there was some depression of fecundity at equilibrium, but he concluded on the basis of his data that cannibalism was the major factor in the regulation of population size.

The relative effectiveness of larval cannibalism on eggs versus adult cannibalism on pupae was experimentally investigated by Young (1970). By interfering with the two types of predation, it was demonstrated that the release of pupae from adult predation resulted in larger population sizes than the release of eggs from larval cannibalism. Moreover, the same results were obtained for both CS and CF.

McDonald (1968b) mixed powdered polyethylene with flour and found that space was more important than amount of food in determining the size of CF populations. The relationship between population size and volume of the medium was not a simple one, however, indicating that other variables in environmental geometry must also be important. The number of eggs recovered from populations in the polyethylene-flour medium was considerably lower than in a flour medium of the same volume, indicating a reduction in fecundity and/or an increase in egg cannibalism.

Two points are illustrated by this phase of the research with *Tribolium*. First, it is naive at best to explain the operation of a complex ecological system by a single limiting factor. It seems clear that the equilibrium population size of *Tribolium* may be affected by all of the following: temperature, humidity, flour type and quantity, type and amount of nutritional supplementation, size and shape of the culture vessel, developmental time, fecundity, viability, feeding behavior, interspecific competition, and a multitude of cannibalistic interactions. The list could, of course, be greatly expanded. Many of the items are not independent. This fact is instructive in itself, because interaction effects would lengthen the number of components by at least an order of magnitude. Moreover, genetic factors have not been included, but the response of a population in terms of most of the listed items is strongly dependent on the genetic structure of the population.

Not all of these factors have a similar degree of influence on the equilibrium size of a population and so the second general point to be emphasized is the predominance of various cannibalistic interactions. In Park's words (if we are any judge of style), "... above all else, a *Tribolium* population must be viewed as a cannibalistic orgy–but one with a pattern." (Park, Mertz, and Nathanson, 1968).

THE GENETICS OF *TRIBOLIUM*

Although flour beetles have been used in ecological studies for over 40 years, geneticists were not as quick to recognize the many virtues of this organism for their own investigations. Aside from a few descriptions of mutant genes discovered in various populations, only one genetic study was carried out prior to 1950, and that one has not been published (Kollros, 1944). Virtually all of the published research in population and quantitative genetics has appeared in the last 10 years.

We have already indicated some of the attributes of *Tribolium* that make it useful to geneticists. The versatility of flour beetles as a tool for use in genetic studies is well illustrated by the diversity of investigations in which they have been employed.

Formal Genetics

The formal genetics of *Tribolium* has received a considerable amount of attention in recent years, primarily through the efforts of A. Sokoloff and his associates. In a recent review, Sokoloff (1966a) has described the known mutants of CF and CS and provided genetic maps for those loci which have been assigned to linkage groups. Additional descriptions of various mutant genes may be found in the Tribolium Information Bulletin.

Population Genetics

Many of the attributes of flour beetles that make them useful for the study of ecological processes are also attractive to population geneticists interested in the process of natural selection within experimental populations. A number of investigations have been directed at analysis of the fate of mutant genes in populations; some of these have also included assessment of the population characteristics traditionally emphasized in ecological studies. The first of these investigations was carried out by Kollros (1944; summarized by Sokoloff, 1966a). Populations of CS, initiated with different frequencies of the *pearl* eye-color mutant and its wild-type allele, were maintained by censusing the population and renewing the medium at 30-day intervals. Although natural selection led to a decrease in the frequency of *pearl*, some evidence of equilibrium was obtained (Sokal and Karten, 1964). Interestingly, a study of various fitness components indicated that heterozygotes were often superior to homozygotes (see Sokoloff, 1966a).

McDonald and Peer (1960, 1961) followed the fate of three mutant genes of CF in populations. One of the mutants used was *Striped* (*St*), a dominant sex-linked gene which is lethal to hemizygous males and reduces the viability and longevity of heterozygous females. Two kinds of populations were studied. In the first, generations were made artificially discrete by extracting a large sample of pupae in each generation, classifying the female adults for *St*, and using the sampled males and females as parents for the next generation. Under this scheme, the decline in frequency of *St* was close to that predicted for selection against a sex-linked lethal that reduces the fitness of heterozygous females to 0.4 (relative to 1.0 for normal homozygotes).

The other study was carried out in population cages, where generations were continuous and overlapping, and longevity of the beetles was also a factor. The *St* gene was eliminated from these populations too, but the decline in frequency of *St* was very different from that in the discrete generation cultures. An analysis of the distribution of the various life history stages indicated that changes in overall birth and death rates in the continuous populations were more useful in describing the process of selection against *St* than the adaptive values estimated from study of individual fitness components. Thus, an adequate description of natural selection required consideration of both ecological and genetic factors. Other population cages were started with a mixture of wild type and *black* (*b*), a semidominant autosomal mutant affecting body color. Using an estimated generation time of 18 weeks, and adopting Levene's method for determining adaptive values (Dobzhansky and Levene, 1951), a theoretical curve for selection against *b* was presented that adequately described the observed change in gene frequency.

In the second paper, McDonald and Peer (1961) reported the results of similar experiments using the *split* (*sp*) mutant, an autosomal recessive that produces shortened elytra, which fail to meet at the midline. The *sp* gene was rapidly eliminated in population cages, and, from other experiments, this was attributed to greatly reduced viability and fecundity of mutant females. The effect was probably intensified by high density in the population cages.

Later, McDonald (1963) again used population cages to investigate the equilibrium characteristics of various CF populations. Two genetically different populations were found to differ significantly in equilibrium size, birth and death rates, and interactions between fecundity and density. Population size regulation was remarkably precise in both kinds of populations; very little fluctuation in numbers was observed after equilibrium size was attained. There was no evidence of total depletion of the medium, indicating that other characteristics might be responsible for regulation of numbers. McDonald suggested that density-dependent population control, which conserves the food supply, might help to insure the ultimate survival of the population, and that the characteristics that lead to regulation of numbers might represent an adaptive complex evolved through a history of habitation in a stable, undisturbed environment.

More recently, McDonald and Davidson (1965) and McDonald and Stoner (1968) have used the *black* mutant to investigate the ability of flour beetle populations to survive through time in unrenewed media. Following Thoday (1953) and Lewontin (1957, 1961), population fitness was defined in terms of ability to survive.

McDonald and Davidson (1965) initiated populations with large (50 pairs) and small (2 pairs) numbers of founders of a wild-type strain and a strain carrying the *black* mutant. The wild-type strain reached the same population size and survived for a longer time than the mutant strain, regardless of the number of founders. The *black* strain produced larger populations when a large number of founders was used, but the larger populations did not survive as long as those initiated with a small number. In a subsequent study, McDonald and Stoner (1968) used the same *black* and wild-type strains and also a hybrid between the two. Populations were initiated with a single pair of adults and were censused at two-week intervals; again the medium was not renewed. In one set of cultures, live adults were discarded every four weeks, whereas nothing was removed from the others. For both experimental treatments, the hybrid populations survived for a longer period of time in the face of the environmental stress. It was postulated that this resulted from a greater efficiency of hybrid individuals in the conversion of flour into beetles. A comparison of replicate populations within each of the six experimental units again showed a negative correlation between population size and survival time. This is not particularly surprising, since a smaller number of beetles would presumably not deplete the available resources as rapidly as a larger number. However, this led to the suggestion that ability to regulate population size may have evolved as a result of interpopulation selection. As we noted earlier, an alternative view of the evolution of population size regulation may be found in Williams (1966).

An attempt to bridge the gap between genetic and ecological investigations of *Tribolium* has been made by Robert Sokal and his associates. A series of studies on the ecological genetics of CS has focused attention on the ecological factors involved in the process of natural selection. In the early studies (Schlager, 1963; Sokal and Huber, 1963), *sooty* (*s*), an autosomal mutant that affects body color, was utilized to investigate the various fitness components of the three genotypes at a single locus at different densities and gene frequencies. The results were then compared to the changes in frequency of *s* in populations. Schlager studied replicate populations initiated using frequencies of 0.25*s*, 0.50*s*, and 0.75*s* in addition to homozygous controls. The frequency of the mutant gene decreased throughout the course of the experiment (252 days) in both the 0.50*s* and 0.75*s* populations. However, in the 0.25*s* populations an apparent equilibrium was established at a frequency of 0.33*s*. This was attributed to superiority of heterozygotes; supporting evidence for this came from an analysis of a number of components of fitness. The experiment was not continued long enough to determine whether the other populations would reach a predicted equilibrium. Many of Schlager's populations were infected with an unidentified "disease," and this may also have affected the results. Changes in gene frequency, particularly during the initial phase of population growth, were found to depend on both density and gene frequency.

An analysis of the effects of density and gene frequency on adult survivorship and weight was carried out with the same strains by Sokal and Huber (1963). Both the environmental (density) and genetic (gene frequency) variables had significant effects on body weight, whereas only density was important in determining survivorship. The three genotypes did not respond uniformly to changes in density or gene frequency, indicating interactions between the genotypes, much like those first described by Lewontin (1955) with *Drosophila.*

Because work with the *sooty* gene is hampered by difficulty in distinguishing heterozygotes from wild-type homozygotes, Sokal turned to another CS body color mutant, *black* (*b*), which produces an easily recognizable heterozygote (Sokal and Karten, 1964; Sokal, 1967a,b; Sokal and Sonleitner, 1968). The first study was similar to the work with effects of density and gene frequency using *sooty.* Again, adaptive values of the three genotypes were found to depend on both density and gene frequency. Furthermore, genotypic interactions were important for both survivorship and body weight. The performance of genotypes when present at very low frequencies was of particular interest. Homozygotes for *black* were especially poor when rare, indicating that conventional selection theory may not be adequate at very low or high gene frequencies.

Sokal and Sonleitner (1968) reported the results of an extensive investigation of genetic and ecological processes determining the fate of the *black* (*b*) mutant in experimental populations. Populations were established with initial frequencies of *b* of 0.25, 0.50, and 0.75, in addition to control populations using the wild-type and mutant strains. The generations were made artificially discrete by discarding parents after a three-day egg-laying period. This procedure removes the complexity of adult survival from the study, and is expected to simplify the interpretation of the genetic and ecological processes involved. These populations were followed over periods ranging from 16 to 35 generations in three different experiments. Measurements on a variety of fitness components were taken during the course of the experiments. The frequency of the mutant gene remained high in virtually all populations. Detailed analyses of egg input, cannibalism, and changes in developmental time during the course of the experiments indicated that these factors were all involved in determining the observed changes in gene frequency.

Analysis of zygotic frequencies among the eggs produced by adults in the mixed populations showed consistent deviations from expectations based on parental frequencies. The results suggested the possibility of frequency dependence of fecundities of the three genotypes; when *black* was present at low frequency the deviations favored *black*, whereas at high frequency the deviations were in the opposite direction, favoring wild type. Frequency-dependent selection has become a central topic in population genetics in recent years. In large part this has been inspired by the work of Petit (1951, 1958), Ehrman (1966, 1968), Spiess and Spiess (1969), and others on frequency-dependent mating success in *Drosophila.* These studies have usually indicated that rare male genotypes are relatively more successful in mating than more frequent genotypes. Wool (1967) carried out a test of the rare-male phenomenon with the same strains used by Sokal and Sonleitner (1968). Although he found no evidence of rare-male advantage, he did find that when *black* females were rare, they mated significantly more often than expected.

More recently, Sinnock (1970), using different wild-type and *black* strains, has produced conclusive evidence for frequency-dependent reproduction in CS. By analyzing numbers of progeny produced from matings in which one of the three genotypes was rare among males, Sinnock found that the rare genotype always had a selective advantage. It was not possible to determine whether the frequency dependence involved mating behavior or fecundity.

Wool (1970) reported the results of an attempt to pin down the factor(s) responsible for the deviations in egg output observed by Sokal and Sonleitner. Fecundity was measured in females taken from the two pure strains (+ and *black*) and in hybrids between the strains. Hybrid females were more fecund, but the differences were not sufficient to account for the results obtained by Sokal and Sonleitner. No significant differences could be detected in egg hatchability or in the fitnesses of gametes or zygotes prior to oviposition. Wool then tested the relative mating success of the three kinds of males by placing virgin males in cultures containing + and *black* females. The males were introduced in a 1:1 ratio, two kinds at a time, and mating success was noted by scoring genotypes of progeny produced by females fertilized during a period of up to two hours. Although mating occurred at random, as measured by the relative frequencies of fertilization of the two kinds of females, hybrid males were much more successful in mating than either *black* or wild-type males, thereby implicating nonrandom mate selection as the major cause of the deviations. Superiority of hybrid males, all other things being equal, would be expected to lead to an increase in the frequency of *black* when it is present at low frequency and to a decrease when at high frequency.

Some caution should be taken in extending the observations made during a two-hour period to the results obtained by Sokal and Sonleitner in cultures where mated adults were allowed to oviposit for three days. Wool found that only 40% to 50% of the females were fertilized during the two-hour period in his experimental apparatus (which was quite different from the population jar used in the selection experiments). It would have been useful to know whether the hybrid males were more successful in fertilizing the remaining females as well. In a small experiment in which single virgin males of an outbred CS strain were placed with 5, 10, 15, and 20 virgin females in 5 g of flour for a 24-hour period, the mean numbers of females fertilized at least once were 5, 8.5, 11.8, and 15.7, respectively, in six replicates (Dawson, unpublished). Since a single male can inseminate a large number of females, the advantage postulated by Wool may in fact exist if the hybrid males are equally successful in fertilizing all females. Another problem must be taken into account; like *Drosophila, Tribolium* females may copulate repeatedly, but the most recent sperm preferentially fertilize the eggs (Schlager, 1960). Therefore the postulated advantage of hybrid males must include not only a greater speed of initial mating, but also a significantly higher frequency of mating throughout the three-day period. Even the demonstration of random mating is not sufficient to rule out the possibility of this kind of selection.

An interesting feature of the selection study was the response of developmental time to the experimental techniques utilized by Sokal and Sonleitner. In all of the selection lines, the larval period was substantially lengthened. Evidence was marshalled to support the hypothesis that this resulted from selection against individuals that pupated early, because of a greatly increased probability of being cannibalized by the remaining larvae. Since individuals of the *black* genotype had a slower developmental rate than the others, this mechanism would lead to selection favoring *black*.

Some discrepancies between the data obtained by Sokal and Sonleitner (1968) and by Sokal and Karten (1964) prompted further study of the effects of density and gene frequency on survival and developmental time (Sokal, 1967a, b; Wool, 1969b). Sokal and Karten had found that black individuals had a higher survival than wild type, yet in the control strains for the selection study the reverse was true. It was found that this could not be accounted for by differences in the length of the egg laying period (Sokal, 1967a), by changes in genetically determined differences between the two strains (Sokal, 1967b), or by differences in the geometry of the culture containers (Wool, 1969b).

Park and his coworkers, studying various strains of CS and CF, had previously reported variation among strains in the major factors involved in regulation of population size. Sokal and Sonleitner (1968) reported a similar difference for their strains. Egg cannibalism was suggested as the predominant factor regulating population size in the wild-type strain. The mutant strain was characterized by low egg cannibalism, yet mortality during the larval and pupal stages very nearly compensated for this reduction.

One important conclusion from this study was that the process of natural selection, at least in *Tribolium* populations, was best characterized by studying the frequencies of genotypes rather than genes. The complexity of the selective forces that operate differentially on different life history stages can be grossly misrepresented by looking at gene frequencies from adults in successive generations. This problem has recently been approached theoretically by Franklin and Lewontin (1970).

The study by Sokal and Sonleitner represents a significant step in experimental population biology. In spite of the complexity introduced by the attempt to simultaneously analyze the genetic and ecological factors involved in the process of natural selection, it clearly points out the direction in which future research must move if we are to understand what is really going on in populations.

The populations studied by Sokal and Sonleitner (1968) also provided material for a test of an hypothesis regarding senescence (Edney and Gill, 1968). One of the predictions of this hypothesis is that longevity would be expected to decrease in populations in which adults are killed after being allowed to reproduce for only a short portion of their normal reproductive period. As noted earlier, Sokal and Sonleitner discarded adults in each generation after a three-day egg laying period. Longevity of wild-type and *black* males and females was studied for the selected strains, which had gone through 40 generations of selection, and mass-mated controls (Sokal, 1970). In these strains, males always lived longer than females. There was no difference between the selected strains and controls with respect to range of longevity, but median longevity was significantly decreased in selected populations. This finding may therefore be taken to support the hypothesis of Edney and Gill.

Other population studies using mutant genes have been carried out by Sokoloff (1965, 1967, 1969a). In the first of these, various combinations of three automsomal genes located in different linkage groups in CS were introduced into populations at frequencies of 0.50. The populations were censused monthly, at which time, following the technique developed by Lerner and Ho (1961), the adults were classified and discarded, and the remaining life stages were returned to fresh medium. It was found that the frequency of the *pearl* gene increased in some populations and declined in others. This result is contrary to that of Kollros (1944); however, both the strains and the techniques used in the two studies were so different that such a comparison is of dubious significance. The *black* gene decreased in frequency in Sokoloff's study, whereas Sokal and Sonleitner (1968) reported an increase in frequency in populations initiated with a gene frequency of 0.50. Again, it is difficult to make meaningful comparisons between these two investigations. The third gene, *microcephalic*, was strongly selected against. Sokoloff (1969a) reported that *melanotic stink glands (msg)*, an autosomal recessive gene, which has been found in both CS and CF, declined rapidly in frequency in populations of both species.

Much of the work done on mutant genes in experimental populations has used the same procedure in order to obtain founding individuals for the populations. This procedure often involves a cross between unrelated mutant and wild-type stocks. In such a case, the subsequent fate of the mutant gene will depend not only on the effects on fitness of the locus being studied, but also on other loci where the two stocks differ. In particular, close linkages with loci affecting fitness may partially explain the frequently observed persistence of mutant genes in *Trobolium* populations. Merrell (1965) has noted that many presumed examples of overdominance in *Drosophila*, deduced from the occurrence of intermediate gene frequency equilibria, may actually have resulted from linkage disequilibrium. An illustration of this problem was provided by the results of population studies involving a balanced lethal system in CS (Dawson, 1970a).

In long-term experiments, differences in the genetic backgrounds of the two original strains would of course be reduced in *hybrid* populations, so that eventually only genes that were very closely linked to the mutant locus being studied would contribute differentially to the fitness of various genotypes. However, the control populations that are used for comparison are the original wild-type and mutant strains, which will clearly not become identical genetically during the course of the experiment.

As will be discussed later, Dawson (1965a) carried out artificial selection for fast and slow development in CS and CF. In the strain of CF selected for fast development, a mutant gene, *thumbed* (thu^s), was found among 18th and 19th generation progeny (Dawson, 1968d). The elytra of mutant individuals are short, usually widely divergent, and raised from the abdomen. Since mutant adults had never been observed among the parents (collected as pupae), the thu^s gene was effectively a recessive lethal in the selection line. Yet, despite complete selection against the mutant genotype, the frequency of thu^s was about 0.3 in this strain, suggesting strong selection favoring heterozygotes. In another population, where selection for fast development was suspended, the frequency of thu^s decreased. An apparently stable gene frequency equilibrium was obtained in a population initiated with progeny from a cross between the mutant and the Berkeley synthetic wild-type strain, which had served as the foundation population for the selection experiment. The initial increase in frequency of thu^s was ascribed to mutation within a linked complex of genes, which, when heterozygous, promotes fast development. In the experimental population, the apparent gene-frequency equilibrium was probably the result of linkage disequilibrium generated from the cross between the selected line and the synthetic strain.

Another interesting case of increase in frequency of a mutant gene as a result of artificial selection occurred in an experiment involving two-way selection for larval weight on two different diets (Yamada and Bell, 1969). One of their selection lines exhibited a unique response to selection, indicating the presence of a peculiar genotype-environment interaction. In this line, selection for decreased 13-day larval weight on a medium containing corn oil did not produce the expected response. Furthermore, this particular strain, in contrast to others developed in this investigation, had a higher larval weight on another medium that lacked corn oil but was otherwise "poor." Subsequent work by Constantino et al. (1966, 1967, 1968) showed that this was due to the presence of an autosomal recessive gene that, when homozygous, produces a sensitivity to corn oil. Mutant individuals have a lowered ability to utilize a number of unsaturated fatty acids, but grow normally on diets lacking corn oil. Thus, artificial selection for smaller size led to an increase in frequency of the mutant gene. It was suggested that the mutant was fixed by about generation 12, and no further response to selection was observed.

An illustration of the problem inherent in trying to define the fitness of populations by extrapolation from Darwinian fitness of individual genotypes was reported by Dawson (1969). During the course of a competition experiment between CS and CF, an eye color mutant appeared in CS in one of 10 replicate cultures. As the frequency of the mutant gene (*chestnut*, c^D, an eye color mutation) increased, the percentage of CS in the culture decreased until the species was eliminated (Fig. 4). CS was the winner in the remaining nine replicates. Mutant homozygotes and/or heterozygotes clearly had a higher fitness

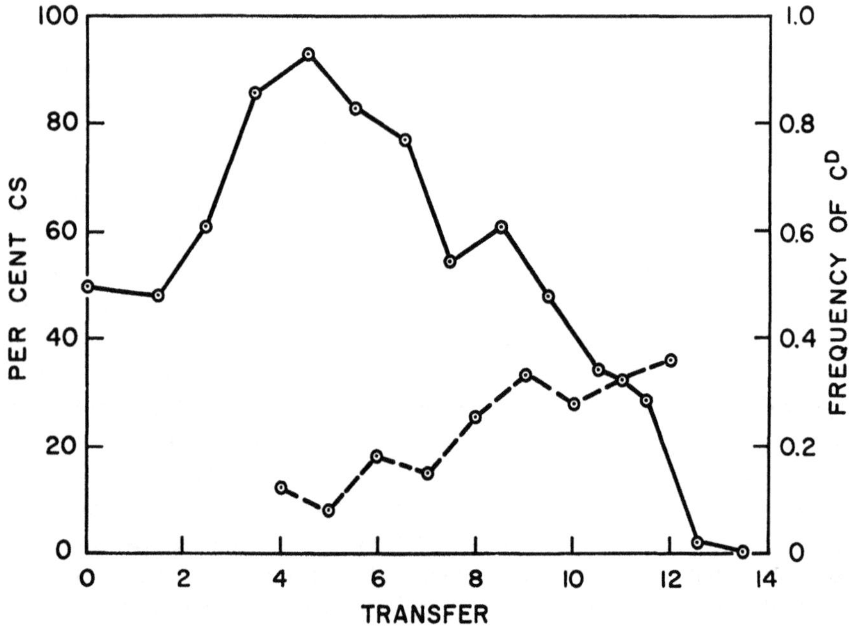

Fig. 4. Percentage of CS (solid line) and the frequency of the C^D gene (dashed line) during the course of competition with CF. The percentage of CS is plotted using a shifting two-point average in order to smooth out fluctuations in the curve which result from the technique of discarding adults at each monthly census (transfer). (Dawson, 1969)

than normal homozygotes. Dawson suggested that the reduced fitness of the CS population in competition with CF resulted from an increased production of genotypes with poor interspecific competitive ability as a result of changes at other loci. That is, the gene pool of a population has an integrated structure, or coadaptation, so that substitution of a new allele at one locus may lead to a general repatterning of the gene pool. Although this explanation is not satisfactory in many respects, others are even less so. At any rate, this observation clearly points out some of the conceptual difficulties associated with fitness considerations.

The problem of genetic load has been perhaps the most controversial one in population genetics for the last 10 years. Levene et al. (1965) utilized a number of populations of CS and CF to test the method devised by Morton, Crow, and Muller (1956) to differentiate between mutational and segregational loads.

Nonlinearity of inbreeding effects and unexplainable differences between the populations indicated that the methods of analysis, particularly "load ratios," were not suitable for making this distinction.

In an extensive series of experiments, Wallace (1958, 1963) has produced evidence that randomly induced mutations in second chromosomes of *Drosophila* improve viability of homozygotes but not heterozygotes. Crenshaw (1965) has utilized similar methods to test the effects of irradiation on inbred lines of CF. Preliminary results obtained with one inbred line indicated that production of live offspring was significantly increased in female progeny of irradiated males as compared to those from nonirradiated males. In fact the increase was much greater than that observed by Wallace. It is of some interest in this regard that Crenshaw's studies were carried out under suboptimal culturing conditions, suggesting that increased heterozygosity may be more important under environmental stress.

Quantitative Genetics

Tribolium has proved to be an exceptionally useful organism to quantitative geneticists. One of its major advantages over *Drosophila* for this purpose is the fact that flour beetles have a relatively large number of chromosomes (10 pairs in CS, 9 pairs in CF). Therefore, one is less likely to encounter the kinds of problems that have become classic in quantitative genetic studies with *D. melanogaster*, which has only 3 pairs of large chromosomes; for example, response to artificial selection is often retarded pending the occurence of recombinational events (e.g., Mather and Harrison, 1949; Thoday and Boam, 1961).

In the last decade or so, A. E. Bell and his associates at Purdue University have used CS to evaluate the relative efficiency of various artificial selection procedures, and to investigate a number of other aspects of quantitative genetic theory. Table 4 summarizes much of the published work that has been done with CS by Bell's group and others on various quantitative traits.

Body Weight. The heritability of body weight in CS has been found to be high (Table 4). There is also a considerable amount of dominance variance, particularly for larval weight, which leads to overestimates of heritability in some of the experimental designs used to obtain variance components. Realized heritabilities, based on response to artificial selection, are generally lower than those obtained from the base populations prior to selection. In spite of these complications, there is a large component of additive genetic variance that can be manipulated by artificial selection, and much of the research in this area has used body weight as the experimental trait.

Several artificial selection experiments involving body weight at different ages have been carried out in order to test various theoretical expectations. Since much of the work done in Bell's laboratory has recently been summarized (Bell, 1969), we will not review it here. Although this work, and that done by others,

Table 4. Quantitative traits of CS which have been studied in selection experiments. (Modified from Bell, 1969)

Trait	Relative Heritability	Degree Heterotic	Genetically Correlated	References
Adult Weight (AW)	high	slight	PW + LW +	Englert and Bell (1963, 1969) Okada and Hardin (1967, 1970)
Pupal Weight (PW)	high	slight	AW + LW + DT +, −	Bell and Moore (1958) Bray et al. (1962) Wilson et al. (1965) Bartlett et al. (1966) Enfield et al. (1966, 1969) Scheinberg et al. (1967) Scheinberg (1969) Englert and Bell (1969, 1970) Wong and Boylan (1970)
Larval Weight (LW)	medium	moderate	PW + DT − AW +	Englert and Bell (1963, 1969, 1970) Hardin and Bell (1967) Scheinberg et al. (1967) Okada and Hardin (1967, 1970) Wilson et al. (1968) Scheinberg (1969) Yamada and Bell (1969)
Development Time (DT)	medium to low	moderate to slight	PW +, − LW −	Dawson (1965a, 1965c, 1966b) Englert and Bell (1963, 1969, 1970) Scheinberg et al. (1967) Scheinberg (1969)
Larval Molts	–	high	LW − DT +	Englert and Bell (1963, 1969) Dawson (1967c)
Chaetal Number	medium to low	slight	–	Lange and Bell (1970)
Insecticide Resistance	–	–	–	Dyte and Blackman (1967) Dyte and Rowlands (1968)

is of considerable interest to quantitative genetics, it is not directly relevant to "the *Tribolium* model." Pertinent references to this work are included in Table 4.

A long-term study of response to selection for increased pupal weight is being carried out by Enfield et al. (1966, 1969) in an attempt to determine the importance of linkage and overdominance with respect to this trait. Selection was initiated in an F_3 population formed from a cross between two highly inbred lines. Data obtained in the F_1 generation indicated a considerable amount of heterosis for pupal weight, as also observed by Englert and Bell (1963) and by Boylan and Wong (1965). Directional dominance of genes for large size was indicated.

An analysis of selected lines (after 12 generations of selection), control lines and the original inbred lines indicated that linkage of genes affecting pupal weight was an important source of variation, and that there were very few, if any, overdominant loci. The latter observation is not surprising for traits that are not major components of fitness (Lerner, 1954). The high heritability of body weight is also expected for such traits. Furthermore, despite 38 generations of full sib mating, at least one of the inbred lines used by Enfield et al. does not appear to differ substantially in weight from the Berkeley synthetic strain (Dawson, 1965a) from which it was derived. Thus, a lack of inbreeding depression is indicated, as as expected for such traits. It is difficult to compare the studies of Dawson and Enfield et al. because the temperature and humidity conditions employed by the latter authors were not stated. Also, the pupal weights were measured at different times; Dawson used first-day pupal weight, whereas Enfield et al. measured weight 21 days after the second of a three-day egg laying period. Moreover, this inbred line was found to possess a considerable amount of genetic variation, raising the possibility of contamination during the inbreeding period. It is of interest to note that Wong and Boylan (1970), who used the same "inbred" line in forming a base population for another selection experiment, also reported the presence of additive genetic variance for pupal weight. This leads to some problems in the interpretation of theoretical models and/or statistical analyses based on the assumption of homozygosity of inbred lines for studies involving this particular strain.

A number of studies have been directed at analysis of genotype-environment interactions using body weight in CS as the experimental trait. Hardin, Rogler, and Bell (1967) studied 14-day larval weight in two populations, one of which had previously been selected for large pupal weight. Interactions between strain and humidity as well as between strain and various nutritional ingredients were observed. Bray et al. (1962) found that interaction between genotype and humidity had a significant effect on response to selection for pupal weight. Corresponding effects of nutritional level were reported by Hardin and Bell (1967) and by Yamada and Bell (1969).

For other traits, Kidwell et al. (1964) demonstrated genotype-environment interactions for number of offspring in two combinations of temperature and humidity, and Dawson (1965b) found similar interactions between genotype and humidity in determination of developmental time in several strains of both CS and CF. The widespread occurrence of genotype-environment interactions emphasizes the previously mentioned fallacy of considering only genetic or environmental factors in the study of *Tribolium* populations.

Another problem in quantative genetics that has been investigated using body weight of flour beetles as an experimental trait is the design of optimum control populations. Although the primary use of control populations is in artificial selection experiments, they have also been used extensively in studies of natural selection in experimental populations. Basically the problem is to maintain a control population in which genetic changes are identical to those in a selected population in all respects other than changes directly related to the selection process. Or, to put it another way, the ideal control population would be one in which random genetic drift is minimized and the mean value of the trait under selection remains stable through time. Bray et al. (1962), Robinson and Bray (1965), and Bray (1966) have reported the results of theoretical and experimental research on this problem. Comparison of a number of different methods for maintaining control populations indicated that the standard procedure of mass mating of controls is more subject to random genetic drift than others when genotype-environment interactions are an important source of variation. Later work involved analysis of the effects of restrictions against inbreeding and unequal family size as measured by stability of the means of control populations. Both restrictions were found to be effective experimentally.

As might be expected, the best control population is one that is closely related to the selected line in origin and in time. This is best accomplished by the method of repeat mating, where the individuals selected in one generation are used first to produce progeny for the next generation of selection and then again to produce the control population for the following generation. Although the longevity of flour beetles is such that this and other methods are feasible, it should be noted that virtually all investigations that employ control lines use the mass mating method.

In many of the selection experiments described above, the selected trait was time-dependent (e.g., 13-day larval weight, 19-day pupal weight, etc.). Under these circumstances the selected character must be thought of as a combined trait that includes not only body weight but also developmental time. This is particularly true for pupal weight, which decreases with pupal age (Englert and Bell, 1963). Thus, response to selection for pupal weight at some specified age measured from time of egg laying could conceivably occur by a shift in developmental time with little or no change in the frequencies of genes affecting pupal weight. On the other hand, selection for first-day pupal weight, even though it might lead to a correlated change in developmental time, would be expected to involve genes concerned with determination of weight.

An additional difficulty encountered in the study of larval weight results from the alternate periods of growth and molting as larvae pass through successive instars. Because of this alternation, weight does not increase monotonically with age (Howe, 1968a). Moreover, selection for weight at a fixed age may be confounded with changes in the number of instars so that the nature of the selected trait may change through time. It is not surprising, therefore, that in several of these experiments the observed responses to selection were considerably different from those predicted on the basis of genetic parameters estimated in the foundation population. Furthermore, because of these differences in technique, it is difficult to compare the results of different selection experiments involving what may superficially appear to be the same trait.

Constantino (1968, 1969) has approached the same problem by incorporating a time parameter into simple models of quantitative inheritance. This parameter modifies the genotypic value of an individual and permits variance components, heritabilities, and other quantitative measurements to be interpreted as a function of the time at which the measurements are made.

Developmental time. Several investigations have involved artificial selection for developmental time. Dawson (1965a) utilized selection for fast and slow development, measured as days from egg to pupa, in a study of the relationship between genotype and interspecific competitive ability. No control or replicate selection lines were maintained, but the main objective of the study was to produce genetically different populations, rather than to analyze the selection process. Nevertheless, it is of interest to note that selection for fast development was successful in both CS and CF. We will return to this point shortly. Response to selection for fast and slow development in CS was highly asymmetrical, with a greater response in the slow direction. This was due in part to greater selection differentials in this line. Realized heritability estimates for the first seven generations were equal for the two lines, but they diverged significantly in later generations. This divergence demonstrated that the response was truly asymmetrical. The results of selection and other supporting evidence indicated that developmental time behaved in a manner expected of a trait that is an important component of fitness in Lerner's (1954) model of genetic homeostasis.

Scheinberg et al. (1967) in their comparison of various selection methods, also were successful in selection for fast development, although the response was considerably lower than that predicted from an analysis of the foundation population. Englert and Bell (1970) carried out two-way selection for fast and slow development in three replicate lines. Asymmetrical response to selection was observed, but in the opposite direction to that reported by Dawson (1965a). That is, selection for fast development was more effective, giving a realized heritability estimate of 0.38, as compared to 0.26 for slow selection. Unfortunately, it is difficult to make a meaningful comparison between these two experiments since Englert and Bell selected only for six generations, and the asymmetry observed by Dawson was evident only in later generations. In fact, had Dawson used only the first six generations to calculate realized heritabilities, there would have been a slightly asymmetrical response in favor of fast development.

The existence of an intermediate optimum for developmental time in *Tribolium* populations is somewhat surprising at first glance, since flour beetles are undoubtedly colonizing species and as such would be expected to show little if any genetic variance in developmental time (Lewontin, 1965). Selection for fast development in *Drosophila* has not been markedly successful despite several attempts (e.g., Sang and Clayton, 1957; Hunter, 1959; Clarke et al., 1961). However, the opposite result has been obtained in one study of *D. pseudoobscura* (Marien, 1958).

The situation in *tribolium* is complicated by the cannibalistic nature of flour beetles, at least in laboratory populations. Since pupae are highly vulnerable to cannibalism, the obvious selective advantages of fast development may, in part, be countered by increased exposure to predation, at least under conditions of high density or limited food. Data on survival of pupae relative to their time of emergence in populations containing adults and larvae confirm this notion (Dawson, unpublished). Sokal and Sonleitner (1968) have also implicated selection of this type as an important factor in flour beetle population biology, and Wool (1969a) has provided evidence for reduced survival of early pupating individuals in cultures where there are large numbers of larvae present. This phenomenon is somewhat similar to the "soft selection" described by Wallace (1968), which operates against early pupating *Drosophila* in glass vials. However, it is less likely that the mechanism in *Drosophila*, which involves a mechanical dislodging of pupae from the side of the vial and their subsequent drowning in a liquid culture medium, operates in natural populations. On the other hand, cannibalism may well operate in *Tribolium* in nature; for example, it has been suggested that CS may be an important natural predator on the eggs of various moth species (Trehan and Rajarao, 1945; Adeyemi, 1968).

Productivity. A number of investigations have been concerned with productivity, measured in a variety of ways. Some of the numerous reports of the effects of various environmental factors on productivity have been referred to earlier. Published genetic studies have been confined to strain comparisons (e.g., Phillips and McDonald, 1958; Sokoloff, Shrode, and Bywaters, 1965) which are not of much interest except to prove that genes affect productivity. Crenshaw and Lerner (1964) analyzed productivity in a series of inbred strains of CF and CS, and found that there was significantly more interstrain variation in CS than CF for three productivity characteristics.

A short term artificial selection experiment for increased fecundity has been carried out in CS (Dawson, unpublished). The objective was to determine whether fecundity in "natural" populations is at a maximum level or if, instead, there is some intermediate optimum for this trait as there is for developmental time and body weight. Selection was successful, but the realized heritability was quite low, as expected for a trait that is a major fitness component.

Lewontin (1965) presented an ingenious analysis of selection on life-cycle components, which led to the prediction that colonizing species should be characterized by low amounts of genetic variance in developmental time, but relatively larger amounts for fecundity. As pointed out earlier, the cannibalistic tendency of flour beetles may partially counteract selection for fast development, leading to an intermediate optimum for this trait. This has not been compensated by a maximization of fecundity, at least in the one population studied. However, a comparison of several laboratory wild-type strains of both species showed greater interpopulation variance in fecundity than in developmental time.

Correlations Among Traits. Many of the investigations referred to above have reported correlated responses to selection and/or correlations among various quantitative traits. Although this is not surprising, there is one pitfall which has generally been ignored and has trapped several workers. This relates to the correlation between developmental time and body size (Howe, 1961, 1966, 1968b). The problem arises because of the distribution of developmental time, which is distinctly nonnormal, often being bimodal for CS. A typical example, given in Fig. 5, employs data from the study presented in Table 3.

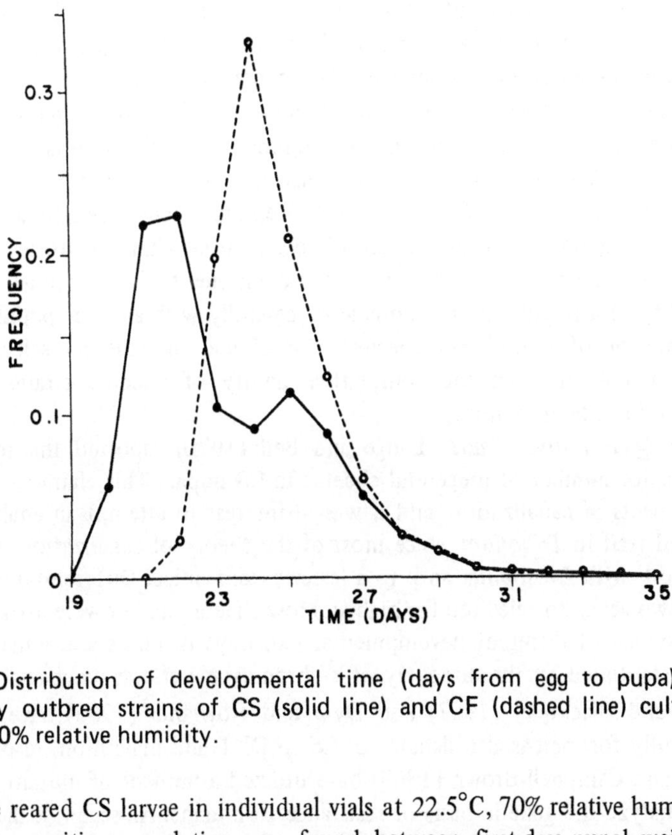

Fig. 5. Distribution of developmental time (days from egg to pupa) in the Berkeley outbred strains of CS (solid line) and CF (dashed line) cultured at 29°C, 70% relative humidity.

Howe reared CS larvae in individual vials at 22.5°C, 70% relative humidity. A very low positive correlation was found between first-day pupal weight and length of the larval period. However, it was possible to separate the population into two fractions, one "quick" developing and the other "slow." When this was done there was a high negative correlation between the two traits within each of the two fractions. These results indicate that a great deal of caution should be exercised in the choice of statistics used to analyze developmental time (Dawson, 1965a) and, in particular, the relationship between developmental time and weight.

A summary of genetic correlations among various traits may be found in Table 4. Recently Englert and Bell (1969) examined 13-day larval weight, first-day pupal weight, and pupation time in a number of the selected populations previously studied in Bell's laboratory. They found that direct selection for each of these traits had produced marked changes in patterns of growth and development in the selected populations. Genetic correlations among these traits and others varied considerably in the different selection lines. An attempt was made to account for the differences in terms of physiological changes in the hormone systems involved in metamorphosis resulting from the genetic changes brought about by directional selection for the various traits. Dawson (1967c, 1968b) had previously proposed that the physiological basis of response to selection for fast and slow development in his experiments might lie in changes in the hormone systems regulating metamorphosis.

Perhaps the most general correlated response to selection for many quantitative traits is a decrease in fitness. In common with the results of artificial selection in *Drosophila* and other organisms, reductions in fitness have been reported for some of the *Tribolium* work. We might argue at length, as seems to be a recent trend among population biologists, over what constitutes a proper measure of fitness. Leaving that aside, one component of fitness must surely be the ability of a population to compete successfully with another population of the same or of a different species. The effects of artificial selection for developmental time on the competitive ability of selected strains will be considered in a later section.

Other Quantitative Traits. Lange and Bell (1970) reported the results of selection for number of pregenital chaetae in CS pupae. This character exhibits some aspects of canalization, and it was of interest to attempt an analysis of a canalized trait in *Tribolium*, since most of the theory of canalization has come from work with *Drosophila* and, to a lesser extent, mice. Differential responses of the two sexes to selection for high and low chaeta number were attributed to the existence of different developmental pathways in males and females. This was substantiated by the discovery of Y-chromosome effects on chaeta number.

Dyte and Blackman (1967) and Dyte and Rowlands (1968) have selected successfully for increased resistance of CS to DDT and malathion, respectively. Champ and Campbell-Brown (1969) have utilized a number of mutant genes in CS to analyze the genetic basis of resistance to lindane. All seven of the linkage groups tested (by crossing resistant and mutant strains and then backcrossing to one or both parental strains) were found to contain factors for resistance. This work suggests a potential use for *Tribolium* in toxicological investigations.

Some genetic work has been done with quantitative behavioral traits in *Tribolium*. However, we will defer discussion of these studies to the next section.

Population Structure

Several observations made during the course of extensive studies with the Berkeley synthetic strains of CS and CF have led to some speculations on the relative homeostatic properties of the two species and their strategies of adaptation to varying environments (Dawson, 1968b, 1970b). Briefly, the observations are as follows: Artificial selection for fast development was more strongly opposed by natural selection in CF. A calculation of the homeostatic strength (Robertson, 1956) of the selected lines showed that the proportionate return of developmental time toward the original level following one generation of suspended selection was greater for CF than CS. Analysis of correlated responses to selection indicated that the reproductive performance of CF was less affected by the genetic changes resulting from selection. Inbreeding depression was found to be much more severe in CS for developmental time, productivity, and survival of the inbred lines. Furthermore both intrastrain and interstrain variability of developmental time was considerably greater in CS. The frequency of occurrence of morphological deviants was measured in several strains of each species, leading to the conclusion that the CF strain possessed a higher degree of developmental homeostasis.

In a later analysis of the role of founder effects in determining the outcome of competition between different synthetic strains of CS and CF, inbreeding in the form of population-size bottlenecks was found to affect CS much more than CF. The competitive ability of CS was considerably reduced in those cultures that had been subjected to the most severe bottlenecks before the start of the experiment (Dawson, 1970b).

Thus observations from two different sets of synthetic populations indicate that CF populations possess higher levels of genetic and developmental homeostasis. Dawson (1968b) suggested that this reflects a greater reliance of this species on superior heterozygous genes or gene complexes. Presumably, these genes have accumulated in response to natural selection for adaptation to a more variable, more extreme, and more unpredictable environment. Similar conclusions were reached by Tantawy and Mallah (1961) from a comparative study of several populations of *Drosophila melanogaster* and *D. simulans* collected in temperate and tropical regions.

Dobzhansky (1962) has distinguished between rigid and flexible inversion polymorphisms in various *Drosophila* species. Both represent adaptations to heterogeneous environment in time or space. Rigid polymorphisms are the result of superiority of heterozygotes over the entire range of environments encountered by a population. Flexible polymorphisms arise when different genotypes have greater fitness in different portions of the environmental range. In the latter case, inversion polymorphism can be maintained even though there is no heterozygote advantage in any single habitat (for theoretical treatments of this problem, see Levene, 1953; Levins and MacArthur, 1966; Prout, 1968; Maynard Smith, 1970).

Adaptation to a heterogeneous environment can also be accomplished by phenotypic plasticity, or individual homeostasis (Lewontin, 1957). Here, individual genotypes can survive and reproduce over a wide range of environmental conditions. The distinction between Lewontin's individual homeostasis and Dobzhansky's rigid polymorphism is that the former does not require the flexible phenotype to be produced by a heterozygous genotype. Lewontin contrasts individual homeostasis with populational homeostasis based on genetic diversity within the population.

Jain and Marshall (1967) examined a number of natural populations of two widely adapted species of wild oats, *Avena fatua* and *A. barbata*, for genetic polymorphism and for genotypic and phenotypic variances for quantitative traits. For the single-gene traits, *A. fatua* populations were found to be highly polymorphic, whereas *A. barbata* populations were virtually monomorphic. For quantitative traits, *A. barbata* exhibited greater phenotypic variation, but had a lower genotypic component. It was suggested that these observed differences reflected different strategies of adaptation to heterogeneous environments. *Avena barbata* appears to rely more on phenotypic plasticity, or individual homeostasis, whereas *A. fatua* relies more on genetic diversity, or populational homeostasis. It was emphasized that this difference is a relative one. Further studies (Marshall and Jain, 1968, 1969) have confirmed the existence of greater phenotypic plasticity in *A. barbata* and led to the conclusion that individual homeostasis is probably the most important adaptive mechanism in both species. The negative correlation between individual homeostasis and genetic variation is in accord with theoretical arguments developed by Levins (1963, 1968).

Somewhat similar relationships exist for at least two quantitative traits in CS and CF. For developmental rate, CS has the greater total phenotypic variance, whereas CF has a higher genotypic component (Dawson, 1965c). The same is true for first-day pupal weight (Dawson, unpublished). There are no comparative data on genetic polymorphism in natural populations. The information presently available might be interpreted to indicate that CF relies to a greater extent on populational homeostasis, and CS more on individual homeostasis. It would appear, however, that the CF populations are superior with respect to both developmental flexibility and populational homeostasis, supporting Lerner's (1954) suggestion that the latter follows as a consequence of the former if heterozygote superiority is an important determinant.

THE BEHAVIOR OF *TRIBOLIUM*

In contrast to the extensive ecological and genetic literature on *Tribolium*, relatively few behavioral studies have been performed with flour beetles. The major behavioral problems that have been investigated deal with prey selection in cannibalistic interactions, habitat selection, and dispersal behavior. We will treat these subjects in the indicated order and conclude with a brief consideration of miscellaneous other behavioral studies.

Before starting this section, a cautionary example of our orientation is perhaps in order. Park (1934a) has presented a detailed description of the process that an individual beetle goes through in manipulating and then eating an egg. Our primary concern in this paper is different. We will consider only that behavior which has ramifications at the population level. Thus we are not interested so much in the means by which an individual manipulates and punctures an egg, as in the behavioral fact that beetles of both species eat eggs (Park, Mertz, et al., 1965). Moreover, we would also like to know whether flour beetles can distinguish their own species' eggs from another species' eggs and might thereby exert differential predation.

Prey Selection

Behavioral data on prey "selection" have been presented by Park et al. 1965, 1968, and 1970. The experiments in the first and third of these studies can not be taken to indicate a direct behavioral preference, since the prey populations of CF and CS were not mixed together, and thus the predator did not have the direct opportunity for choice. However, several relevant observations were made, for instance, both CF and CS adults exhibit different rates of exploitation on their own eggs in comparison to eggs of the other species (Park, Mertz, et al., 1965). The form of the difference in this case was somewhat surprising; both species ate approximately 10% more CF eggs than CS eggs. In contrast, larvae of both species fed at higher rates on CS eggs than on CF eggs. It was also determined that CS larvae preyed more heavily on CF pupae than on CS pupae. CF larvae showed relatively little pupal cannibalism, but there was a slight tendency for rates to be higher on CS than on CF pupae.

The predation of adults on pupae has been studied in greater detail than any other cannibalistic interaction. Park, Mertz, et al. (1965) reported that the adults of two of the four CF strains they studied fed on CS pupae at higher rates than on CF pupae, and all four of the CS strains exploited CF pupae at higher rates than CS pupae. These relationships hold even when adults of the two species are combined Park, Nathanson, et al. (1970).

Park, Mertz, and Nathanson (1968) exposed mixtures of CF and CS pupae to cannibalism by either CF or CS adults. These experiments may therefore be used to look for preferential cannibalism, and the results clearly demonstrate that prey preferences exist for both species. Moreover, the preferences were in the expected direction: CF adults preferred CS pupae, and CS adults preferred CF pupae.

Ryan et al. (1970) have recently reported experiments in which they impregnated neutral pith discs with aqueous extracts of either CS or CF pupae. These discs were then placed in petri dishes along with a number of adults of one species or the other. In the first experiment starved adults were used, and each dish contained one CF disc and one CS disc. Choices were scored by recording at 15-minute intervals the number of beetles touching each disc. The experiment was also performed with larvae and with nonstarved adults in dishes containing two additional discs, one with flour extract and one with no extract. In each case, the CF adults and larvae were shown to orient preferably to CS discs. The opposite preference was demonstrated when CS adults or larvae were studied.

It therefore seems clear that both adults and larvae of both species not only have the ability to discriminate between CF and CS prey, but do so actively and preferentially.

Finally, we come to the evolution of cannibalistic behavior, as well as to the significance of this behavior as a regulatory mechanism in natural populations of flour beetles. Unfortunately, very little is known about the ecology of *Tribolium* outside of the laboratory. There have been suggestions that CS may be an important natural predator on the eggs of several species of moths (Trehan and Rajarao, 1945; Adeyemi, 1968), and in the absence of the crucial evidence we may assume that cannibalistic behavior is not merely an artifact of laboratory husbandry. On the basis of this assumption, Dawson (1968a) has speculated on the evolution of such behavior.

Suppose that in a population of noncannibalistic beetles, one female is an egg eater. Further, suppose that this female cannot discriminate between her own eggs and those of other females. Then all eggs have an equal chance of being eaten, and the cannibal is not placed at a selective disadvantage. Now suppose that there is some benefit gained as a result of egg eating, such as a higher fecundity or increased longevity. Natural selection would then favor the cannibal, and any genetic component of cannibalistic behavior would have a selective advantage.

The crucial supposition in this argument is the advantage accorded the cannibal. A number of studies have given some support to this idea. Stanley (1938) showed that CF females maintained on a diet of eggs had higher fecundity rates than females reared on a standard medium. Ho and Dawson (1966) reported that CS females that had been fed eggs during the larval stage had increased fecundity, as compared to larvae raised without egg supplementation. This experiment had only two replicates, but the results are suggestive. Rich (1956) and Sonleitner (1961) demonstrated that females of the two species are much more cannibalistic than males. Although the reason for this sex difference is not known, it is consistent with a selective advantage to cannibalistic behavior that is expressed in increased fecundity or longevity of the female predators or increased survival of progeny.

The possibility of a nutritive gain from egg-eating is also supported by the observation that predation is much more intense on nutritionally deficient media. Numerous studies of interspecific competition between CF and CS performed at 29°C, 70% relative humidity, have shown that CF is rapidly eliminated on such media as corn flour (Sokoloff et al., 1965a; Inouye and Lerner, 1965; Dawson, 1967b), rice flour or soy flour supplemented with yeast (Sokoloff et al., 1965b), and others (Dawson, unpublished). Although Sokoloff et al. (1965b) state that this "must be explained by an inhibition of reproduction of CF by CS," it is more likely the result of increased predation.

We have obtained evidence (unpublished) that the rates of both egg and pupal cannibalism are significantly higher for CF and CS in corn flour than in wheat flour. Furthermore, CS adults living on corn flour are more fecund if they are provided with eggs to eat (Dawson, unpublished).

An increase in cannibalistic behavior through natural selection favoring the more voracious genotypes poses further questions. Does natural selection maximize cannibalism, or is there an intermediate optimum level maintained by stabilizing selective forces? Although we have seen that the rate of cannibalism is density-dependent, active searching behavior could presumably have evolved if natural selection maximizes cannibalism. There is at least one strain of CF that occasionally becomes extinct under laboratory conditions through "cannibalistic suicide" (Park, Leslie, and Mertz, 1964). What factor(s) keep this from happening in other strains? These questions and others about the nature of cannibalistic behavior deserve further study.

The role of cannibalistic predation in determining the outcome of interspecific competition experiments between CF and CS has been discussed by Park et al. (1964, 1965) and Dawson (1968a). Dawson pointed out that in order for this behavior to be of significance in competition, there must be either prey selection or a differential advantage to the "eaters.". As we have seen, the studies of Park et al. (1965, 1968, 1970) clearly show that there is a whole complex of preferential predatory behaviors. A differential advantage is supported by the finding of Ho and Dawson (1966) that CS but not CF females laid more eggs when they had been fed eggs during the larval stage.

Habitat Selection

The fitness of individuals within a population is determined by the separate and interacting effects of their genetic constitution and the environment. Since the physical characteristics of the place in which they live constitute an important category of environmental elements, it would be surprising if most motile organisms failed to display a strong and consistent habitat selection. Unfortunately, such behavior is difficult to measure for many animals, because of their size or because it is difficult to establish a meaningful series of habitats. Thus, most of our knowledge of habitat selection behavior is inferred from the actual distribution of organisms, under the assumption that distribution equals choice or preference. However, it is relatively easy to devise environmental gradients for beetles living in stored grain products.

Amos (1969) studied habitat selection by two species of *Carpophilus* in humidity gradients, and Amos and Dennler (1969) investigated three different strains of *Oryzaephilus surinamensis* on a combined temperature and humidity gradient. Some habitat selection was indicated in each of these studies, although the sample sizes and number of replicates were quite small.

Amos and Waterhouse (1969) studied the distribution of CS in narrow troughs varying in temperature and moisture content. Wheat flour (supplemented with

10% yeast) was used for the medium, and gradients of two different steepnesses were studied. There were only two replicates, and only 15 pairs of beetles were employed in each, so replication was minimal. No measures of variability were presented in this study, which concluded that the steepness of the gradient affected the distributions of eggs (produced in a two-day period), pupae, and adults. No differences between the two gradients were reported for the developmental period or survival pattern of adults.

A more detailed study of the biology of CS and CF on environmental gradients will be presented in the last section of this paper.

Environmental conditioning by the odorous secretions of *Tribolium* has been examined in several behavioral investigations. A brief review of some of these studies is presented by Loconti and Roth (1953), in the context of secretions as defensive weapons. The same authors reported an interesting experiment that they performed with CS. Ethylquinone, which had been extracted from beetles, was mixed with small amounts of fresh flour and placed in the center of a petri dish. Other dishes were set up containing fresh flour mixed with synthetic ethylquinone, and fresh flour alone. Two hundred starved beetles were then placed at the perimeter of each dish. In the untreated control dishes, the starved CS quickly entered the flour, but the treated flour was largely undisturbed after a two-hour period. Thus the secretion acted as a repellant to the beetles.

Ghent (1963) devised a simple two-celled choice chamber and placed fresh flour in one side and conditioned flour in the other. The avoidance response of CS confirmed the earlier findings of Loconti and Roth (1953). In contrast, however, CF individuals were strongly attracted to the conditioned flour, indicating a fundamental difference in behavior between the two species. Ghent suggested two possible selective advantages for the striking preference of CF for conditioned flour. One is that the positive response acts to concentrate the species, the other is that CF may be essentially a secondary invader adapted to escape competition with other species, such as CS, which prefer fresh flour. Dawson (1968a), suggested that CF may have evolved its unusual behavioral preference for conditioned flour as an escape from the pressure of cannibalistic predation.

The principal factor involved in the differential response of the two species to conditioned flour was thought to be quinones until Ogden (1969) showed that flour conditioned by quinones is actually repellent to CF adults. Although adult beetles were strongly attracted to media conditioned by other adults, by larvae (which do not secrete quinones), and by frass (larval and pupal exuviae), quinones alone were repellent. Thus quinones do not seem to be the determining factor in the attraction of CF to conditioned flour.

Karten (1965) investigated the effects of conditioned flour on three strains of CS. He used a wild-type strain, an unrelated mutant strain homozygous for *black* (*b*), a semidominant gene, and the heterozygote formed from a cross between the first two strains. Larval survival, developmental time, and fecundity were measured for each strain on various concentrations of flour conditioned by wild type and black adults. In general, the performance of the three strains was influenced by both the source and the concentration of the conditioned flour. Similar responses of *Drosophila* larvae to the metabolic products of other genotypes have been reported by Weisbrot (1966) and by Dawood and Strickberger (1969).

One problem with using *Tribolium* in behavioral studies is that the beetles are beneath the surface of the flour most of the time and not accessible to view. However, this problem can be turned into an advantage, as illustrated by the studies of Ghent (1966) and Wool (1969c). Ghent developed a device to determine the depth distribution of adult and preadult life history stages. This device was composed of five plastic rings stacked on top of one another to form a "fractionable shell vial," comparable to the glass shell vials commonly employed in *Tribolium* research. When CS adults were placed in the vial, they quickly entered the flour, and both sexes became uniformly distributed throughout the four regions of the flour column. CF females also entered the flour soon after their introduction to the vial. In contrast, CF males remained on the surface for a much longer period. When both species were placed in the same vial, CF entered the flour more rapidly than when alone, and the distribution of CS shifted towards the bottom of the vials. But, about five days after the experiment was initiated, the CS distribution again shifted so that CS was more frequent on the surface than CF. This change in distribution was attributed to the avoidance behavior that CS displays towards conditioned flour.

Wool (1969c), used similar factionable shell vials to study the depth distribution of wild type and *black* (*b*) strains of CS. At low densities (five beetles/g), wild type was most abundant in the lower part of the vial after three days, whereas the *black* mutant was more frequent in the upper region. However, at high densities (20 beetles/g), both phenotypes were more frequent in the upper part of the vial. These results suggest a genetic factor in habitat selection, but from the high density experiment it is clear that there is also a large environmental component. There was little change in the distribution of the two phenotypes when placed in the same vial, and thus the interaction which Ghent (1966), observed between CS and CF did not appear in Wool's intraspecific system.

Dispersal

Movements of several types occur within *Tribolium* cultures. Some of these relate to dispersal of the various life history stages (e.g., Ghent, 1966; Wool, 1969c) or to simple density responses (e.g., Ghent, 1963, Wool, 1969c). Pimentel (1958) has described a daily rhythm of surface numbers of CS which imposes microclimatic temperature changes on the culture. Sonleitner (1961) also observed daily fluctuations in surface numbers at high densities, but not at low densities, where the number of beetles on the surface was negligible. The presence of large numbers of beetles on the surface in the latter study was causally related to decreased rates of oviposition.

A different type of dispersal may occur if flour beetles are given the opportunity to migrate. A series of experiments of this type have been performed by Naylor (1959, 1961, 1965), Prus (1963, 1966), Żyromska-Rudzka (1964, 1966a, b), Legay and Chassé (1964), Ernst and Mutchmor (1969), and Ogden (1970a).

In his first two studies, Naylor used choice chambers constructed by suspending replicate vials from a plastic ring. Each vial contained about 1 g flour, so that the top of the flour was level with the floor of the chamber. Males and/or females of CF (Naylor, 1959) or CS (Naylor, 1961) were then placed in the chamber and their distribution was noted after 22 or 46 hours. Females tended towards a uniform distribution, whereas males of both species were more highly aggregated than predicted on the basis of a random distribution. The first of these observations indicates that females tend to select areas of low density, or to avoid areas of high density, for feeding and egg laying. Naylor suggested that this behavior confers a survival value on the species, because fewer suitable sites would be overcrowded by the hatching larvae with a subsequent reduction in survival. We prefer to phrase the selective advantage of this behavior in terms of the fitness of individual females. Furthermore, it could well be that the dispersal behavior of females is an adaptation to reduce the loss of progeny by cannibalism, rather than to decrease the crowding of developing larvae.

The aggregation of CF males is in accord with the reports of CF's attraction to conditioned flour. The same phenomenon was observed for CS, even though it was shown that CS males are repelled by the smell of adults. Naylor also found that females spent more time below the surface of the flour than did males, and that as density increased, fewer beetles of either sex entered the flour.

Later, Naylor (1965) investigated the dispersal of CF adults in response to larvae. Adult females tended to avoid flour inhabited by larvae when unoccupied habitats were available. Moreover, the intensity of the avoidance appeared to increase as the larval density increased. Again, these results are probably best interpreted as indicating exploitation of the environment in a pattern that maximizes individual reproductive fitness.

Several studies have been performed with a simpler apparatus that gives more information on the stimuli evoking dispersal, but less information on the pattern of dispersion. This apparatus, designed by Prus (1963), consists of two culture vials connected by a length of polyvinyl tubing. Flour and experimental animals are placed in one vial, and a thread is run from the flour surface through the tubing into the second vial. This thread serves to define a route by which the beetles may leave the culture vial and migrate to the second vial, where they are trapped if they fall. Figure 6 depicts one of several minor variants of the apparatus.

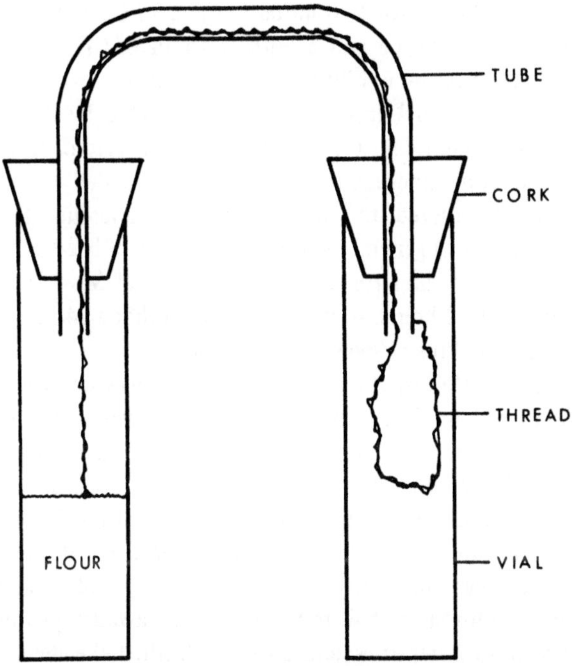

Fig. 6. A variation of the dispersal apparatus designed by Prus (1963).

This apparatus was used by Prus (1966) to study dispersal behavior in 12 strains of *Tribolium*, including the four strains each of CS and CF used by Park and coworkers in studies discussed earlier. Marked differences among strains were found, suggesting a genetic component for this trait. In general, the emigration rate (number of beetles leaving the initial vial in a 24-hour period) was higher for CS than for CF, and higher for males than for females of either species. The beetles used in these experiments were maintained in unrenewed media and tested after 2, 10, 20, and 30 days. As we saw earlier, Ghent (1966) found that CS adults rapidly entered fresh flour, whereas CF remained on the surface for a longer time. This differential response to fresh flour was reflected in changes in emigration rate with time as the flour became conditioned.

Prus also studied surface numbers, and again found significant differences between sexes and among strains. Surface numbers were lower for females, as previously found by Naylor (1959, 1961); sex differences were more pronounced for CS. Thus, males of both species show higher general activity, at least with respect to emigration rate and surface numbers, confirming earlier observations of Surtees (1963) on CS. Trends in surface numbers with time were similar to those observed by Sonleitner (1961). There was a positive correlation between emigration rate and surface numbers. In other studies, Dawson (1964b) reported that four highly inbred strains of CF differed in their tendency to migrate to and remain on the surface in fresh flour. Similar differences were noted by Sokoloff (1966b) for two mutant strains of CS and by McDonald and Fitting (1965) for a wild-type and a mutant strain of CF. McDonald (1968a) reported other differences in mobility of his two strains.

The effectiveness of Prus' apparatus was investigated by Żyromska-Rudzka (1966a) in a study of adult density, flour type (conditioned or fresh), number of emigratory paths, and experimental duration. CF tended to remain on the thread, whereas CS entered the receptacle vial. Increasing the number of receptacle vials did not in general increase the number of beetles that emigrated, but the type of medium did strongly influence dispersal. Most of the emigration took place in the first 12 hours, so it was concluded that a 24-hour experimental period, such as used by Prus (1966), was adequate.

Żyromska-Rudzka (1966b) showed that dispersal was density-dependent, at least with a flour medium. The two strains of CS that were studied had different rates of emigration in response to density; only one strain of CF was used. As might be expected, the emigration rates were higher for CS in self-conditioned flour than in fresh flour. In contrast, flour conditioned by CF reduced the CF dispersal rates. In a totally atrophic medium of baked sawdust, the percentage of emigrating beetles was uniformly high (~90%) and independent of initial density. However, addition of 25% fresh flour to the sawdust greatly reduced the emigatory rates at low densities and so re-established the density dependent dispersal pattern observed in fresh flour.

Ogden (1970a), used the same type of device to investigate the different effects of homotypically (by the same species) and heterotypically (by the other species) conditioned media on the dispersal of CS and CF. CS displayed a strong avoidance response to either type of conditioning. In contrast, CF showed little tendency to leave homotypically conditioned media but more than doubled its dispersal rate when exposed to heterotypically conditioned media. This observation is at variance with Ghent's (1963) speculation that CF is adapted to secondarily invade habitats occupied by other species. However, under either type of conditioning the dispersal of CF was lower than in a fresh medium. Other observations by Ogden confirmed earlier results of Prus (1963, 1966) and Żyromska-Rudzka (1966a, b).

Ogden (1970b) also carried out an artificial selection program for high and low dispersal in wild-type strains of both CS and CF. In common with Dawson's (1965a) experiments on selection for developmental rate, this study lacked suitable control lines and replicated selection lines. Furthermore, the "controls" that were used cannot be considered to be comparable to the selection lines, since the beetles tested in the dispersal apparatus were obtained in a very different manner. Nevertheless it is probable that some separation was obtained in five generations of selection, confirming the existence of a genetic component in dispersal behavior.

An interesting, although undocumented, observation made in this study was that selection did not change the responses of the two species to fresh, as compared to conditioned, flour. Thus, more CS beetles of both lines dispersed in the 24-hour test period when placed in conditioned flour; the opposite was true for both CF lines.

Miscellaneous Behavior Studies

Several behavioral studies that have been performed with flour beetles do not conveniently fit into any of the headings used in this section. These include work on tunneling behavior, sexual behavior, flight behavior, and maze running. Indeed, one of the first studies of *Tribolium* might be termed behavioral because, as mentioned earlier, Chapman (1918) tested the response of CF to different types and coarseness grades of flour.

The work of John Stanley at McGill University has been cited on several occasions in this review. Stanley's contributions to the *Tribolium* literature now span almost 40 years and embody a unique approach to the study of flour beetles. Consider, for example, Stanley's (1966) paper on tunneling behavior. Stanley views the complex of tunnels in a flour mass as determined by new tunneling activity, traversal through old tunnels, conversion of new to old tunnels, and destruction of both new and old tunnels by parallel tunneling activity. In order to analyze this system, Stanley formulated or derived 172 mathematical equations and concluded that the tunnel complex in a *Tribolium* culture represents a sort of "moving steady state."

M.S.M. Stanley and Grundmann (1965) described the mating and oviposition behavior of CF. One interesting aspect of this study was the finding that although females preferred a deep medium for oviposition, a fairly large proportion of the eggs were attached to the sides or bottom of the experimental container.

In the most extensive genetic investigation of *Tribolium* behavior, Lerner and Inouye (1968) carried out artificial selection for speed of running through a T-shaped maze. Rapid responses to selection were observed in both CS and CF; the results with CS are illustrated in Fig. 7 (Lerner, 1968). The low degree of repeatability of this trait and the presence of a large amount of variation between replicate matings make this response all the more remarkable. All of the fifth generation progeny of CS were retested by running them through the maze a second time. Improved performance of males in the second run suggested the possibility that there might be a learning component of behavior in *Tribolium*. Other experiments indicated that the physiological basis of the selection response might involve an increased sensitivity to conditioned flour. Although Lerner and Inouye suggested that this response might involve quinones, the experiments of Ogden (1969) described earlier would indicate that this hypothesis needs further study.

Lerner and Inouye have also selected for flying ability (Lerner, 1968). When starved, flour beetles will occasionally fly; presumably they have retained an ability to do this as an adaptation to survival in a patchy environment. Howe

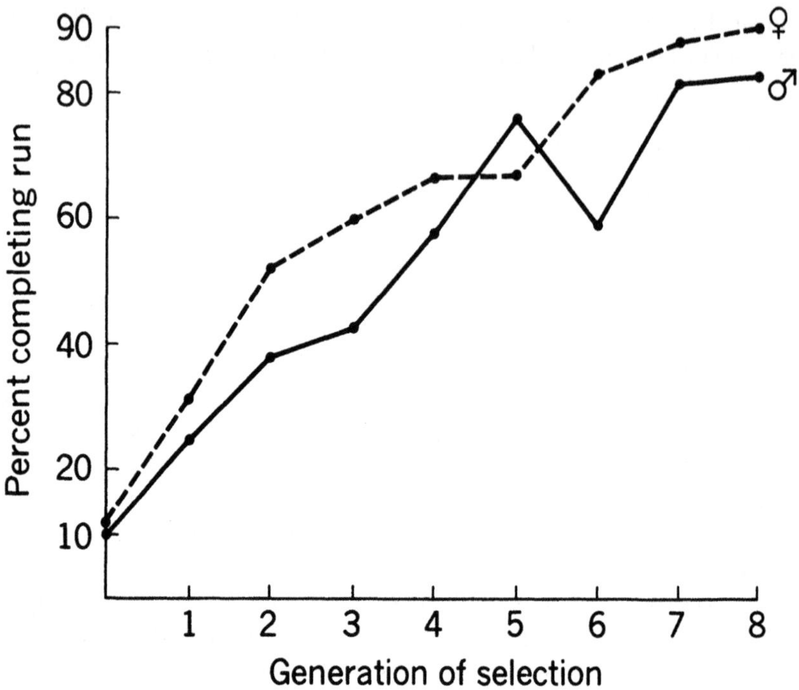

Fig. 7. Selection progress for speed of running through a T-shaped maze by CS. The ordinate shows the percentage of beetles completing the run within a two-hour period in each generation of selection. (Lerner, 1968, copyright by The Johns Hopkins Press).

(1952) has reported that under natural conditions in Nigeria CS spreads actively by flight at dusk, and has been caught in flight far from storage centers. Lerner reported that in a period of only six generations, the percentage of beetles that "flew" out of a petri dish in a 24-hour test period was increased from one or two to fifty. This indicates that flying ability is a highly heritable trait.

THE *TRIBOLIUM* MODEL

Having examined in some detail various ecological, genetic and behavioral aspects of populations of CS and CF, we now turn to a consideration of what may well be the best available illustration of the case for population biology. This is "the *Tribolium* model" of interspecific competition. By the late 1950's, after some 30-odd years of laboratory investigation with flour beetles, the major ecological problems had been defined and studied in some detail. However, some of the conclusions of these studies raised questions and problems that transcended the field of ecology. These areas remained unexplored until the early 1960's, when the geneticists "discovered" *Tribolium*.

Before going any further we need to define the term competition as it is used in the *Tribolium* literature. The many connotations of "competition" have made it one of the most ambiguous terms in ecology, and attempts to define it almost invariably lead to heated debates. One commonly used definition was proposed by Birch (1957): "Competition occurs when a number of animals (of the same or of different species) utilize common resources the supply of which is short; or if the resources are not in short supply, competition occurs when the animals seeking that resource nevertheless harm one or the other in the process." In the standard experimental design for *Tribolium* studies, food resources are, for all practical purposes, unlimited. Competitive ability is therefore determined by factors such as cannibalism, rather than by differential success in exploiting common limiting resources. In fact, the *Tribolium* system has been referred to as a double predator-prey interaction. We will return to this point later; in our view, there is no reason why the interaction should not be referred to as competition as long as it is understood how the term is being used.

As we saw earlier when we discussed the ecology of CF and CS, virtually all of our detailed knowledge of both intra- and interspecific competition has come from the intensive investigations of Thomas Park and his associates. Although the two species have also been used in competition with various other insects, we will concentrate on studies involving CS and CF. It is convenient to divide the discussion into two parts, the first of which can be labelled "The Ecological Era" and the second "The Genetic Era." Following these two divisions, we will consider some investigations that have attempted to synthesize the different approaches.

The Ecological Era

The ecological era began with the first of the major studies of *Tribolium* competition by Park (1948). Since a knowledge of the basic technique developed by Park is important to what follows, we will describe it in some detail. For the most part, the experiments are carried out in small vials containing a prescribed amount (usually 8 g) of a flour-yeast mixture. Two pairs of young adults of both species are introduced into each of a number of replicate vials, which are then placed into an incubator under controlled temperature and humidity conditions. Single-species control cultures are initiated with four pairs of adults. At monthly intervals, the cultures are censused as illustrated in Fig. 8. The various life history stages are separated by use of sieves of various pore size, and returned to fresh flour after counting. In the competition cultures this process is repeated until one of the two species is eliminated.

In the first of Park's studies of competition between the "Chicago" strains of CS and CF (Park, 1948), 92 competition cultures were established in an incubator maintained at 29°C, 70% relative humidity. Three sets of cultures, not equally replicated, contained 8 beetles in 8 g of medium, 40 beetles in 40 g, and

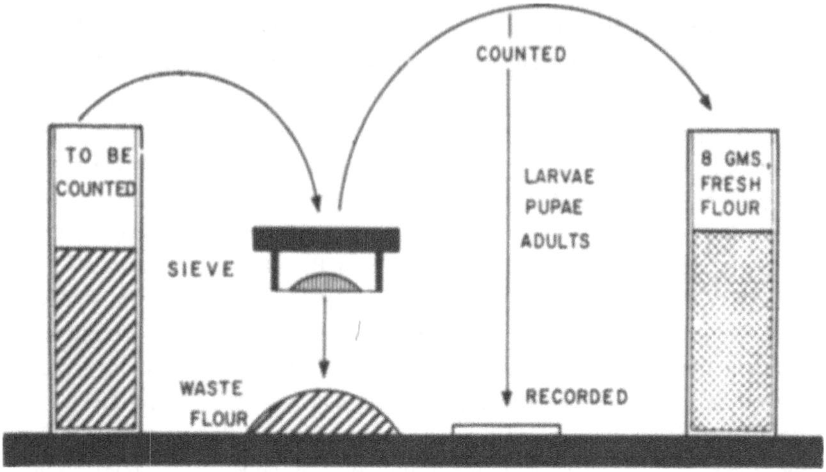

Fig. 8. Schematic diagram showing the technique of censusing a *Tribolium* population. (Park, 1962, copyright 1962 by the American Association for the Advancement of Science).

80 beetles in 80 g, respectively. The experimental design also involved varying the initial ratio of the two species; however, many of the cultures were parasitized by the sporozoan *Adelina tribolii*, which causes much greater mortality in CS than in CF and complicates the process of interspecific competition. Park also established 18 parasite-free cultures using 2 pairs of adults of each species in 8 g of flour. CF was eliminated in 12 and CS in 6.

No cases of coexistence were observed in this study. In fact, as might be expected in a uniform, homogeneous environment, coexistence has never been found in experiments with *Tribolium*. However, long periods are sometimes required for complete elimination of the losing species. Park (1948) reported that up to 840 days were required, and in other experiments it has taken over 4 years for the outcome to be decided. Miller (1967) has argued that this extended coexistence of CS and CF is a significant feature of the competition process. Although the requisite data cannot be extracted from Park's publications, our experience with competition cultures handled in the same manner indicates that this may be an "artificial coexistence." By this we mean that although the losing species may persist for a long period of time, persistence is not necessarily accompanied by recruitment. There is greater justification for arguing that a more meaningful measure of the time to elimination is the last census at which actual recruitment occurs (as indicated by an increase in the number of adults–live plus dead–of the eventual loser). Although the continued persistence of the loser after this time may, through predatory behavior, have important effects on the abundance of the other species, the absence of further recruitment surely signifies "competitive death" in the same way that a dominant sterility gene signifies genetic death.

Although CS was usually the losing species in the parasitized cultures, it is of interest to our subsequent discussion to note that the only cases of elimination of CF occurred in cultures initiated with a small number of adults. Of a total of 45 cultures started with 8 adults, CS was eliminated in 37 and CF in 8; when 40 or 80 adults were used, CS lost in all 29 cultures. A test for equality of the percentages (Sokal and Rohlf, 1968) of CS winners in the "small number" and "large number" cultures leads to the conclusion that the percentages are not equal ($P < 0.001$).

Before moving on to the next of Park's major works, a brief consideration will be given to the relative performance of the two species in parasite-free pure cultures maintained in 8 g of flour (Park, 1948; Park and Frank, 1950). The mean number of individuals (larvae, pupae and adults) in equilibrium cultures was higher for CS than for CF because the proportion of immatures was consistently higher in the CS cultures. The mean number of adults was approximately equal for the two species. Over the entire period of observation (140 days) the difference in number of adults was significantly higher for CS; however, this difference may be misleading since the time span includes the pre-equilibrium period.

Since the particular combination of temperature and humidity employed by Park (1948) has been used extensively in studies to be considered later, we present in Table 5 a summary of the results of competition between the

Table 5. Outcome of competition between the "Chicago" strains of CS and CF in cultures maintained at 29°C, 70% relative humidity

Number of Replicate Cultures	Number in Which CS Won	CF Won	Reference
18	12	6	Park (1948)
20	19	1	Kennington (1953)
28	24	4	Park (1954)
10	9	1	Park and Lloyd (1955)

"Chicago" strains of CS and CF in four separate investigations. When these data are treated as a 4 × 2 contingency table, they are found to be homogeneous, suggesting a remarkable stability of the competition process (Park, Leslie, and Mertz, 1964).

The second major investigation (Park, 1954) involved an analysis of the effects of temperature and humidity, as representative examples of physical parameters of the environment, on intra- and interspecific competition. Three temperatures (24°, 29°, and 34°C) and two relative humidities (30% and 70%) were used to form six treatment combinations. A summary of the extensive data from this experiment is given in Table 6.

Table 6. Results of intra- and interspecific competition observed by Park (1954). For each temperature and humidity combination, the first and second entries give the equilibrium numbers (*K*) per gram flour of all stages (except eggs), and the equilibrium numbers of adults for both CS and CF in single species cultures. The third and fourth entries give the number of mixed species cultures and the percentage of replicates won by each species

Relative Humidity (%)			30			70	
Temperature (°C)		24	29	34	24	29	34
	CS	3	19	10	45	50	38
K-all stages		) a	) a	) a	) a	) a	
	CF	26	30	24	28	33	41
	CS	1	9	5	18	22	22
K-adults		) a	) a	) a			
	CF	14	17	12	18	22	23
No. replicates		20	30	29	28	28	29
	CS	0	13	10	29	86	100
% won by							
	CF	100	87	90	71	14	0

[a] Significant difference between species.

Looking first at the data from control populations, it is clear that temperature and humidity had a pronounced effect on population size. As pointed out earlier, the equilibrium population size of CF is much less affected by variations in the environment than that of CS. In all comparisons at the lower humidity, CF maintained significantly larger populations than CS. At the higher humidity, there were no significant differences in the mean numbers of adults of the two species, but there were more larvae and pupae in the CF populations at 24° and 29° and therefore a greater total population size. It should be noted that the CS populations at 24° and 30% relative humidity, all died out; thus these conditions were too extreme for survival of CS. Park suggested that "adult beetles not raised in these conditions produce new generations, but their progeny cannot." As we will see later, this phenomenon is not a general one for CS populations.

In the competition cultures at the cold-dry extreme (24°C, 30% R.H.), CS was always eliminated by CF. However, as noted above, all of the CS control populations also died out under these conditions, although at a slower rate. Thus the presence of CF merely hastened the elimination of CS. At the hot-wet extreme, CS won in all replicates. In all four intermediate environments. the outcome was "indeterminate" in that each species won in some of the replicates. It is clear from the comparison of the control data and the outcome of competition that it is not possible to predict the latter from a knowledge of the former, except perhaps to say that when the equilibrium number of CF adults is significantly greater than that of CS adults in control cultures, CF will usually be the superior species in competition. However, later work has shown that this relationship does not always hold.

The results obtained in this study led to the formulation of a stochastic model to explain the indeterminate outcome of competition (Neyman et al., 1956). Briefly, the major points of the model are as follows: (1) for a given temperature-humidity combination, it is possible to specify a proportion of cultures in which a given species will win (as pointed out earlier, the data in Table 5 attest to the remarkable stability of the overall outcome). (2) In individiual replicates, the outcome is determined by random events. An illustration of the fate of one hypothetical replicate is given in Fig. 9. The two limits (L_1 and L_2) are determined by the environmental conditions in which the

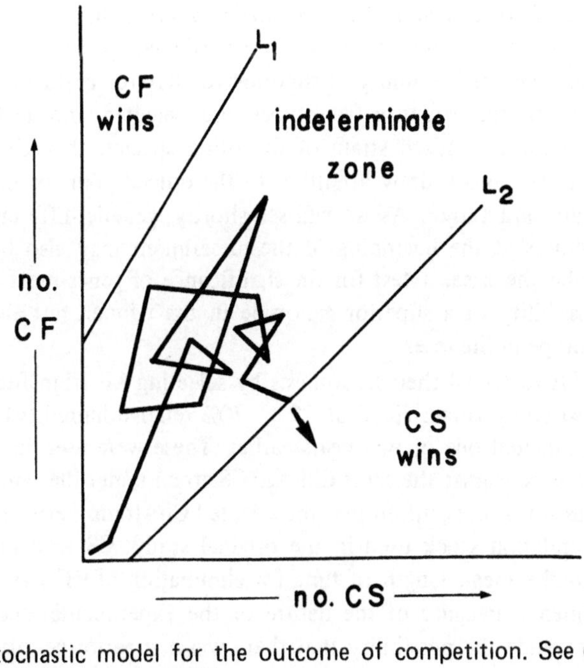

Fig. 9. A stochastic model for the outcome of competition. See discussion in text. (After Neyman et al., 1956)

culture is kept. The plotted variable is the ratio of numbers of adults of the two species, which is initially 1. During the process of competition, the path traced by the ratio is described as a "random walk" as long as the ratio remains within the limits set by L_1 and L_2. However, once the ratio crosses either L_1 or L_2, the outcome has been determined for that culture. That is, there is an underlying random process that causes the ratio to wander until it encounters one of the limits, L_1 or L_2. At that point the process is irreversibly determined and chance events no longer decide the outcome of competition. In Fig. 9, we illustrate a situation in which CS is the eventual winner.

In this model it is assumed that genetic variability within species is not important in determining competitive outcome. Or, to put it another way, the outcome is not influenced by the genetic properties of the founder individuals of the replicate cultures. Later work in Park's laboratory shows clearly that he had not dismissed genetic factors from consideration. It is clear, however, that he failed to consider the possibility of difference in the "genetic input" into replicate cultures at the time of initiation of the experiments.

This is illustrated by the approach to a test of the importance of genes in determining competitive outcome (Park and Lloyd, 1955), which was undertaken following a suggestion by Professor Th. Dobzhansky. If one assumes that an indeterminate outcome of competition can be traced to differences in the genetic properties of the individuals in different replicate cultures *that have arisen as a result of selection for competitive ability*, then it is reasonable to suppose that for those cultures in which a particular species wins, the average genotype for competitive ability is superior to that for cultures in which this species loses. If the superior genotypes are selected and tested again in competition with a standard strain of the other species, they should perform better. The italics above draw attention to the crucial point of the hypothesis tested by Park and Lloyd. As we will see shortly, genetic differences between replicate cultures at the beginning of the experiment may also be important. Should this be the case, a test for the significance of genes must consider not only the possibility of a superior genotype in the winner, but also that of an inferior genotype in the loser.

Park and Lloyd tested their hypothesis by selecting CF from four cultures of Park's (1954) study (maintained at 29°C, 70% relative humidity) in which CS had been eliminated one or two years earlier. These were used in a total of 10 competition tests against the same Chicago CS strain under the same conditions. Selection was not successful, in that the selected CF strains performed no better than the foundation stock used in the original study. CS won in 9 of the 10 cultures, and the mean length of time for elimination of CF was the same for both experiments. Because of the nature of the experimental design, it is not possible to exclude the possibility that there was a superior genotype in one of the four CF cultures. Although Park and Lloyd do not provide the necessary information, the culture in which CF won was one of either two or three replicates started from one of the four original cultures. At any rate, with reservations based on the long time period between successful elimination of CS by CF in the four original cultures and the start of this set of experiments, Park and Lloyd concluded that "natural selection resulting from competition between the two (populations) is not likely to have had appreciable effects in determining the outcome."

Park's results were for some time cited as the classical example of indeterminism in biology. In Park's own words, "The outcome of the struggle appears to be determined by something comparable to the toss of a coin" (Park, 1955). It was this point that stimulated the geneticists to initiate experiments on the *Tribolium* model.

The Genetic Era

The importance of genetic factors in determining the outcome of competition between CS and CF was first demonstrated by I. Michael Lerner and his associates at the University of California. Lerner became interested in the *Tribolium* model because of the reported indeterminate outcome and his experiments were designed to investigate this aspect of the problem (Lerner and Ho, 1961; Lerner and Dempster, 1962).

Prior to carrying out competition experiments, Lerner established new strains of CS and CF by systematic matings among six or seven strains of each species obtained from a variety of sources. These synthetic strains, previously referred to as the "Berkeley" strains, were highly heterogeneous, and have been maintained since that time as closed populations under a husbandry procedure designed to maintain genetic variation in the populations. The Berkeley CS strain was marked with *sooty*, an autosomal recessive gene affecting body color. This marker facilitates rapid discrimination between adults of the two species without microscopic examination.

In the first set of experiments, 40 replicate competition cultures were established at 29°C, 70% relative humidity. The techniques for censusing the cultures were identical to those of Park (see Fig. 8) except that (1) the cultures were initiated with 10 rather than 2 pairs of adults of each species (single species control cultures received 20 pairs of adults); and (2) in 20 cultures the adults were discarded at each monthly census. This latter procedure was tried as a means of reducing the length of time required for determination of competitive outcome.

The results of this experiment are shown in Fig. 10. CS was the winner in all 36 replicates (four replicates were discarded prior to completion of competition because of a sporozoan infection). Thus the outcome was completely determinate under conditions that had previously been reported as leading to indeterminacy (see Table 5). The only difference in the *techniques* used by Park and by Lerner in the first 20 cultures was the number of founding adults in each culture. However, since different *strains* of CS and CF were used, it is possible that Lerner merely had a superior CS strain and/or an inferior CF strain, and that the results would have been the same if only two pairs of adults had been used to start the cultures.

The modified technique introduced by Lerner to shorten the duration of the competition process produced the desired result (Fig. 10). Although the outcome was the same, the length of time required for elimination of CF was

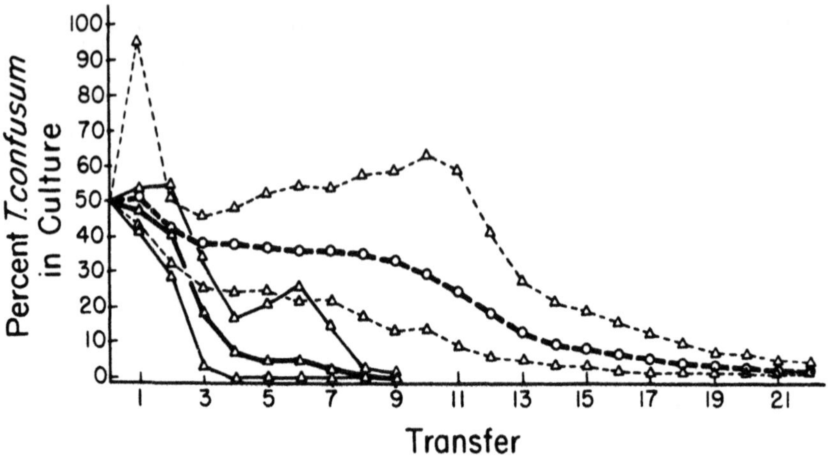

Fig. 10. The proportion of CF adults at each census. The dashed lines represent the first experiment in which all life stages are transferred to fresh vials every month. The solid lines represent the second experiment in which the adults are discarded at every transfer. The heavy lines are unweighted averages of 20 cultures (except for the last four transfers in the first experiment in which 19, 18, 17, and 16 cultures remained, respectively). The light lines show the individual cultures in which CF was respectively the most and the least successful. (Lerner and Ho, 1961, copyright by the University of Chicago).

reduced from over two years to only nine months. This technique was therefore adopted as a time-saving device in all future experiments in Lerner's laboratory.

Lerner and Ho (1961) also attempted to select for competitive ability, but were unsuccessful just as Park and Lloyd (1955) had been. This failure can be explained if genetic variation in competitive ability is primarily nonadditive. Mass selection is expected to be successful when genetic variance for the trait under selection is largely additive. If, as we suspect, competitive ability is a major component of fitness, the genetic variance in competitive ability would be expected to have a small additive component, with dominance and epistatic deviations assuming greater importance.

At the time that the new synthetic strains of CS and CF were established, Lerner also set up a series of inbred lines. These lines were maintained by full sib mating and, at various times following the initiation of inbreeding, were used in competition experiments. The results of one set of experiments, using lines that had been inbred for 16 to 18 generations, are given in Table 7. For each species, two inbred lines and the hybrid formed from a cross between the two inbred lines were competed in all nine possible combinations at 29°C, 70% relative humidity. Each combination was replicated 10 times. The experiment involving CS 5 and CF 11 in Table 7 had only eight replicates. The other two cultures were discarded because of sporozoan infection; CS was winning in both at the time of termination.

Table 7. Outcome of competition between strains of CS and CF (observed by Lerner and Dempster, 1962) given as the number of replicates won by each species

CS Strains	Replicates Won By	CF Strains: CF1	CF11	CF (1×11)
CS 3	CS	0	0	0
	CF	10	10	10
CS 5	CS	10	8	0
	CF	0	0	10
CS (3 × 5)	CS	10	10	10
	CF	0	0	0

If this entire set of experiments is treated as a single set of replicates, CS wins in 48 and CF wins in 40 of the cultures. However, for competition between any pair of populations of CS and CF, the outcome is completely determinate. The differences between sets must be genetic, so that by identifying genotypes and reducing the amount of genetic variation within sets, the random effect in determining the outcome is eliminated.

In another set of experiments with different inbred lines, initiated after 13 generations of inbreeding, there was still some indeterminacy present. Out of a total of 88 noninfected cultures, there were two cases of reversal of the usual outcome. This suggests that the outcome of competition becomes more determinate as genetic variation is reduced by inbreeding, and thus indicates a major genetic component in competitive ability.

How, then, to explain Park's results? Lerner and Dempster (1962) suggested that the indeterminacy observed by Park could be traced to the fact that he used only two pairs of adults of each species to start each replicate culture. This procedure results in a founder effect, which is a form of random genetic drift. In essence the idea is that two pairs of virgin adults do not carry a representative sample of the gene pool of the population from which they were derived, and that this sampling process results in genetic differences between cultures at the time the experiments are started. By using ten pairs of adults, initial genetic heterogeneity among replicate cultures is reduced, leading to a more uniform result. Lerner and Dempster did not rule out the possibility that stochastic, or random, events occurring within different replicates could influence the outcome of competition, particularly when the competitors were closely matched. In their experiments, however, such random phenomena, if present, were not of sufficient importance to counteract the effects of controlling genetic variation by the use of inbred strains.

It may be recalled at this point that Park (1948) had already provided evidence to support the founder effect proposed by Lerner and Dempster.

Dawson and Lerner (1962, 1966) and Dawson (1970b) have provided direct experimental evidence in support of the hypothesis of Lerner and Dempster. In the first study, progeny of adult beetles, taken from the competition experiments between inbred lines of CF and CS, were used in a selection program for decreased developmental time. Selection was successful in two inbred lines of CF extracted from the competition cultures initiated after 13 generations of inbreeding. However, no response to selection was found in any of the four inbred lines from the experiment summarized in Table 7. These competition cultures were started following 16 to 18 generations of inbreeding. As noted earlier, the outcome of competition in the latter experiment was completely determinate, whereas some indeterminacy was observed in the former. Thus the selection results provided independent confirmation of the presence of genetic variation when the outcome of competition was indeterminate. By the same criterion, they demonstrated an absence of genetic variation in the inbred lines used in the experiment with determinate outcome.

Dawson and Lerner (1966) compared the outcomes of paired experiments initiated with two and ten pairs of adults from various strains of CS and CF. A greater amount of indeterminacy was found in experiments where two pairs of adults were used. In addition, variability in the percentage of CS in ten replicate cultures was greater for the two pair series.

Since these experiments were initiated with different numbers of beetles, it can be argued that differences in initial density may be of importance in determining both the pattern and the outcome of competition. Hence, another set of experiments was established in which the founder effect occurred one generation before the start of competition (Dawson, 1970b). Seven experimental series, each containing 10 replicates, were started with 10 pairs of adults of each species. These beetles were, in turn, the progeny of 1, 2, 4, 8, 16, 32, or 64 pairs of adults taken from new, highly heterogeneous synthetic populations. The hypothesis to be tested was that there should be greater variability between replicate cultures in the competition series where small numbers of founder adults were used. The hypothesis was confirmed.

Another way to test the importance of genetic factors in determining competitive outcome would be to alter the genetic composition of a population and measure the correlated response in competitive ability. This approach was used by Dawson (1965a; 1966a; 1967a, b) who, starting with the Berkeley synthetic populations of CS and CF, produced genetically different strains by means of artificial selection for developmental rate. The selection results have been described in a previous section; here we will give a brief summary of the competition experiments.

At various times during the selection program, the selected strains were placed in competition with the synthetic stock of the opposite species. The most general observation was a decrease in competitive ability as a correlated response to selection for either fast or slow development. This was interpreted on the basis of a genetic model relating competitive ability and developmental rate (Dawson, 1966a).

One exception to the general trend was found in the CS strain that had been selected for fast development for 17 generations (Dawson, 1967a). When placed in competition with the CF stock, this CS strain eliminated CF more rapidly than did the CS stock. Further analysis indicated that the improvement in the selected CS strain was probably due to increased cannibalistic activity of larvae (and perhaps adults).

Clearly then, alteration of the genetic composition of a population by artificial selection can lead to changes in competitive ability. Similar results were obtained by Sokoloff (1969b) using strains of CS selected for high and low body weight by Dr. I. Franklin.

As noted earlier, Park and Lloyd (1955) and Lerner and Ho (1961) were unsuccessful in attempts to select for increased competitive ability in CF. A third failure was reported by Inouye and Lerner (1965). However, Dawson (1966a, and unpublished) has been able to select for *reduced* competitive ability in a CS strain that had previously been artificially selected for slow development.

As mentioned earlier, Lerner and Dempster (1962) did not rule out the possibility that unique events occurring in individual cultures could lead to indeterminate outcome of competition. Dawson (1969) provided an example of this type of event. In one of ten replicate cultures, an eye color mutant occurred in CS. As already discussed, the mutant gene increased in frequency at the same time CS was being eliminated in competition, although CS was the winning species in the other nine replicates.

Rudiments of the Synthesis

A final confirmation of Lerner's hypothesis may be found, as noted earlier, in Park, Leslie, and Mertz (1964). Four strains of each species, all but one produced by a haphazard inbreeding program of variable duration, were chosen on the basis of a productivity criterion, and competed in all 16 possible combinations at 29°C, 70% relative humidity. Ten replicate cultures for each combination were started with two pairs of adults of each species. As shown in Table 8, a striking reduction in the indeterminacy was obtained. Eleven of the 16 combinations produced a completely determinate outcome, in spite of the fact that only two pairs of founders were used for each culture. On the basis of the results obtained by Lerner, one would predict that with more inbreeding, the indeterminacy would be reduced still further.

An interesting feature of this study is that the outcome of competition for any particular combination of strains depended very strongly on the strain of CS being used (Table 8). Similar results can be seen in Table 7 for the inbred lines used by Lerner and Dempster (1962). In Park, Leslie, and Mertz (1964), CS strains II and III were almost always eliminated, regardless of the CF strain they competed against. Conversely, CS strains I and IV were almost always victorious in competition.

It is of some interest to note, in connection with the earlier discussion of inbreeding effects, that the latter two CS strains were the least inbred of the

Table 8. Outcome of competition between strains of CS and CF (observed by Park, Leslie, and Mertz 1964) given as the number of replicates won by each species

		CF Strains			
CS Strains	Replicates Won By	I	II	III	IV
	CS	10	10	10	10
I	CF	0	0	0	0
	CS	1	0	0	4
II	CF	8	10	10	6
	CS	0	0	0	0
III	CF	10	9	10	10
	CS	9	9	9	8
IV	CF	1	0	1	2

four. Strain I was an outbred stock that had been in the laboratory for a number of years. Strains II, III, and IV had been "inbred" for 7, 5, and 3 generations, respectively (Park, Mertz, and Petrusewicz, 1961). Furthermore, the presence of considerable genetic heterogeneity in strain IV, as indicated by later observations that this strain had "changed" (Leslie et al., 1968; Mertz, 1969), is in accord with the relatively high amount of indeterminacy associated with this strain. As shown in Table 8, three of the five combinations with indeterminate outcome involved CS strain IV.

In the earlier discussion, evidence was presented to show that CS is more severely affected by inbreeding than CF. Previous work, by Park and his associates (see Table 5) had shown that under this set of conditions CS usually wins in competition. The fact that CS strains II and III, which had undergone the most "inbreeding," exhibited greatly reduced competitive ability, is consistent with the earlier evidence.

The "genetic strains" used in this set of experiments have been the subject of a number of studies by Park and his students and associates. Many of these have been referred to in earlier sections. Park, Leslie, and Mertz (1964) utilized the results of some of these studies and the data on performance of the various strains in single-species populations to obtain estimates of parameters needed for construction of another stochastic model. The model, developed by Leslie (1962) for the *Tribolium* system, generated a series of expectations of competitive outcome that could be compared to the experimental results. In general, the agreement between predicted and observed outcome was rather good. It would be of considerable interest to apply this model to experiments in which the problem of genetic heterogeneity between cultures at the onset of competition was not an important factor.

The relationship between performance in single species cultures, as measured by the equilibrium number of beetles (K), and competitive ability is an intriguing one. Comparing outbred populations of the two species under the "standard" conditions of 29°C, 70% relative humidity, Lerner and Ho (1961) found that under their experimental procedures, CF maintained larger populations in pure cultures but was the losing species in competition. Park's (1954) technique yielded opposite results. Comparing strains within species, Lerner and Ho (1961) reported a positive correlation between K and competitive ability. Crenshaw and Lerner (1964) and Crenshaw (1966) provided further evidence from detailed analyses of productivity of the various inbred and hybrid strains used in Lerner's experiments.

On the other hand, these relationships did not hold in the study of Park, Leslie, and Mertz (1964). For CS, the two most successful strains in competition maintained the largest and smallest numbers of beetles in pure cultures. All four CF strains were more or less equally successful in competition, despite rather striking differences in performance as single populations. There was a marked difference among the various strains in rate of cannibalism (Park, Mertz, et al., 1965), which was shown to be an important factor in determining the competitive ability of the various strains. The inescapable conclusion, following Park, Mertz, and Petrusewicz (1961) and Dawson (1967a), is that competitive ability is the sum total of numerous factors that interact in exceedingly complex ways.

It should also be pointed out that qualitatively different selective processes would be expected to operate in cultures subjected to the two different husbandry procedures. When adults are discarded at each census, the mean density of adults through time is reduced. To the extent that adult density affects the competitive process, we would expect the adaptive structure of two competing populations to reflect r selection. That is, selection would favor those genotypes with maximum early reproductive ability, regardless of adult efficiency in utilizing environmental resources or in regulating population size. On the other hand, K selection should predominate in equilibrium populations in which adults are not discarded. This type of selection would favor genotypes that maintain high fecundity in the presence of large adult numbers, even though the reproductive rate of these genotypes might be relatively low in the absence of adult density restraints.

Of course, the preadult stages are subject to density restraints under both husbandry procedures. To the extent that preadult density determines the pattern of selection, the two procedures are expected to lead to similar results. Nevertheless, it would be surprising if the techniques used in the laboratories of Park and Lerner did not produce differences in growth rates, developmental times, population sizes (of all life history stages), and rates of cannibalism.

All the studies described up to this point were carried out using a medium consisting of whole wheat flour supplemented with brewer's yeast. More recently, competition experiments have been conducted in a number of other kinds of flour, using the Lerner and Ho (1961) technique of discarding adults. When competition cultures were established in corn flour (Sokoloff et al.,

1965a), CF was rapidly eliminated. Subsequently the numbers of CS decreased to a very low level. It was suggested that corn flour is nutritionally deficient for CS, but that CS individuals can obtain the necessary nutrient(s) from predation of CF eggs and/or pupae. Therefore, it was concluded that CS can survive on corn flour as long as CF is present, but once CF has been eliminated, CS has effectively committed "near-suicide." These experiments were terminated after 10 months. However, the evidence indicates that CS populations can survive on corn flour for at least 22 months (Inouye and Lerner, 1965) and, furthermore, that the mean number of adults increases with time, suggesting that CS is capable of adapting to the corn flour medium (Dawson, 1968a).

Sokoloff and Lerner (1967) have attempted to refute the notion that CS and CF are ecologically similar and—on the basis of experiments carried out at different times using different procedures, without controls, and in two cases with only a single replicate which was started with the survivors of at least 20 months of previous competition—have argued that competition may occur in corn flour, but does not occur in yeast-supplemented wheat flour. Their approach was to accept competition as occurring *only* when the competitors jointly exploited resources that are present in short supply (the first part of Birch's 1957 definition, quoted earlier in the present section). Interference competition (the second part of Birch's definition) was rejected as meaningless because of its breadth. Given these constraints, it may be argued that competition does not occur between CS and CF if (1) the two species do not share common resources (that is, they are not ecologically similar in a given habitat), or (2) if resources are not present in limiting supply. Having constructed this framework, Sokoloff and Lerner then proceeded to list some of the differences in behavior, productivity, survivorship, and nutrition between CS and CF that we have previously reviewed. On this basis, they concluded that "CS and CF can no longer be considered as having similar ecology." Secondly, they argued that in yeast-supplemented wheat flour, the numbers in even single-species cultures are held below the carrying capacity of the medium because of cannibalism. Thus when both species are present, a mutual prey-predator relationship, but not competition, is presumed to occur in this medium. In contrast, it was suggested that certain required nutrients are in limited supply in corn, thereby explaining the increased cannibalism rates in corn flour. Because of this, it was concluded that cannibalism in corn flour may represent competition for nutrients.

Several important points are overlooked by these arguments. First, interference competition is not only a "real" phenomenon, but we know more about it than we do about exploitation competition (see Birch, 1957; Miller, 1967). Second, any two groups of organisms distinct enough to be termed separate species will also differ ecologically. As Kohn and Orians (1962) have pointed out, "Ecological differences may be striking . . . between very closely related sympatric species which are difficult to distinguish morphologically." Moreover, by the analysis of Sokoloff and Lerner, strain differences of the magnitudes reported for both CS and CF might be enough to disqualify strains of the same species from being termed "ecologically similar."

Cannibalism occurs in both environments, although at different rates, and so it is difficult to see the distinction between cannibalism as a prey-predator interaction in one habitat and as a competitive interaction in the other. In both habitats, it seems clear that cannibalism not only has an individual fitness value, but is of significance in the regulation of population size (Park, Mertz, et al., 1965; Dawson, 1968a). Moreover, it is appropriate to reiterate the findings of Park et al. (1965, 1968) that both CF and CS preferentially select "prey" of the other species. Thus, although it does not seem profitable to quibble over the semantics of the interaction between CS and CF, it is important to recognize that environmental changes primarily affect the *quantitative*, rather than the qualitative, form of the interaction.

A number of other aspects of interspecific competition have been investigated using CS and CF, but lack of space prevents their full discussion here. Numerous further reports of competition studies involving genetically different strains and a variety of different kinds of flour can be found in the various issues of the Tribolium Information Bulletin, as well as in Sokoloff (1969b), Dawson (1967b), Inouye and Lerner (1965) and others. The effects of varying the initial proportions of the two species have been investigated by Park (1957) and by Leslie et al. (1968).

In addition to the stochastic models developed by Neyman et al., (1956) and Leslie (1962; used in Park, Leslie, and Mertz, 1964), a number of other theoretical and computer simulation models have been developed for the *Tribolium* system (e.g., Landahl, 1955; Taylor, 1968; Niven, 1969). These models share the common failure of the ecological model described in the Prologue; they assume that all individuals of each species are genetically identical.

CS and CF have also been used in interspecific competition experiments involving other *Tribolium* species (Sokoloff and Ho, 1966, 1969), other granivorous beetles (Park, Gregg, and Lutherman, 1941; Crombie, 1945, 1946; Polnik, 1960; Lefkovitch, 1968) and moths (Adeyemi, 1968). Although these studies will not be considered in detail, it is of interest to note that in a high proportion of the combinations tested, CS and/or CF are superior to the other insects in competition. The overall results suggest that the more voracious predatory habits of CS and CF may provide these species with a significant advantage in competition.

Before closing this section it is appropriate to interject a word of caution concerning the techniques used in handling competition cultures. The monthly census interval utilized by Park and most other investigators roughly approximates the generation time of both CF and CS under conditions of 29°C, 70% relative humidity on a yeast-supplemented wheat flour medium. The modified technique introduced by Lerner and Ho (1961) yielded results similar (in pattern) to those of Park when a similar environment was used. However, this technique has also been employed, as we have seen, for many other studies using diverse environmental conditions. In many of these environments there are marked differences in developmental time of the two species, and the practice of discarding adults can lead to severe bottlenecks in population size, particularly in deficient media.

Perhaps an illustration will clarify these points. As part of an investigation of ours on the behavior of *Tribolium* populations in heterogeneous environments (to be discussed in the last section), populations were also established in the 16 uniform environments resulting from a factorial combination of four temperatures and four flour types (see Fig. 13). Competition between CS and CF was carried out in jars containing 20 g of medium, which were started with 20 pairs of adults of each species. CF was the winner in all parasite-free cultures (there were four replicates at each combination, and four of the 64 cultures had to be discarded with both species still present, although in all four cases the numbers of CS were very low).

The marked superiority of CF was somewhat surprising to us, even though the relative humidity (which was not precisely controlled) was fairly low in some incubators. Therefore we initiated a series of competition experiments using the Lerner and Ho technique, with 10 replicates in each of 12 of the 16 environments. These experiments are still in progress, but an extremely interesting trend has been observed. At 30°, CF has won in all 29 of the cultures where the outcome has been decided, and is substantially ahead in the remaining 11. On the other hand, a considerable amount of indeterminacy has been found at both 26° and 34°. For example, in 26° wheat CF has won in 3 and CS in 6 of the 9 completed cultures; in 34° wheat the results are exactly reversed. In many of these experiments, particularly at 26°, the populations of both species have gone through one or more extreme bottlenecks in numbers.

These preliminary results indicate to us that caution should be exercised in the use of the Lerner and Ho procedure in studies of interspecific competition under conditions different than those which have been shown to lead to similar patterns when compared with the procedures utilized by Park.

Thus the analysis of competition between species of *Tribolium* has recently proceeded along both ecological and genetic lines. However, questionable conclusions have been reached by members of each group because of their failure to jointly consider both environmental and genetic factors. The moral of the story seems obvious.

All of the studies reviewed in this section have been performed in uniform, homogeneous environments. As we pointed out earlier, coexistence of two competing species is not to be expected in this type of environment. In the final section, we will look at the results of an attempt to extend the *Tribolium* model to a complex, heterogeneous environment.

THE DYNAMICS OF *TRIBOLIUM* IN COMPLEX ENVIRONMENTS

> The longer we study flour beetles, the less we view them as beads in an urn. Rather, we come increasingly to the opinion that they are organisms with a complicated, and sometimes capricious, personality!
>
> Park, Mertz, et al. (1965)

Environmental variation is a basic characteristic of any natural habitat, yet most laboratory investigations of the relationships between populations and their environments have been performed under designs that seek to eliminate rather than measure the effects of environmental variation. Our purpose in the present section is twofold: first, we wish to sketch several aspects of our experiments with *tribolium* in complex environments, and, second, we will indicate our view of promising directions for future experimental research in population biology with flour beetles.

To our knowledge, only one previous attempt has been made to investigate competition in complex environments, and that one was quite different from our own. Lerner et al. (1966) constructed a population cage with a mixture of several types of flour in a large central chamber, and a single flour type in each of several projecting side chambers. There were only slight differences in distribution of the two species in individual cages, and coexistence was not obtained in mixed species cages.

The experiments to be described were performed in heterogeneous environments using one physical variable, temperature, and one biological variable, food. A temperature gradient varying from 20° to 36°C was produced by a modified counter-current heat exchange device (Fig. 11). At right angles to this was a nutritional "gradient" consisting of four discrete steps: 100% wheat flour (hereafter referred to as "wheat"); 2/3 wheat flour and 1/3 corn flour ("2/3-wheat"); 1/3 wheat flour and 2/3 corn flour ("2/3-corn"); 100% corn flour ("corn"). However, as the beetles milled through the flour, the edges between the different flour types became less discrete, and thus more closely approximated a true gradient. Populations were maintained in 40 cm square plexiglass boxes, termed "arenas," each containing 1280 g flour. The arenas were placed on the heat exchange devices as shown in Fig. 11. The heat exchangers were located in a constant temperature room kept at 25°C, 60% relative humidity.

At the time of each census, the arena was divided into 16 cells (Fig. 12) with a metal grid. The hypothetical "centers" of each of these 16 cells were used for experiments performed in uniform environments. For temperature, the four "centers" were 22°, 26°, 30°, and 34°. These numbers are also used in identifying the cells of the arenas (Fig. 12).

There are three major components of the present study:

(1). Arena populations: Sixteen arenas were established. Of these, four contained CS, four contained CF, and eight contained both species. The eight single-species arenas, which will be referred to as CS or CF "control arenas,"

Fig. 11. Arrangement of the heat exchangers within a constant temperature and humidity room. Several arenas are shown on the heat-exchange devices.

were started with 2500 adult beetles. For the eight "competition arenas," 1250 adults of each species were used. These populations were used to measure growth rates, distributional patterns, and competition in spatially variable environments, and also to study the patterns of adaptation to a complex environment.

(2). Single cell populations: sixteen individual environments were established as indicated above. There were four replicates each of CS and CF, and four replicates of the two species together. Screw-top jars (with silk bolting cloth replacing the center of the top) were used to achieve the same depth of flour as in the arenas. Populations were initiated with 40 pairs of adults (20 pairs of each species for the competition cultures) in 20 g flour. These experiments were designed to provide data on the behavior of populations in uniform environments for comparison with the arena results.

(3). Estimation of population parameters: The 16 individual environments were again used to measure a number of parameters of interest in this project. Among these are age-specific fecundity and survivorship, survival of the immature life stages, rates of egg and pupal cannibalism, and developmental times under the various conditions.

All of the beetles used in these investigations were taken from new, highly heterogeneous, synthetic strains that were established by systematic crosses

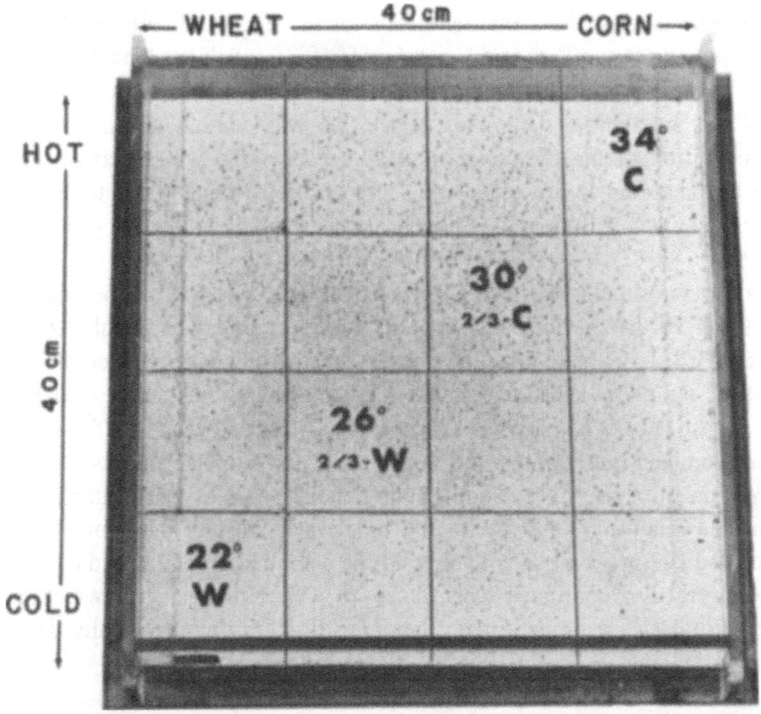

Fig. 12. Photograph of an arena with an overset to show the temperature and nutritional gradients, and the nomenclature of the 16 cells.

among several strains of each species. A body color mutant, *sooty*, was incorporated into the CS strain to aid in separating the two species in mixed cultures. These strains have been reared under a husbandry method designed to maintain maximum genetic variation in the stock cultures.

The flour used in all aspects of this study was obtained from the same sources even though part of the work has been performed at Yale University and part at the University of Illinois and at Oregon State University. Standard sieves were used, and although censusing techniques varied because of the amount of flour involved, every effort was made to keep them as similar as possible.

Both the arenas and single cell populations were censused (transferred) every four weeks. The contents of each cell (jar) were sifted through sieves of various pore sizes in order to remove adults, pupae, larvae, and eggs. Live adults were counted and weighed (separately by species where appropriate) and the pupae were counted (and also weighed in the arena work). Larval counts were obtained during some of the transfers. Dead adults were recorded separately and discarded. All live stages were then placed in fresh flour, thoroughly mixed, and returned to the cell (jar).

Unfortunately, a sporozoan parasite (*Nosema* sp.) was introduced into some of the stock cultures, the arenas, and some of the single cell cultures. As reported by Park (1948) and Park and Frank (1950), CF is much less sensitive than CS to the effects of parasitism. Partly for this reason, and partly because our analysis of CS is at a more preliminary stage, we will confine our discussion to a consideration of the CF populations. Moreover, only two aspects of the study will be presented in any detail: (1) effects of environmental heterogeneity on fitness characteristics, and (2) behavioral adaptations to the environmental mosaic.

One point should be made with regard to the CS single cell populations. As discussed in the previous section, Park (1954) found that populations of the Chicago CS strain could not survive under an environmental regimen of 24°, 30% relative humidity and yeast-supplemented wheat flour. In contrast, our nonparasitized CS populations survived for almost two years at 22°C on all four flour combinations (which did not have yeast supplementation). With the exception of corn flour, equilibrium population sizes were established that remained relatively stable up to the time when most of the populations were discarded (22 months). It is likely that the combination of 22° and corn flour is too severe for continued persistence of CS; although the two remaining populations are declining in size, they are still recruiting new adults after some 32 months.

Fitness

Many different parameters have been suggested as measures of population fitness. These include rate of population growth (r), equilibrium population size (K), and mean body weight. Data have been collected on each of these and will be presented briefly here, and in more detail elsewhere.

The r values obtained in this study for CF are presented in Fig. 13. These values are based on weekly censuses of survivorship and fecundity in each of the 16 environments. An analysis of variance on the three individual replicates (each started with four pairs of 1- to 2-day old adults and combined to calculate the rates presented in Fig. 13) reveals that both temperature and flour have highly significant effects on the rate of population growth, but do not appear to interact (Table 9). The growth rates increase with increasing temperature and, less regularly, with increasing percentage of wheat flour.

By the 18th transfer, the numbers of adults in both the arena and single-cell populations had reached an approximate stablility. The equilibrium population size for each replicate, K, was estimated as the average number of live adults recovered in transfers 18, 19, 20, and 21. An analysis of variance was performed on these averages (Table 9) and it may be concluded that both temperature and flour have significant main effects in both the arena and single-cell populations. Additionally, a significant interaction between the two was found in the arenas, but not in the single cells.

CF - GROWTH RATES (r)

	wheat	2/3-W	2/3-C	corn	mm
34°	4.83	4.68	4.35	3.19	4.26
30°	4.29	4.70	4.42	3.04	4.11
26°	2.70	3.03	2.45	1.28	2.37
22°	0.93	0.79	0.89	0.53	0.79
mm	3.19	3.30	3.03	2.01	

Fig. 13. Growth rates (*r*) estimated for 16 different combinations of temperature and flour. mm = marginal means.

Table 9. Factorial analysis of growth rates and of adult and pupal numbers and adult weights in arena (A) and single cell (SC) populations of CF. Entries are mean squares.

			Mean No. Adults		Mean No. Pupae		Mean Wt. Adults	
Factor	d.f.	Growth Rates	A	SC	A	SC	A	SC
Temperature (T)	3	42.626[c]	50266[c]	10869[c]	27835[c]	783.7[c]	0.054[c]	0.220[c]
Flour (F)	3	4.458[c]	28504[b]	5456[c]	1004	344.6[c]	0.053[c]	0.042[c]
T X F	9	0.412	19471[b]	177	1406	116.6	0.001	0.005[a]
Error	45	0.112	5059	207	814	59.5	0.002	0.002

[a]$p \leq 0.05$.
[b]$p \leq 0.01$.
[c]$p \leq 0.001$.

Figure 14 gives the estimated adult *K* values for both experimental designs. In order to compare the arenas and single cells directly, the total numbers of adults per cell have been reduced to the number of individuals per gram of flour. The correlation between the two sets of measures is, of course, quite high (> 0.9) and the conclusions from the statistical analysis are the same for both "per cell"

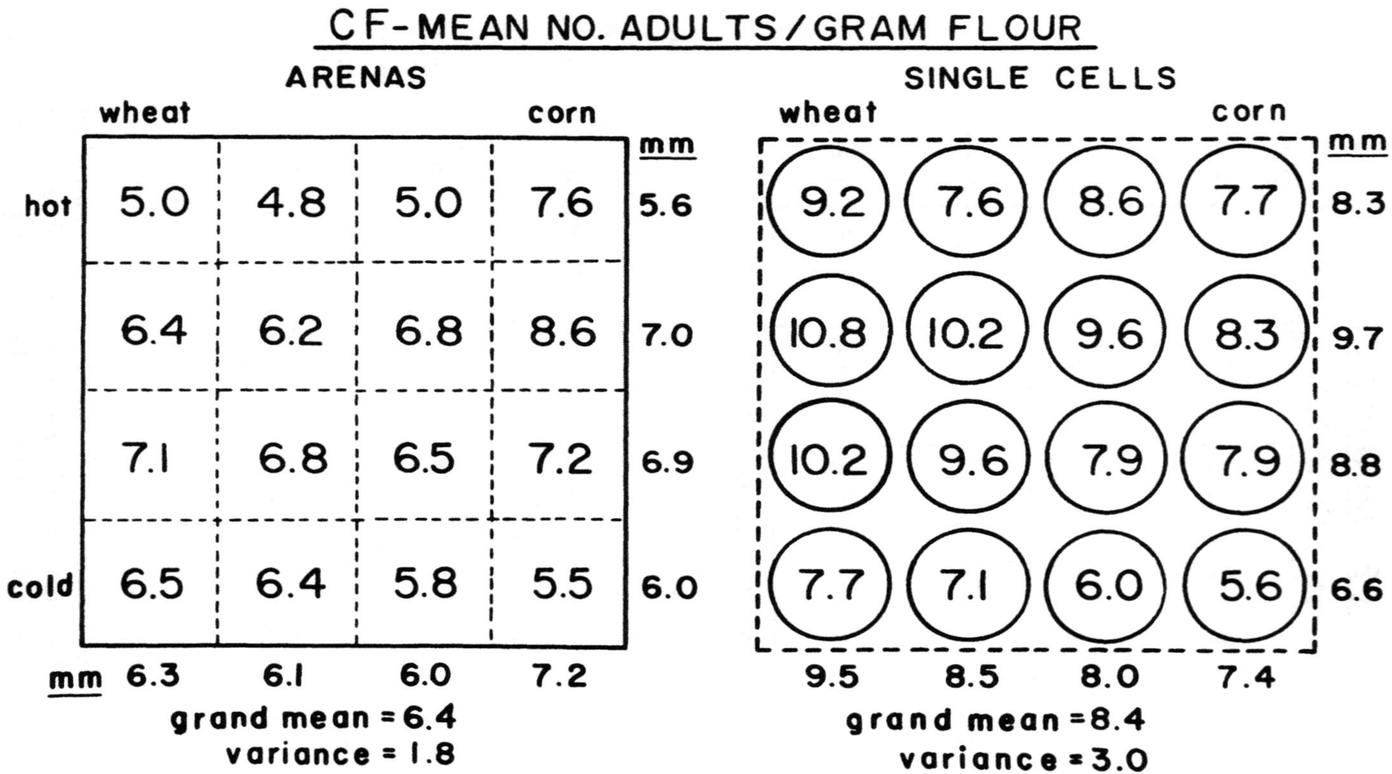

Fig. 14. Mean numbers of adults per gram flour for four replicate populations in both the arena and single cell populations. The numbers in each replicate were averaged over transfers 18 to 21. mm = Marginal means; the grand means and variances are presented at the bottom of the figure.

and "per gram" densities. However, the need for some caution is indicated because of the different sizes of the two environments; each cell in the arenas had 80 g flour, whereas only 20 g were used in the single cell populations. Thus the differences between comparable values in Fig. 14 may reflect not only the structures of the environments, but also the different quantities of flour.

Three major differences between the single cell and arena populations are evident in Fig. 14; these differences are in the equilibrium population density, the variance in equilibrium population density, and the distribution of *K*.

(1). Population density: The overall mean population density in the single cells is approximately 25% higher than in the arenas and, moreover, the sign of the difference holds in 15 of the 16 cells. It may be that the lower equilibrium densities in the arenas reflect the presence of the sporozoan parasite. Park (1948) and Park and Frank (1950) found that about 20% more adults (and 20% fewer larvae and pupae) occurred in nonparasitized CF populations than in populations parasitized with the sporozoan *Adelina tribolii.* Thus the presence of the sporozoan in the arenas and absence of the parasite in the reported single cells may explain the differences in adult population size. However, in contrast to Park's results, the pupal numbers are also much higher in the parasite-free cultures. This may indicate that there are also effects on population sizes directly attributable to the arena environment.

(2). Variance in equilibrium population density: Equilibrium population densities between cells in the arenas are less variable than the equilibrium densities between the single cells. In Fig. 14, the variance of the cell means over transfers 18 to 21 is shown to be about 40% larger for the single cell populations than for the arena populations. The reasons for this relationship (unlike those for the differences in total population size) are quite clear.

In the first few transfers after initiation of the arena populations, the adults tended to concentrate in the central temperature bands. However, as new adults were recruited to the population, the individuals dispersed and the overall distribution tended towards greater uniformity. These movements are undoubtedly more complicated than those observed by Naylor (1959) for CF in a spatially complex series of uniform environments. For instance, it is quite clear from the *r* values presented earlier that both flour and temperature significantly affect the fitness characteristics of survivorship and/or fecundity. However, these measures were made under optimum, or near optimum, densities and thus do not satisfactorily express fitness under the restraints of *K* selection (King and Anderson, 1971). We might therefore expect an individual with maximum fitness in one cell under optimal density conditions to emigrate to a different cell if the advantage gained by escaping the density restraint is great enough to overcome the loss of fitness in the new suboptimum physical environment. This process, accompanied by natural selection, should lead to a more uniform distribution of population sizes than under single-cell conditions, where emigration is not physically possible. However, as we shall see, this is only a partial explanation for the variance in the distribution of adults.

(3). Distribution of equilibrium population sizes: As pointed out earlier, both temperature and flour have significant effects on the distribution of K values in the two types of environments (Table 9). However, the patterns of adult distribution do not show a positive correlation (Table 10), primarily because of the different responses to flour. In the single-cell experiment, the optimum environment is clearly in the wheat flour, whereas in the arenas it is in corn flour. (The significant interaction between temperature and flour in the arenas [Table 9] reflects the weak reversal of this pattern at 22°C.) Both types of environments lead to maximum K's in the central, moderate temperature bands.

Table 10. Correlation coefficients for growth rates and for adult and pupal numbers and adult weights in arena (A) and single cell (SC) populations of CF

Correlation Coefficients

		Growth Rate	*Mean No. Adults* A	*Mean No. Adults* SC	*Mean No. Pupae* A	*Mean No. Pupae* SC	*Mean Wt. Adults* A	*Mean Wt. Adults* SC
Growth Rate		–						
Mean No. Adults	A	−0.148	–					
	SC	0.561c	0.207	–				
Mean No. Pupae	A	0.572c	0.162	0.345b	–			
	SC	−0.467c	−0.200	−0.116	−0.549c	–		
Mean Wt. Adults	A	−0.500c	−0.058	−0.398b	−0.381b	0.097	–	
	SC	−0.514c	0.098	0.004	−0.548c	0.618c	0.249a	–

a$p \leqslant 0.059$.
b$p < 0.01$.
c$p \leqslant 0.001$.

One might expect the numbers of adults and pupae to be negatively correlated because of pupal cannibalism. However, the correlation coefficient for this relationship is not significant within either the arena or single cell populations (Table 10). Although this is somewhat surprising for the single cell populations, the explanation for the arenas is clear and reveals one of the more exciting results of our study.

The distributions of pupal numbers are presented in Fig. 15. Again, the effects of both major variables are highly significant (Table 9), but the most striking aspect of the pupal distributions in the arenas is the strong skew toward the warm end. Over 96% of the pupae recovered in transfers 18 to 21 were in either the 30° or the 34° band. In marked contrast, over 60% of the pupae in the single cells were found in the 22° and 26° populations. These distributions are reflected in the highly significant negative correlation between the pupae in the arena and single cell populations (Table 10). The explanation for this striking

PUPAE/GRAM FLOUR

ARENAS

	wheat			corn	mm
hot	0.57	0.66	1.07	1.51	0.95
	0.99	0.82	1.06	0.78	0.91
	0.08	0.04	0.02	0.05	0.05
cold	0.03	0.01	0.02	0.01	0.02
mm	0.42	0.38	0.54	0.59	

grand mean = 0.48
variance = 0.33

SINGLE CELLS

wheat			corn	mm
1.42	1.41	0.98	0.64	1.11
1.37	1.17	1.08	0.79	1.10
1.85	1.72	1.73	0.87	1.54
2.41	2.23	1.96	1.37	1.99
1.76	1.63	1.44	0.91	

grand mean = 1.44
variance = 0.28

Fig. 15. Mean numbers of pupae per gram flour for four replicate populations in both the arena and single cell populations. The numbers in each replicate were averaged over transfers 18 to 21 mm = Marginal means; the grand means and variances are presented at the bottom of the figure.

difference is quite simple: larvae migrate to the warm end of the arena to pupate. To support this finding, the distributions of larval sizes were recorded in several of the arenas. The vast majority of the larvae in the cold end were very small. The mean weight of the larvae increased with increasing temperatures as the larvae migrated towards the warm end. Finally, the quiescent last larval instar (the "prepupa") was seldom encountered in the cold end of the arenas.

One advantage of this behavior is quite clear. The duration of the pupal period is two to three times longer at 22° than it is at 34° (14.5 versus 5.3 days). Since the risk of pupal death through cannibalism is a positive function of the length of the pupal stage, the migratory behavior of the larvae must decrease the risk of death through cannibalism by adults.

Temperature-dependent developmental rates also largely explain the skewed distribution of pupae in the single cell experiment. Migration is not possible under the single cell conditions and so, because of the extended developmental period, pupae tend to "pile up" in the colder cells. But again, these conclusions must be examined in light of other environmental correlates. For instance, an analysis of pupal cannibalism by adults indicates that the rate of increase of cannibalism with temperature roughly parallels the rate of decrease of developmental period, so these patterns may interact to determine the pupal distribution.

Finally, two additional behavioral patterns should be mentioned at this point, even though our analysis of these patterns is still incomplete. The first of these observations is that the adults in the arenas have become divided into "subpopulations" that may or may not be reproductively isolated from each other. Adults collected in one cell of the arena, for instance 22°-wheat, were marked and placed in a different cell, for instance 34°-corn at the time of census. Four weeks later, at the next census, a larger proportion of the marked adults had returned across the arena to the 22°-wheat cell than could be expected on the basis of change. This result has been consistently obtained for CF in each of the four diagonal cells labeled in Fig. 12, thus eliminating "corner effects" as a possible explanation. When single-cell adults from wheat and corn were tested in flour gradients, it was clear that they had not developed any preferences. Thus, in the arenas, "discrete subpopulations" were developed by active habitat selection, but in the single cells, where there was no opportunity to choose, no divergence occurred.

Second, indirect evidence indicates that females temporarily leave their home area to oviposit. This result may also be interpreted in terms of maximizing individual fitness in the midst of a "cannibalistic orgy" because it suggests that the areas of egg-laying and egg-eating are spatially discrete.

Obviously, strong selection has operated in both the arena and single-cell populations. For several of the characteristics that we have mentioned, the structure of the complex environment has led to different patterns of adaptation than found in the single-cell environments.

However, in some respects the two designs led to similar results. An example is provided by the development of the arena adult "subpopulations." Each of these presumptive units in an arena occupies a relatively small portion of the total environmental mosaic, perhaps comparable to a unit in the single-cell experiment. It is noteworthy that both temperature and flour have highly significant effects on adult weight in the two types of environments (Table 9). Moreover, the distributions of weights are similar in the arenas and single cells (as indicated by the significant positive correlation presented in Table 10). Thus, at least in this respect, a common pattern is obtained in spite of the different means of environmental exploitation.

CONCLUDING VIEWS

This study initiates a new phase in research with flour beetles. Certainly, the observed patterns belie much of the confidence we may take in our understanding of the population biology of *Tribolium.* They also raise a number of questions that have significance beyond the investigation of flour beetles *per se.* For instance, the nature of the interaction between r and K is one of the major foci of current *theoretical* research in this area of biology, yet it is almost totally devoid of experimental treatment. An intriguing observation that is relevant to this issue can be gleaned from Table 10. The r values obtained in this study show a significant positive correlation with the adult K values in the single-cell populations. However, r and K are negatively correlated (though not significantly so at the 5% level) in the arenas. It would be interesting to know whether these relationships are determined more directly by the patterns of environmental exploitation, or by the effects of natural selection. Experimental analyses of this problem are currently being pursued.

An extension of this work with *Tribolium* to the problem of extinction would be very interesting. MacArthur and Wilson (1967) and Levins (1970) have developed models that examine the relationship between various ecological properties of a population and the probability of its extinction. In addition, Levins has extended the analysis of this problem to complex environments and to cases in which the probability of extinction is a function of the gene pool of the population.

Genetic factors in the adaptation of *Tribolium* to complex environments could be readily investigated using the electrophoretic allozyme analyses that have been applied with such striking results to *Drosophila.* (Preliminary tests of our new synthetic populations of CS and CF indicate considerable variation in both esterase and MDH patterns, but little or no variation in LDH.) An even more intriguing possibility is a study of the relationship between genetic variation and behavioral variation among the presumptive subpopulations in the arenas. For instance, the question could be posed: "Is the variation in pupation site observed between corn and wheat flours, and 30° and 34° temperature bands, related to genetic variation?"

And finally, we return to the problem of competition in a complex environment. Our original goal in setting up the arena work was to investigate the influence of environmental variation on the interaction between CS and CF in complex environments. Although we did not obtain coexistence because of the parasitic infection mentioned earlier, the problem remains amenable to experimental treatment with *Tribolium.* Studies such as this may give us some feeling for how complex an environment must be in order to lead to competitive coexistence rather than competitive exclusion. It is this type of investigation which will help bridge the increasing gap between the analysis of field and laboratory populations.

EPILOGUE

This is the first paper to broadly review population studies on flour beetles since the boundaries of the field expanded beyond ecology. At the current rate of increase of the *Tribolium* literature, it seems safe to predict that this paper will also be the last one with such a review. We have not done complete justice to the work of all of the cited authors or included all of the papers written on *Tribolium* because, even now, a complete review would require a book of considerable length. However, we hope that by considering the entire *range* of population problems investigated with flour beetles we can stimulate others, who would not take the time to read such a book, to recognize the relevance of *Tribolium* to their own research interests.

Acknowledgments

We gratefully acknowledge the comments of Wyatt W. Anderson, who read the manuscript, and the support of NSF grants GB 8191 and GB 16131, under which some of the data presented in the last section were obtained.

REFERENCES

Adeyemi, S. A. O. 1968. A laboratory study of competition between *Tribolium castaneum* (Hbst.) (Coleoptera, Tenebrionidae) and three moth species. Bull. Entom. Res., 58:31-45.

Alexander, P., and D. H. R. Barton. 1943. The excretion of ethylquinone by the flour beetles. Biochem. J., 37:463-465.

Amos, T. G. 1969. Reactions of *Carpophilus* spp. (Coleoptera, Nitudulidae) to humidity with reference to gradient steepness. Anim. Behav., 17:9-13.

——and S. N. Dennler. 1969. *Oryzaephilus surinamensis* (L.) (Col., Silvanidae) on a temperature-moisture gradient. J. Stored Prod. Res., 5:173-175.

——and F. L. Waterhouse. 1969. The distribution and biology of *Tribolium castaneum* (Coleoptera, Tenebrionidae) on temperature gradients varying in steepness. Entom. Exp. Appl., 12:44-52.

Anderson, W. W. 1971. Genetic equilibrium and population growth under density-regulated selection. Amer. Natural. (in press)

——and C. E. King. 1970. Age-specific selection. Proc. Nat. Acad. Sci. U.S.A., 66:780-786.

Andres, A. 1931. Catalogue of the Egyptian Tenebrionidae. Bull. Soc. Entom. Egypte, 15:74-125.

Andrewartha, H. G., and L. C. Birch. 1954. The Distribution and Abundance of Animals. Chicago, University of Chicago Press.

Ashford, R. W. 1968. Sporozoan parasites of *Tribolium castaneum* (Herbst): *Triboliocystis garnhami* Dissanaike a synonym of *Farinocystis tribolii* Weiser. J. Protozool., 15:466-470.

Bartlett, A. C., A. E. Bell, and V. L. Anderson. 1966. Changes in quantitative traits of *Tribolium* under irradiation and selection. Genetics, 54:699-713.

Bell, A. E. 1969. The nature of selection responses in *Tribolium.* Jap. J. Genet., 44 (Suppl. 1):299-309.

——and C. H. Moore. 1958. Further comparisons of reciprocal recurrent selection with conventional methods of selection for the improvement of quantitative characteristics. Proc. X Int. Cong. Genet., 2:20-21.

Birch, L. C. 1947. The ability of flour beetles to breed in wheat. Ecology, 28:322-324.

——1948. The intrinsic rate of natural increase of an insect population. J. Anim. Ecol., 17:15-26.

——1953a. Experimental background to the study of the distribution and abundance of insects. I. The influence of temperature, moisture and food on the innate capacity for increase of three grain beetles. Ecology, 34:698-711.

——1953b. Experimental background to the study of the distribution and abundance of insects. II. The relation between innate capacity for increase in numbers and the abundance of three grain beetles in experimental populations. Ecology, 34:712-726.

——1953c. Experimental background to the study of the distribution and abundance of insects. III. The relation between innate capacity for increase and survival of different species of beetles living together on the same food. Evolution, 7:136-144.

——1957. The meanings of competition. Amer. Natural., 91:5-18.

——T. Park, and M. B. Frank. 1951. The effect of intraspecies and interspecies competition on the fecundity of two species of flour beetles. Evolution, 5:116-132.

Boyce, J. M. 1946. The influence of fecundity and egg mortality on the population growth of *Tribolium confusum* Duval. Ecology, 27:290-302.

Boylan, W. J., and W. C. Wong. 1965. Heterosis in *Tribolium castaneum.* Canad. J. Genet. Cytol., 7:84-87.

Bray, D. F. 1966. Variability and restrictions against inbreeding and unequal family size in control populations of *Tribolium.* Genet. Res., 7:122-129.

——A. E. Bell, and S. C. King. 1962. The importance of genotype by environment interaction with reference to control populations. Genet. Res., 3:282-302.

Brereton, J. L. 1962a. A laboratory study of population regulation in *Tribolium confusum.* Ecology, 43:63-69.

——1962b. Evolved regulatory mechanisms of population control. *In* Lepar, G. W., ed., The Evolution of Living Organisms, pp. 81-93. Melbourne, Melbourne University Press.

Champ, R. B., and M. Campbell-Brown. 1969. Genetics of lindane resistance in *Tribolium castaneum* (Herbst) (Coleoptera, Tenebrionidae). J. Stored Prod. Res., 5:399-406.

Chapman, R. N. 1918. The confused flour beetle (*Tribolium confusum* Duval). Minnesota State Entom. Report, 17:73-94.

——1928. Quantitative analysis of environmental factors. Ecology, 9:111-122.

——and W. Y. Whang. 1934. An experimental analysis of the cause of population fluctuations. Science, 80:297-298.

Chittenden, F. H. 1896. Insects affecting cereals and other dry vegetable foods. *In* The Principal Household Insects of the United States. U. S. Dept. Agric. Entom. Bull., 4:112-131.

Clarke, J. M., J. Maynard Smith, and K. C. Sondhi. 1961. Asymmetrical response to selection for rate of development in *Drosophila subobscura.* Genet. Res., 2:70-81.

Constantino, R. F. 1968. The genetical structure of populations and developmental time. Genetics, 60:409-418.

——1969. Effects of a time-dependent model of genotypic value on the covariance between relatives, variance components and estimates of heritability. Genetics, 63:521-529.

——A. E. Bell, and J. C. Rogler. 1966. Genetic control of lipid metabolism in *Tribolium.* Nature (London), 210:221-222.

——A. E. Bell, and J. C. Rogler. 1967. Genetic analysis of a population of *Tribolium.* I. Corn oil sensitivity and selection response. Heredity, (London), 22:529-539.

—— J. C. Rogler, and A. E. Bell. 1968. Genetic analysis of a population of *Tribolium.* II. Metabolic pattern of corn oil sensitive anomaly. Heredity (London), 23:477-483.

Crenshaw, J. W. 1965. Radiation induced increases in fitness in the flour beetle *Tribolium confusum.* Science, 149:426-427.

—— 1966. Productivity and competitive ability in *Tribolium.* Amer. Natural., 100:683-687.

—— and I. M. Lerner. 1964. Productivity of inbred strains of *Tribolium confusum* and *Tribolium castaneum.* Ecology, 45:697-705.

Crombie, A. C. 1942. The effect of crowding upon the oviposition of grain-infesting insects. J. Exp. Biol., 19:311-340.

——1943. The effect of crowding upon the natality of grain-infesting insects. Proc. Zool. Soc. London, [A], 113:77-98.

——1945. On competition between different species of graminivorous insects. Proc. Roy. Soc. London, [B], 132:363-395.

——1946. Further experiments on insect competition. Proc. Roy. Soc. London, [B], 133:76-109.

——1947. Interspecific competition. J. Anim. Ecol., 16:44-73.

Dawood, M. M., and M. W. Strickberger. 1969. The effect of larval interaction on viability in *Drosophila melanogaster.* III. Effects of biotic residues. Genetics, 63:213-220.

Dawson, P. S. 1962. Aberrant segregation ratios in *Tribolium castaneum.* Genetics, 47:950.

——1964a. Age at sexual maturity in female flour beetles, *Tribolium castaneum* and *T. confusum.* Ann. Entom. Soc. Amer., 57:1-3.

——1964b. An interesting behavioral phenomenon in *Tribolium confusum.* Tribolium Info. Bull., 7:50-52.

——1965a. Genetic homeostasis and developmental rate in *Tribolium.* Genetics, 51:873-885.

——1965b. Genotype-environment interactions and developmental rate in the flour beetles *Tribolium castaneum* and *T. confusum* (Coleoptera, Tenebrionidae). Ann. Appl. Biol., 56:21-25.

——1965c. Estimation of components of phenotypic variance for developmental rate in *Tribolium.* Heredity (London), 20:403-417.

——1966a. Developmental rate and competitive ability in *Tribolium.* Evolution, 20:104-116.

——1966b. Correlated responses to selection for developmental rate in *Tribolium.* Genetica, 37:63-77.

——1967a. Developmental rate and competitive ability in *Tribolium.* II. Changes in competitive ability following further selection for developmental rate. Evolution, 21:292-298.

——1967b. Developmental rate and competitive ability in *Tribolium.* III. Competition in unfavorable environments. J. Stored Prod. Res., 3:193-198.

——1967c. Interspecific competition, egg cannibalism and the length of larval instars in *Tribolium.* Evolution, 21:857-858.

——1968a. Xenocide, suicide, and cannibalism in flour beetles. Amer. Natural., 102:97-105.

——1968b. Developmental and genetic homeostasis in two species of flour beetles. Evolution, 22:217-227.

——1968c. Genetic evidence for an hypothesis concerning evolution in *Tribolium.* J. Hered., 59:188-190.

——1968d. A mutant gene maintained at high frequency in spite of complete selection against the homozygote. Genetics, 59:537-542.

——1969. A conflict between Darwinian fitness and population fitness in *Tribolium* "competition" experiments. Genetics, 62:413-419.

——1970a. Linkage and the elimination of deleterious mutant genes from experimental populations. Genetica, 41:147-169.

——1970b. A further assessment of the role of founder effects in the outcome of *Tribolium* competition experiments. Proc. Nat. Acad. Sci. U.S.A., 66:1112-1118.

—— and I. M. Lerner. 1962. Genetic variation and indeterminism in interspecific competition. Amer. Natural., 96:379-380.

—— and I. M. Lerner. 1966. The founder principle and competitive ability of *Tribolium.* Proc. Nat. Acad. Sci. U.S.A., 55:1114-1117.

Dissanaike, A. S. 1955. A new Schizogregarine *Triboliocystis garnhami* n. g., n. sp. and a new microsporidian *Nosema buckleyi* n. sp., from the fat body of the flour beetle *Tribolium castaneum.* J. Protozool., 2:150-156.

Dobzhansky, Th. 1962. Rigid vs. flexible chromosomal polymorphisms in Drosophila. Amer. Natural., 96:321-328.

—— and H. Levene. 1951. Development of heterosis through natural selection in experimental populations of *Drosophila pseudoobscura.* Amer. Natural., 85:247-264.

Dyte, C. E., and D. G. Blackman. 1967. Selection of a DDT-resistant strain of *Tribolium castaneum* (Herbst) (Coleoptera, Tenebrionidae). J. Stored Prod. Res., 2:211-228.

—— and D. G. Rowlands. 1968. The metabolism and synergism of malathion in resistant and susceptible strains of *Tribolium castaneum* (Herbst) (Coleoptera, Tenebrionidae). J. Stored Prod. Res., 4:157-173.

Edney, E. B., and R. W. Gill. 1968. Evolution of senescence and specific longevity. Nature (London), 220:281-282.

Edwards, D. K. 1958. Altered vitality associated with a black mutation and influenced by sex in *Tribolium confusum* Duval (Coleoptera: Tenebrionidae). Physiol. Zool., 31:323-329.

Ehrman, L. 1966. Mating success and genotype frequency in *Drosophila.* Anim. Behav., 14:332-339.

—— 1968. Frequency dependence of mating success in *Drosophila pseudoobscura.* Genet. Res., 11:135-140.

Enfield, F. D., R. E. Comstock, and O. Braskerud. 1966. Selection for pupa weight in *Tribolium castaneum.* I. Parameters in base populations. Genetics, 54:523-533.

—— R. E. Comstock, R. Goodwill, and O. Braskerud. 1969. Selection for pupa weight in *Tribolium castaneum.* II. Linkage and level of dominance. Genetics, 62:849-857.

Englert, D. C., and A. E. Bell. 1962. Sex ratio in *Tribolium castaneum* Herbst as influenced by larval starvation. Oikos, 13:118-124.

—— and A. E. Bell. 1963. Genetic differences in the growth curve of *Tribolium castaneum.* Growth, 27:87-99.

—— and A. E. Bell. 1969. Components of growth in genetically diverse populations of *Tribolium castaneum.* Canad. J. Genet. Cytol., 11:896-907.

—— and A. E. Bell. 1970. Selection for time of pupation in *Tribolium castaneum.* Genetics, 64:541-552.

Erdman, H. E. 1962. Beginning of reproduction determined by age of the female flour beetle, *Tribolium confusum* (Coleoptera: Tenebrionidae). Naturwissenschaften, 18:428.

—— 1963. The differential sensitivity of flour beetles, *Tribolium confusum* and *T. castaneum* to X-ray alteration of reproductive abilities, induced dominant lethals, biomass and survival. J. Exp. Zool., 153:141-147.

—— 1964. Sexual precocity of the male flour beetle, *Tribolium castaneum* Herbst, and the influence of temperatures on reproduction during early adult life. Canad. Entom., 96:656-659.

—— 1965. Modifications of productivity in flour beetles, *Tribolium castaneum* Herbst, due to X-ray dose, hypothermia, and the sex exposed. Radiation Res., 25:341-351.

—— 1966. Modification of fitness in species and strains of flour beetles due to X-ray and DDT. Ecology, 47:1065-1072.

Ernst, S. A., and J. A. Mutchmor. 1969. Dispersal of three species of grain beetles as a function of thermal acclimation, temperature, and larval size. J. Stored Prod. Res., 5:407-412.

Franklin, I., and R. C. Lewontin. 1970. Is the gene the unit of selection? Genetics, 65:707-734.

Gause, G. F. 1931. The influence of ecological factors on the size of population. Amer. Natural., 65:70-76.

Ghent, A. W. 1963. Studies of behavior of the *Tribolium* flour beetles. I. Contrasting responses of *T. castaneum* and *T. confusum* to fresh and conditioned flours. Ecology, 44:269-283.

——1966. Studies of behavior of the *Tribolium* flour beetles. II. Distribution in depth of *T. castaneum* and *T. confusum* in fractionable shell vials. Ecology, 47:355-367.

Good, N. E. 1933. Biology of the flour beetles *Tribolium confusum* Duv. and *T. ferrugineum* Fab. J. Agric. Res., 46:327-334.

——1936. The flour beetles of the genus *Tribolium.* U.S. Dept. Agric. Tech. Bull., 498:1-58.

Gray, H. E. 1948. The biology of flour beetles. The Northwestern Miller, 26:3a-14a, 18a.

Hardin, R. T., and A. E. Bell. 1967. Two-way selection for body weight in *Tribolium* on two levels of nutrition. Genet. Res., 9:309-330.

——J. C. Rogler, and A. E. Bell. 1967. Genetic and environmental interactions in growth of *Tribolium castaneum.* Canad. J. Zool., 45:139-144.

Hinton, H. E. 1942. Secondary sexual characters of *Tribolium.* Nature (London), 149:500-501.

——1948. A synopsis of the genus *Tribolium* MacLeay, with some remarks on the evolution of its species-groups (Coleoptera, Tenebrionidae). Bull. Entom. Res., 39:13-56.

Ho, F. K. 1960. Discrimination between the pupae of *Tribolium confusum* Duv. and *T. castaneum* (Hbst.) (Coleoptera: Tenebrionidae). Ann. Entom. Soc. Amer., 53:280-281.

——1967. Indentification of *Tribolium* larvae by their setal characteristics (Coleoptera: Tenebrionidae). Ann. Entom. Soc. Amer., 60:729-732.

—— and P. S. Dawson. 1966. Egg cannibalism by *Tribolium* larvae. Ecology, 47:318-322.

Holdaway, F. G., and H. F. Smith. 1933. Alteration of sex ratio in the "flour beetle" *Tribolium confusum* Duval, following starvation of newly hatched larvae. Aust. J. Exp. Biol. Med. Sci., 11:34-43.

Howe, R. W. 1952. Entomological problems of food storage in Northern Nigeria. Bull. Entom. Res., 43:111-144.

——1956. The effect of temperature and humidity on the rate of development and mortality of *Tribolium castaneum* (Herbst) (Coleoptera, Tenebrionidae). Ann. Appl. Biol., 44:356-368.

——1960. The effects of temperature and humidity on the rate of development and mortality of *Tribolium confusum* Duval (Coleoptera, Tenebrionidae). Ann. Appl. Biol., 48:363-376.

——1961. Developmental time and weight in *Tribolium castaneum.* Tribolium Info. Bull., 4:21-22.

——1966. Developmental period, and the shape of the curve representing it in stored products beetles. J. Stored Prod. Res., 2:117-134.

——1968a. Changes in weight during development in two stored products beetles. J. Stored Prod. Res., 4:213-220.

——1968b. A further consideration of the heterogeneity of the developmental period of *Tribolium castaneum* (Herbst) (Col., Tenebrionidae) in constant environmental conditions. J. Stored Prod. Res., 4:221-231.

Hunter, P. E. 1959. Selection of *Drosophila melanogaster* for length of larval period. Z. Vererb., 90:7-28.

Inouye, N., and I. M. Lerner. 1965. Competition between *Tribolium* species (Coleoptera, Tenebrionidae) on several diets. J. Stored Prod. Res., 1:185-191.

Jain, S. K., and D. R. Marshall. 1967. Population studies in predominantly self-pollinating species. X. Variation in natural populations of *Avena fatua* and *A. barbata.* Amer. Natural., 101:19-33.

Karten, I. 1965. Genetic differences and conditioning in *Tribolium castaneum.* Physiol. Zool., 38:69-79.

Kennington, G. S. 1953. The effects of reduced atmospheric pressure on populations of *Tribolium castaneum* and *Tribolium confusum.* Physiol. Zool. 26:179-203.

Kidwell, J. F., A. E. Freeman, L. H. Haverland, and G. M. H. Rolfes. 1964. Genotype × environment interaction in *Tribolium castaneum.* Genet. Res., 5:335-340.

King, C. E., and W. W. Anderson. 1971. Age-specific selection. II. The interaction between *r* and *K* during population growth. Amer. Natural., 105:137-156.

Kohn, A. J., and G. H. Orians. 1962. Ecological data in the classification of closely related species. Syst. Zool., 11:120-127.

Kollros, C. L. 1944. A study of the gene "pearl" in *Tribolium castaneum* Herbst. Ph. D. Dissertation, University of Chicago.

Laird, M. 1959. Gregarines from laboratory colonies of flour beetles, *Tribolium castaneum* Herbst and *T. confusum* Duval at Montreal. Canad. J. Zool., 37:378-381.

Landahl, H. D. 1955. A mathematical model for the temporal pattern of a population structure, with particular reference to the flour beetle. II. Competition between species. Bull. Math. Biophys., 17:131-140.

Lange, E. L., and A. E. Bell. 1970. Genetic variation in chaetae number of *Tribolium.* 65:669-679.

Lefkovitch, L. P. 1968. Interaction between four species of beetles in wheat and wheat feed. J. Stored Prod. Res., 4:1-8.

Legay, J. M., and J. L. Chassé. 1964. Etude experimental de la sortie d'un enciente chez *Tribolium confusum.* C. R. Acad. Sci. [D] (Paris), 358:717-719.

Lerner, I. M. 1954. Genetic Homeostasis. Edinburgh, Oliver and Boyd.

——1968. Heredity, Evolution and Society. San Francisco, William H. Freeman.

——and E. R. Dempster. 1962. Indeterminism in interspecific competition. Proc. Nat. Acad. Sci. U.S.A., 48:821-826.

——and F. K. Ho. 1961. Genotype and competitive ability of *Tribolium* species. Amer. Natural., 95:329-343.

—— and N. Inouye. 1968. Behavior genetics, with special reference to maze-running of *Tribolium. In* Dronamraju, K. R., ed., Haldane and Modern Biology. Baltimore, John Hopkins Press.

—— A. Sokoloff, and F. K. Ho. 1966. Genotype-environment interactions in the outcome of "competition" in *Tribolium.* I. Population cage studies. Tribolium Info. Bull., 9:06-105.

Leslie, P. H. 1962. A stochastic model for two competing species of *Tribolium* and its application to some experimental data. Biometrika, 49:1-25.

——and T. Park. 1949. The intrinsic rate of natural increase of *Tribolium castaneum* Herbst. Ecology, 30:469-477.

—— T. Park, and D. B. Mertz. 1968. The effect of varying the initial numbers of the outcome of competition between two *Tribolium* species. J. Anim. Ecol., 37:9-23.

Levene, H. 1953. Genetic equilibrium when more than one niche is available. Amer. Natural., 87:331-333.

—— I. M. Lerner, A. Sokoloff, F. K. Ho, and I. R. Franklin. 1965. Genetic load in *Tribolium.* Proc. Nat. Acad. Sci. U.S.A., 53:1042-1050.

Levins, R. 1963. Theory of fitness in a heterogeneous environment. II. Developmental flexibility and niche selection. Amer. Natural., 97:75-90.

—— 1968. Evolution in Changing Environments. Princeton, Princeton University Press.

—— 1970. Fitness and optimization. Biomathematics, 1:389-400.

——and R. MacArthur. 1966. The maintenance of genetic polymorphism is a spatially heterogeneous environment: Variations on a theme by Howard Levene. Amer. Natural., 100:585-589.

Lewontin, R. C. 1955. The effects of population density and composition on viability in *Drosophila melanogaster.* Evolution, 9:27-41.

——1957. The adaptations of populations to varying environments. Sympos. Quant. Biol., 22:395-408.

——1961. Evolution and the theory of games. J. Theoret. Biol., 1:382-403.

——1965. Selection for colonizing ability. *In* Baker, H. G., and G. L. Stebbins, eds., The genetics of Colonizing Species. New York, Academic Press.

—— 1970. The units of selection. Ann. Rev. Ecol. Syst., 1:1-18.

Lipa, J. J. 1968. On two microsporidians: *Nosema rohitei* Weiser from *Tribolium confusum* and *Nosema roeiseri* sp. n. from *Rhizopertha dominica.* Acta Protozool., 5:375-380.

Lloyd, M. 1968. Self regulation of adult numbers by cannibalism in two laboratory strains of flour beetles (*Tribolium castaneum*). Ecology, 49:245-259.

—— and T. Park. 1962. Mortality resulting from interactions between adult flour beetles in laboratory cultures. Physiol. Zool., 35:330-347.

Loconti, J. D., and L. M. Roth. 1953. Composition of the odorous secretion of *Tribolium castaneum.* Ann. Entom. Soc. Amer., 46:281-289.

MacArthur, R. H. 1962. Some generalized theorems of natural selection. Proc. Nat. Acad. Sci. U.S.A., 48:1893-1897.

—— and E. O. Wilson. 1967. The Theory of Island Biogeography. Princeton, Princeton University Press.

MacLagan, D. S. 1932. The effect of population density upon rate of reproduction with special reference to insects. Proc. Roy. Soc. London, [B], 111:437-454.

—— and E. Dunn. 1935. The experimental analysis of the growth of an insect population. Proc. Roy. Soc. Edinburgh, 55:126-139.

Marien, D. 1958. Selection for developmental rate in *Drosophila pseudoobscura.* Genetics, 43:3-15.

Marshall, D. R., and S. K. Jain. 1968. Phenotypic plasticity of *Avena fatua* and *A. barbata.* Amer. Natural., 102:457-467.

—— and S. K. Jain. 1969. Genetic polymorphism in natural populations of *Avena fatua* and *A. barbata.* Nature (London), 22:276-278.

Mather, K., and B. J. Harrison. 1949. The manifold effects of selection. Heredity, 3:1-52, 131-162.

Maynard Smith, J. 1970. Genetic polymorphism in a varied environment. Amer. Natural., 104:487-490.

McDonald, D. J. 1963. Natural selection in experimental populations of *Tribolium.* III. Characteristics of *Tribolium confusum* populations in equilibrium. Amer. Natural., 97:383-396.

—— 1968a. Mobility in *Tribolium confusum.* Ecology, 49:770-771.

—— 1968b. The response .of *Tribolium confusum* to variations in the volume of its environment. Ecology, 49:1166-1168.

—— and L. J. Davidson. 1965. Population fitness in *Tribolium.* I. An approach to the problem. Amer. Natural., 99:463-470.

—— and L. Fitting. 1965. Activity differences in black and wild-type strains of *Tribolium confusum.* Tribolium Info. Bull., 8:124-125.

—— and N. J. Peer. 1960. Natural selection in experimental populations of *Tribolium.* I. Preliminary experiments with population cages. Genetics, 45:1317-1333.

—— and N. J. Peer. 1961. Natural selection in experimental populations of *Tribolium.* II. Factors affecting the fitness of the mutant "split" of *Tribolium confusum.* Heredity, 16:317-330.

—— and L. Stoner. 1968. Population fitness in *Tribolium.* II. Population characteristics influencing the capacity for survival. Amer. Natural., 102:323-336.

McFadden, J. T. 1963. An example of inaccuracies inherent in interpretation of ecological field data. Amer. Natural., 97:99-116.

Merrell, D. J. 1965. Competition involving dominant mutants in experimental populations of *Drosophila melanogaster.* Genetics, 52:165-189.

Mertz, D. B. 1961. Identification of *Tribolium* species by pupal characteristics. Ecology, 42:811.

—— 1969. Age distribution and abundance in populations of flour beetles. I. Experimental studies. Ecol. Monogr., 39:1-31.

—— and R. B. Davies. 1968. Cannibalism of the pupal stage by adult flour beetles: An experiment and stochastic model. Biometrics, 24:247-275.

—— T. Park, and W. J. Youden. 1965. Mortality patterns in eight strains of flour beetles. Biometrics, 21:99-114.

Miller, L. W. 1944. Investigations of the flour beetles of the genus *Tribolium.* II. Effect of different mill fractions on the larval development and survival of *T. castaneum* (Herbst) and *T. confusum* (Duv.). J. Agric. Victoria, 42:365-373, 377.

Miller, R. S. 1967. Pattern and process in competition. Advances Ecol. Res., 4:1-74.

Morton, N. E., J. F. Crow, and H. J. Muller. 1956. An estimate of the mutational damage in man from data on consanguineous marriages. Proc. Nat. Acad. Sci. U.S.A., 42:855-863.

Naylor, A. F. 1959. An experimental analysis of dispersal in the flour beetle, *Tribolium confusum.* Ecology, 40:453-465.

——1961. Dispersal in the red flour beetle, *Tribolium castaneum.* Ecology, 42:231-237.

——1965. Dispersal responses of female *Tribolium confusum* to presence of larvae. Ecology, 46:341-343.

Neyman, J., T. Park, and E. L. Scott, 1956. Struggle for existence. The *Tribolium* model: biological and statistical aspects. Proc. Third Berkeley Sympos. Math. Stat. Prob., 4:41-79;

Nicholson, A. J. 1933. The balance of animal populations. J. Anim. Ecol., 2:132-178.

Niven, B. S. 1969. Simulation of two interacting species of *Tribolium.* Physiol. Zool., 42:248-255.

Nowosielski-Slepowrow, B. J. A., F. L. Waterhouse, and D. E. A. Strevens. 1968. Sub-zero mortality responses of *Tribolium confusum* Duval (two stocks) and *T. castaneum* (Herbst) (from stocks) analyzed by weighted regression lines based on individual temperature LD50's. Physiol. Zool., 41:440-446.

Ogden, J. C. 1969. Effect of components of conditioned medium on behavior in *Tribolium confusum.* Physiol. Zool., 42:266-274.

——1970a. Aspects of dispersal in *Tribolium* flour beetles. Physiol. Zool., 43:124-131.

——1970b. Artificial selection for dispersal in flour beetles (Tenebrionidae: *Tribolium*). Ecology, 51:130-133.

Okada, I., and R. T. Hardin. 1967. An experimental examination of restricted selection index, using *Tribolium castaneum.* I. The results of two-way selection. Genetics, 57:227-236.

——and R. T. Hardin. 1970. An experimental examination of restricted selection index, using *Tribolium castaneum.* II. The results of long-term one-way selection. Genetics, 64:533-539.

Park, T. 1932. Studies in population physiology: The relation of numbers to initial population growth in the flour beetle *Tribolium confusum* Duv. Ecology, 13:172-181.

——1933. Studies in population physiology. II. Factors regulating initial growth of *Tribolium confusum* populations. J. Exp. Zool., 65:17-42.

——1934a. Observations on the general biology of the flour beetle *Tribolium confusum.* Quart. Rev. Biol., 9:36-54.

——1934b. Studies in population physiology. III. The effect of conditioned flour upon the productivity and population decline of *Tribolium confusum.* J. Exp. Zool., 68:167-182.

——1935. Studies in population physiology. IV. Some physiological effects of conditioned flour upon *Tribolium confusum* Duv. and its populations. Physiol. Zool., 8:91-115.

——1936a. Studies in population physiology. V. The oxygen consumption of the flour beetle *Tribolium confusum* Duv. J. Cell. Comp. Physiol, 7:313-323.

——1936b. Studies in population physiology. VI. The effect of differentially conditioned flour upon the fecundity and fertility of *Tribolium confusum* Duv. J. Exp. Zool., 73:393-404.

——1937. Experimental studies of insect populations. Amer. Natural., 71:21-33.

——1938a. A note on the size and composition of old *Tribolium confusum* populations. Amer. Natural., 72:24-33.

——1938b. Studies in population physiology. VIII. The effect of larval population density on the post-embryonic development of the flour beetle, *Tribolium confusum* Duv. J. Exp. Zool., 79:51-70.

——1939. Analytical population studies in relation to general ecology. Amer. Midl. Natural., 21:235-255.

——1941. The laboratory population as a test of a comprehensive ecological system. Quart. Rev. Biol., 16:274-293, 440-461.

——1942. Integration in infra-social insect populations. Biological Symposia, 8:121-138.

——1945. Life tables for the black flour beetle *Tribolium madens* Charp. Amer. Natural., 79:436-444.

——1948. Experimental studies of interspecies competition. I. Competition between populations of the flour beetles *Tribolium confusum* Duv. and *Tribolium castaneum* Herbst. Ecol. Monogr., 18:265-308.

——1954. Experimental studies of interspecies competition. II. Temperature, humidity and

competition in two species of *Tribolium.* Physiol. Zool., 27:177-238.
—— 1955. Experimental competition in beetles with some general implications. *In* Cragg, J. B., and N. W. Pirie, eds., The Numbers of Man and Animals, pp. 69-82. London, Oliver and Boyd.
—— 1957. Experimental studies of interspecies competition. III. Relation of initial species proportion to competitive outcome in populations of *Tribolium.* Physiol. Zool., 30:22-40.
—— 1962. Beetles, competition, and populations. Science, 138:1369-1375.
—— and W. Burrows. 1942. The reproduction of *Tribolium confusum* Duv. in a semi-synthetic wood dust medium. Physiol. Zool., 15:476-484.
—— and M. B. Davis. 1945. Further analysis of fecundity in the flour beetles, *Tribolium confusum* Duval and *Tribolium castaneum* Herbst. Ann. Entom. Soc. Amer., 38:237-244.
—— P. P. H. DeBruyn, and J. A. Bond. 1958. The relation of X-irradiation to the fecundity and fertility of two species of flour beetles. Physiol. Zool., 31:151-170.
—— and M. B. Frank. 1948. The fecundity and development of the flour beetles *Tribolium confusum* and *Tribolium castaneum* at three constant temperatures. Ecology, 29:368-374.
—— and M. B. Frank. 1950. The population history of *Tribolium* free of sporozoan infection. J. Anim. Ecol., 19:95-105.
—— E. V. Gregg, and C. Z. Lutherman. 1941. Studies in population physiology. X. Interspecific competition in populations of granary beetles. Physiol. Zool., 14:395-430.
—— P. H. Leslie, and D. B. Mertz. 1964. Genetic strains and competition in populations of *Tribolium.* Physiol. Zool., 37:97-162.
—— and M. Lloyd. 1955. Natural selection and the outcome of competition. Amer. Natural., 89:235-240.
—— D. B. Mertz, W. Grodzinski, and T. Prus. 1965. Cannibalistic predation in populations of flour beetles. Physiol. Zool., 38:289-321.
—— D. B. Mertz, and M. Nathanson. 1968. The cannibalism of pupae by adult flour beetles. Physiol. Zool., 41:228-253.
—— D. B. Mertz, and K. Petrusewicz. 1961. Genetic strains of *Tribolium*: their primary characteristics. Physiol. Zool., 34:62-80.
—— E. V. Miller, and C. Z. Lutherman. 1939. Studies in population physiology. IX. The effect of imago population density on the duration of the larval and pupal stages of *Tribolium confusum* Duv. Ecology, 20:365-373.
—— M. Nathanson, J. Ziegler, and D. B. Mertz. 1970. Cannibalism of pupae by mixed-species population of adult *Tribolium.* Physiol. Zool., 43:166-184.
—— and N. Woollcott. 1937. Studies in population physiology. VII. The relation of environmental conditioning to the decline of *Tribolium confusum* populations. Physiol. Zool., 10:197-211.
Pearl, R., T. Park, and J. R. Miner. 1941. Experimental studies on the duration of life. XVI. Life tables for the flour beetle *Tribolium confusum* Duval. Amer. Natural., 75:5-19.
—— and L. J. Reed. 1920. On the rate of growth of the population of the United States since 1790 and its mathematical representation. Proc. Nat. Acad. Sci. U.S.A., 6:275-288.
Petit, C. 1951. Le rôle de l'isolement sexuel dans l'evolution des populations de *Drosophila melanogaster.* Bull. Biol. France Belgique, 85:392-418.
—— 1958. Le déterminisme génétique et psycho-physiologique de la compétition sexuelle chez *Drosophila melanogaster.* Bull. Biol. France Belgique, 92:248-329.
Phillips, A. L., and D. J. McDonald. 1958. A comparison of productivity in a mutant and wild strain of *Tribolium castaneum* Herbst. Amer. Natural., 92:376-378.
Pimentel, D. 1958. Alteration of microclimate imposed by populations of flour beetles (*Tribolium*). Ecology, 39:239-246.
Polnik, A. 1960. Effects of some intraspecies processes on competition between two species of flour beetles, *Latheticus oryzae* and *Tribolium confusum.* Physiol. Zool., 33:42-57.
Prout, T. 1968. Sufficient conditions for multiple niche polymorphism. Amer. Natural., 102:493-496.
Prus, T. 1963. Search for methods to investigate mobility in *Tribolium.* Ecology, 44:801-803.
—— 1966. Emigrational ability and surface numbers of adult beetles in 12 strains of *Tribolium confusum* Duval and *T. castaneum* Herbst (Coleoptera, Tenebrionidae). Ekol. Pol., [A], 14:547-588.

Rich, E. 1956. Egg cannibalism and fecundity in *Tribolium.* Ecology, 37:109-120.
Robertson, A. 1956. The effect of selection against extreme deviants based on deviation or on homozygosis. J. Genet., 54:236-248.
Robinson, P., and D. F. Bray. 1965. Expected effects on the inbreeding coefficient and rate of gene loss of four methods of reproducing finite diploid populations. Biometrics, 21:447-458.
Ryan, M. F., T. Park, and D. B. Mertz. 1970. Flour beetles: responses to extracts of their own pupae. Science, 170:178-180.
Sang, J. H., and G. A. Clayton. 1957. Selection for larval development time in *Drosophila.* J. Hered., 48:265-270.
Scheinberg, E. 1969. Experimental evidence for factors affecting the relative efficiency of indirect to direct selection in *Tribolium castaneum.* Canad. J. Genet. Cytol., 11:86-96.
—— A. E. Bell, and V. L. Anderson. 1967. Genetic gain in populations of *Tribolium castaneum* under uni-stage tandem selection and under restricted selection indices. Genetics, 55:69-90.
Schlager, G. 1960. Sperm precedence in the fertilization of eggs in *Tribolium castaneum.* Ann. Entom. Soc. Amer., 53:557-560.
——1963. The ecological genetics of the mutant *sooty* in populations of *Tribolium castaneum.* Evolution, 17:254-273.
Sinha, R. N. 1966. Development and mortality of *Tribolium castaneum* and *T. confusum* on seed-borne fungi. Ann. Entom. Soc. Amer., 59:192-201.
——1968. Adaptive significance of mycophagy in stored-product Arthropoda. Evolution, 22:785-789.
——1969. Reproduction of stored-grain insects on varieties of wheat, oats and barley. Ann. Entom. Soc. Amer., 62:1011-1015.
Sinnock, P. 1970. Frequency dependence and mating behavior in *Tribolium castaneum.* Amer. Natural., 104:469-476.
Slobodkin, L. B. 1962. Growth and Regulation of Animal Populations. New York, Holt, Rinehart and Winston.
Smith, H. S. 1935. The role of biotic factors in the determination of population densities. J. Econ. Entom., 28:873-898.
Smith, S. G. 1952. The cytology of some tenebrionid beetles (Coleoptera). J. Morph., 91:325-364.
Sokal, R. R., 1967a. The effect of synchrony of egg hatches on fitness characters and competition in *Tribolium castaneum.* Tribolium Info. Bull., 10:135-141.
——1967b. A comparison of fitness characters and their response to density in stock and selected cultures of wild type and *black Tribolium castaneum.* Tribolium Info. Bull., 10:142-147.
——1970. Senescence and genetic load: Evidence from *Tribolium.* Science, 167:1733-1734.
—— and I. Huber. 1963. Competition among genotypes in *Tribolium castaneum* at varying densities and gene frequencies (the *sooty* locus). Amer. Natural., 97:169-184.
—— and I. Karten. 1964. Competition among genotypes in *Tribolium castaneum* at varying densities and gene frequencies (the black locus). Genetics, 49:195-211.
—— and F. J. Rohlf. 1968. Biometry. San Francisco, W. H. Freeman.
—— and F. J. Sonleitner. 1968. The ecology of selection in hybrid populations of *Tribolium castaneum.* Ecol. Monogr., 38:345-379.
Sokoloff, A. 1965. Studies of natural selection in *Tribolium castaneum* Herbst (Coleoptera) in two different media. Proc. XII. Int. Congr. Entom., 273-274.
——1966a. The Genetics of *Tribolium* and Related Species. Advances Genet., Suppl. 1.
——1966b. A behavioral difference of two mutant strains in *Tribolium castaneum.* Tribolium Info. Bull., 9:109-110.
——1967. Preliminary population studies with mutants of *Tribolium castaneum* Herbst. I-IV. Tribolium Info. Bull., 10:147-160.
——1969a. Genotype-environment interactions in the outcome of "competition" in *Tribolium.* II. Genetic influence in the survival of *Tribolium castaneum* and *Tribolium confusum* in mixed cultures. Genetika, 5:85-98.
——1969b. Effects of selection for high and low body weight on survival of *T. castaneum*

and *T. confusum* in mixed cultures. Tribolium Info. Bull., 11:90-93.
—— and F. K. Ho. 1966. "Competition" between *Tribolium anaphe, T. castaneum, T. confusum, T. destructor* and *T. madens* by pairs. I. Tribolium Info. Bull., 9:110-119.
—— and F. K. Ho. 1969. "Competition" between *Tribolium anaphe, T. castaneum, T. confusum, T. destructor* and *T. madens* by pairs. II. The final results of "competition" between *T. confusum* and *T. madens.* Tribolium Info. Bull., 11:95-98.
—— and I. M. Lerner. 1967. Laboratory ecology and mutual predation of *Tribolium* species. Amer. Natural., 101:261-276.
—— I. M. Lerner, and F. K. Ho. 1965a. Self-elimination of *Tribolium castaneum* following xenocide of *T. confusum.* Amer. Natural., 99:399-404.
—— I. M. Lerner, and F. K. Ho. 1965b. Sudden elimination of *T. confusum* by *T. castaneum* in different media. Tribolium Info. Bull., 8:151-153.
—— R. R. Shrode, and J. H. Bywaters. 1965. Productivity in *Tribolium castaneum.* Physiol. Zool., 38:165-173.
—— H. M. Slatis, and J. Stanley. 1960. The black mutation in *Tribolium castaneum.* J. Hered., 51:131-135.
Solomon, M. E., and B. E. Adamson. 1955. The powers of survival of storage and domestic pests under winter conditions in Britain. Bull. Entom. Res., 46:311-355.
Sonleitner, E. J. 1961. Factors affecting egg cannibalism and fecundity in populations of adult *Tribolium castaneum* Herbst. Physiol. Zool., 34:233-255.
Southwick, C. H. 1955. The population dynamics of confined house mice supplied with unlimited food. Ecology, 36:212-225.
Spiess, L. D., and E. B. Spiess. 1969. Minority advantage in interpopulational matings of *Drosophila persimilis.* Amer. Natural., 103:155-172.
Stanley, J. 1932. A mathematical theory of the growth of populations of the flour beetle, *Tribolium confusum.* Canad. J. Res., 6:632-671.
—— 1938. the egg-producing capacity of populations of *Tribolium confusum* Duv., as affected by intensive cannibalistic egg-consumption. Canad. J. Res., 16:300-306.
—— 1942. A mathematical theory of the growth of populations of the flour beetle *Tribolium confusum* Duval. V. The relation between the limiting value of egg-populations in the absence of hatching and the sex-ratio of the group of adult beetles used in a culture. Ecology, 23:24-31.
—— 1961. Two techniques of use in the control of *Triboliocystis garnhami.* Canad. J. Zool., 39:121-122.
—— 1966. A mathematical theory of the growth of populations of the flour beetle *Tribolium confusum*, Duval. IX. The persistence of tunnels in the flour mass. J. Theoret. Biol., 13:379-411.
Stanley, M. S. M., and A. W. Grundmann. 1965. Observations on the morphology and sexual behavior of *Tribolium confusum* (Coleoptera: Tenebrionidae). J. Kansas Entom. Soc., 38:10-18.
Surtees, G. 1963. Laboratory studies on dispersion behaviour of adult beetles in grain. III. *Tribolium castaneum* (Hbst.) (Coleoptera, Tenebrionidae) and *Cryptolestes ferrugineus* (Steph.) (Coleoptera, Cucujidae). Bull. Entom. Res., 54:297-306.
Tantawy, A. D., and G. S. Mallah. 1961. Studies on natural populations of *Drosophila.* I. Heat resistance and geographical variation in *D. melanogaster* and *D. simulans.* Evolution, 15:1-14.
Taylor, N. W. 1965. A theoretical study of population regulation in *Tribolium confusum.* Ecology, 46:334-340.
—— 1968. A mathematical model for two *Tribolium* populations in competition. Ecology, 49:843-848.
Thoday, J. M. 1953. Components of fitness. Sympos. Soc. Exp. Biol., 7:96-113.
—— and T. B. Boam. 1961. Regular responses to selection. I. Description of responses. Genet. Res., 2:161-176.
Trehan, K. N., and S. A. Rajarao. 1945. A note on the predatory habits of *Tribolium* beetles. Curr. Sci., 14:209-210.
Verhulst, P. F. 1839. Notice sur la loi que la population suit dans son accroissement. Corr. Math. Phys., 10:113-121.

Wallace, B. 1958. The average effect of radiation-induced mutations on viability in *Drosophila melanogaster.* Evolution, 12:532-556.

—— 1963. Further data on the overdominance of induced mutations. Genetics, 48:633-651.

—— 1968. Polymorphism, population size, and genetic load. *In* Lewontin, R. C., ed., Population Biology and Evolution. Syracuse, Syracuse University Press.

Watt, K. E. F. 1955. Studies on population productivity. Three approaches to the optimum yield problem in populations of *Tribolium confusum.* Ecol. Monogr., 25:269-290.

Weisbrot, D. R. 1966. Genotypic interactions among competing strains and species of *Drosophila.* Genetics, 53:427-435.

Williams, G. C. 1966. Adaptation and Natural Selection: A Critique of Some Current Evolutionary Thought. Princeton, Princeton University, Press.

Wilson, S. P., P. V. Blair, W. H. Kyle, and A. E. Bell. 1968. The influence of selection and mating systems on larval weight in *Tribolium.* J. Hered., 59:313-317.

Wong, W. C., and W. J. Boylan. 1970. Intrapopulational selection and correlated response in crossbreds of *Tribolium castaneum.* Genetics, 64:69-78.

Wool, D. 1967. Some observations on mating frequencies in *Tribolium castaneum* strains. Tribolium Info. Bull., 10:182-186.

—— 1969a. The effects of larval age range on survival of two *Tribolium castaneum* strains in mixed cultures through pupal cannibalism. Res. Pop. Ecol., 11:40-44.

——1969b. Differences in population parameters of two *Tribolium castaneum* strains in environments of different shapes. Res. Pop. Ecol., 11:45-56.

——1969c. Depth distribution of adults and immatures of two *Tribolium castaneum* strains in pure and mixed cultures. Res. Pop. Ecol., 11:137-149.

—— 1970. Deviations of zygotic frequencies from expectations in eggs of *Tribolium castaneum.* Genetics, 66:115-132.

Yamada, Y., and A. E. Bell. 1969. Selection for larval growth in *Tribolium* under two levels of nutrition. Genet. Res., 13:175-195.

Young, A. M. 1970. Predation and abundance in populations of flour beetles. Ecology, 51:602-619.

Żyromska-Rudzka, H. 1964. Migration tendency of two genetic strains of *Tribolium*, preliminary observations. Ekol. Pol., [A], 12:379-388.

——1966a. An appreciation of the laboratory set for investigating the emigratory activity of *Tribolium.* Ekol. Pol., [A], 14:127-134.

——1966b. Abundance and emigrations of *Tribolium* in a laboratory model. Ekol. Pol., [A], 14:491-518.

6

Ecological Opportunities and Dominant Families, as Illustrated by the Neotropical Tyrannidae (Aves)

ALLEN KEAST
Department of Biology, Queen's University
Kingston, Ontario, Canada

INTRODUCTION

Each continental fauna has its series of dominant families, characterized by an unusual diversity of genera and species, degree of morphological diversity and, in terms of ecology, by the wide range of ecological roles and niches occupied. Such groups raise many interesting evolutionary questions. How are such radiations to be explained? What range of adaptive zones are occupied? What environmental circumstances, and what attributes of the groups themselves, underlie such success? Involved here are broader questions such as the role played by the associated fauna in influencing patterns of evolutionary development and radiation in a group. In endeavoring to answer this it is basic to investigate how, and by what groups, the adaptive zones of the dominant family are occupied, or utilized, on other continents.

The objective of the present study is to investigate these questions by analysis of an avian family, the Neotropical Tyrannidae. Birds are the most suitable group for this kind of evolutionary study. They possess a range of readily measureable morphological characters, the adaptive significance of which are reasonably well known, and many of the major ecological attributes of species can readily be determined by direct field observation. Included are: habitat specialization and utilization, feeding zones, feeding and perching habits, seasonal movements and, to some extent, foods (discussion in Keast, 1970). The Neotropical Tyrannidae were chosen by the group for study because: (1) with

103 genera and 302 species (de Schauensee, 1966) it is the largest single continental bird family–one tenth of the 2900 species of South American land birds are tyrannids (Smith and Vuilleumier, 1971); (2) The group is very diverse in body structure and size (individual species range in body length in 3 to 14 inches [7.5 to 36 cm]); and (3) it fills an unusual diversity of ecological roles and "adaptive zones" (in the usage of Simpson, 1944). For an overall view of the morphological types comprising the Neotropical tyrannids see de Schauensee (1964), Johnson (1967), and Haverschmidt (1968). Data on ecology are to be found in these works and in Wetmore (1926), Skutch (1960), and Slud (1964). Recently, a study of feeding sites, feeding behavior, and speciation patterns in the high mountain *Muscisaxicola* (chat tyrants), and related genera, has been published (Smith and Vuilleumier, 1971).

METHODS OF ANALYSIS

The review is specifically directed at five aspects. These are: (1) Defining, or "contouring," the morphology of the Neotropical Tyrannidae, the adaptive zones occupied; and, at the species level, how these are split up into individual "niches."

(2) Investigating the mechanisms whereby so many members of a single family can occur on one continent. This breaks down into: (a) a consideration of the numbers of distinct morphological and ecological "types" in the family, and (b) a study of patterns of space and habitat utilization within regional faunas. Major structural types are obviously adapted for distinct roles and hence are not in direct competition. The Tyrannidae are particularly rich in this category. Most regional avifaunas in South America are characterized by large numbers of tyrannid species. Costa Rica, for example, with an area of only 200 square miles (n500 km^2) has 69 species of Tyrannidae. Colombia has 150. Mechanisms operating to ensure ecological exclusion in such assemblages must be surveyed.

(3) A consideration of how the various adaptive zones and ecological roles occupied by the Neotropical Tyrannidae are filled on other continents, that is, how they are split up, and what groups occupy them in the *absence* of tyrannids. In view of the close correlation between structure and ecology in birds this aspect can be handled most efficiently by supplementing general ecological observation by structural studies, and specifically, comprehensive series of body measurements. In this way genera and species of comparable sizes and with comparable structures can be pinpointed.

The avifaunas of Africa and Australia were chosen as the basis of this comparative analysis, since they can be expected to contain the same broad range of adaptive types. The three continents, it will be noted, occupy similar latitudes, have broadly comparable climates and vegetation belts, and have retained a large degree of spatial isolation.

(4) Explaining the "evolutionary success" of the tyrannids. This is looked at both in the contemporary as well as historic contexts, and, specifically, from the

viewpoint from the presence or absence of potential competitors in the adaptive zones occupied by them.

(5) In a final synthesis to try to assess the significance of the associated avifauna in influencing, or limiting, the evolution and radiation of individual groups. This is approached by a comparative analysis of the ways in which the "insectivorous" adaptive zones in particular, and ecological opportunities in general, are subdivided on the three continents of South America, Africa, and Australia. It supplements a more comprehensive comparative analysis of avian evolution in these three avifaunas being published elsewhere (Keast, 1971).

The morphological features on which measurements were carried out are the wing, bill (length, width, depth), tarsus, and hind-toe. These parts of the body are highly plastic in the evolutionary sense and change with minor shifts in feeding and perching habits. Wing length is a good criterion of overall size in birds (Amadon, 1943), and is commonly used for this purpose when body weights are not available. The bill, as the delicate instrument by means of which the bird catches and picks up its prey is a good general indicator of diet. It may vary in length, width, depth, and degree of curvature. There is obviously a good correlation between bill form and the type of size of prey items that can be handled, although much more research is needed in this area. Amongst the major variations to be found in the bills of small insectivorous birds are: (a) a broad, flat bill (along with prominant rictal bristles and, commonly, a wide gape), in species that catch most of their food in the air (e.g., swallows, flycatchers); and (b) a deep (and hence strong) bill in species that take larger bodied insects from the trees and ground. This trend commonly culminates in a heavy bill with a hooked tip (see Fig. 3) such as characterizes the shrikes (Laniidae), which may eat, in addition to large insects, small lizards, and even small mammals. This latter type of bill is seen in the predatory shrike-tyrant, *Agriornis livida*, of the high Andes. A minority of bird species have massive, conical bills, produced by the simultaneous increase both in width and depth. The New Guinea muscicapid flycatcher, *Peltops*, has a bill of this type.

The form of the leg and toes in birds is a good indicator of perching habits, as has been shown by such studies as those of Dilger (1956), Osterhaus (1962), as discussed in Keast (1968, 1970). In the case of South American flycatchers of the genus *Muscisaxicola* long legs apparently may be associated with running, making long hops (as on broken ground), and permitting the owner to stand high and thus see farther over the grassy habitat (Smith and Vuilleumier, 1971, p. 240). Much remains to be learned on the subject of leg form in passerine birds, however. Evolutionary changes in the leg are most marked in the distal segments (Dilger, 1956), hence the tarsus and the toe are the most revealing parts for measurement. When a group becomes more terrestrial in habits this is characteristically reflecting in an increase in leg length. Hence, the relative length of the leg (i.e., tarsus) in two genera is an indication of the relative degree to which they are specialized for terrestrial feeding (see plottings in Fig. 9).

The measurements made were as follows: *wing length* (angle of folded wing to the tip, straightened along a rule); *bill length* (end of feathering on top of bill to the tip); *bill width* (width in the horizontal plane, taken at end of feathering on top of bill); *bill depth* (vertical height of bill at this point); *tarsal length* (tarsometatarsal from articulation with tibiotarsus to junction of hind toe); *hallux length* (from base of hind toe to tip of claw.

All measurements are based on five adult males of each species. Only perfect specimens (i.e., those in which the two halves of the mandible were correctly pressed together) were used in this. The points plotted on the graphs (Figs. 5-9) are based on the means of the various characters.

The African and Australian families that are potential counterparts of the Tyrannidae on those continents, and for which detailed sets of measurements are given, are as follows: *Africa*—Muscicapinae, Sylviinae, Turdinae, Timaliinae, Paridae and Remizidae, Eurylaimidae, Laniidae, Dicruridae, Motacillidae, and Meropidae. *Australia*—Muscicapinae, Falcunculinae, Pachycephalinae, Sylviinae, Acanthizinae, Malurinae, Epthianurinae, Campephagidae, Dicruridae, Dicaeidae, Motacillidae, Meropidae, Cracticidae, and Grallinidae. The Muscicapinae (the so-called "Old World Flycatchers"), which are the counterparts of the tyrannids in the "flycatcher" or "aerial feeding insect-eater" adaptive zone, are covered in the greatest detail, measurements being given for 15 (Africa), and 13 (Australia) genera and species, respectively. In the case of the other families and subfamilies, measurements of enough genera are given in each case to broadly "contour" the range of morphological variation to be found in the group, or at least those members of it that could be counterparts, or partial counterparts of members of the Tyrannidae. The numbers of genera measured range from one to eight, depending on the degree of morphological diversity present in the group. Several of the families occurring in Australia have only one or two species on the continent, hence their meager representation in the figures.

THE SOUTH AMERICAN TYRANNIDAE

This family contains 103 genera and 302 species in continental South America (de Schauensee, 1966). The group is obviously Neotropical in origin (Mayr, 1946). Only 11 genera and 31 species occur in North America (American Ornithologists' Union Checklist), where they fill the role of generalized flycatchers, comparable to the muscicapid flycatchers of the Old World. There has been limited Nearctic radiation, e.g., in *Empidonax*. Most of the species in the list, however, are only marginal in the Nearctic region, and the smaller "warbler-like" tyrannids, which make up a large segment in central America, are largely excluded. This is probably partly because they are tropically adapted, partly because the "warbler" adaptive zone is occupied by parulids and vireos in Nearctica (see discussion later).

The Tyrannidae make up an important segment of regional avifaunas from north to south throughout South America (Table 1). Thus, in Costa Rica there are 40 and 69 breeding genera and species, respectively, representing 21.3% and

Table 1.[a, b]

Total Numbers of Genera and Species

Continent as a Whole	Costa Rica	Colombia	Central Peru	Argentina	
				Northern	Southern
Total Passerines	189(330)	328(816)	366(572)	189(274)	74(115)
Tyrannidae	40(69)	78(150)	69(151)	55(72)	23(36)
103(302)	*21.2(20.9)*	*23.8(18.4)*	*19.9(26.4)*	*29.0(26.3)*	*31.1(31.3)*
Dendrocolaptidae	9(16)	11(26)	11(22)	10(14)	nil
13(45)	*4.2(4.8)*	*3.4(3.2)*	*3.0(3.8)*	*5.3(5.1)*	
Furnariidae	13(17)	26(68)	26(77)	21(35)	16(28)
56(202)	*6.9(5.2)*	*7.9(8.3)*	*7.1(13.5)*	*11.1(12.8)*	*21.6(24.3)*
Formicariidae	16(29)	42(123)	34(78)	13(17)	nil
51(227)	*8.5(8.8)*	*12.8(15.1)*	*9.3(13.6)*	*6.9(6.2)*	
Hirundinidae	3(3)	5(12)	6(10)	5(7)	5(5)
11(23)	*1.6(0.9)*	*1.5(1.5)*	*1.6(1.7)*	*2.6(2.6)*	*6.8(4.3)*
Troglodytidae	9(20)	9(27)	9(15)	2(2)	2(2)
9(98)	*4.8(6.1)*	*2.7(3.3)*	*2.5(2.6)*	*1.1(0.7)*	*2.7(1.7)*
Turdinae	3(11)	4(17)	5(17)	2(6)	1(3)
6(32)	*1.6(3.3)*	*1.2(2.1)*	*1.4(3.0)*	*1.1(2.2)*	*1.4(2.6)*
Sylviinae	3(4)	3(5)	3(3)	1(2)	nil
3(8)	*1.6(1.2)*	*0.9(0.6)*	*0.8(0.5)*	*0.5(0.7)*	
Motacillidae	nil	1(2)	1(5)	1(3)	nil
1(7)		*0.3(0.2)*	*0.2(0.9)*	*0.5(1.1)*	
Vireonidae	2(7)	3(10)	3(6)	2(2)	nil
4(23)	*1.1(2.1)*	*0.9(1.2)*	*0.8(1.0)*	*1.1(0.7)*	
Icteridae	9(12)	18(32)	13(23)	14(19)	5(6)
27(60)	*4.8(3.6)*	*5.1(3.9)*	*3.6(4.0)*	*7.4(6.9)*	*6.8(5.2)*
Parulidae	7(12)	5(21)	4(11)	3(5)	nil
6(33)	*3.7(3.6)*	*1.5(2.6)*	*1.1(1.9)*	*1.6(1.8)*	
Thraupidae	18(43)	40(123)	27(77)	14(23)	1(1)
47(182)	*9.5(13.0)*	*12.2(15.1)*	*7.4(13.5)*	*7.4(8.4)*	*1.4(0.9)*
Cotingidae	13(18)	23(48)	19(38)	8(11)	nil
34(88)	*6.9(5.5)*	*7.0(5.9)*	*5.2(6.6)*	*4.2(4.0)*	

[a] Neotropical Passerine Bird Families, relative abundance of major insectivorous groups (plus tanagers and cotingas) from north to south.

Migrants from the Nearctic Region are excluded.

Numbers of genera and (in parentheses) species are shown, plus the percentages (in italics) that they make up of the total passerine avifauna of each region.

[b] The Costa Rican figures are from Slud (1964), those for Colombia from de Schauensee (1964). The Peruvian figures have kindly been supplied by Dr. L. Short and pertain to the Departments of Lima, Ancash, Huánuco, Junín, Cusco, Huancavelica and Ayacucho, and parts of Libertad, San Martín, Loreto, Apúrimac, Ica, and Arequipa, i.e., they represent the avifauna of a broad tract across central Peru. The figures for northern Argentina, also supplied by Dr. L. Short, are for the provinces of Misiones, Corrientes, and Entre Rios: they thus represent an area that is almost entirely less than 3,000 ft. in height. The figures for southern Argentina are from an unpublished list of Dr. P. J. Humphreys. They refer to the provinces of Rio Negros, Chubut, and Santa Cruz.

20.9% of the avifauna. For Colombia the figures are 78 genera and 150 species (23.8% and 18.4%, respectively). Figures for central Peru are comparable. In northern Argentina there are 55 genera and 72 species (making up 29.0% and 26.3% of the avifauna). In southern Argentina the figures are 23 genera and 36 species (31.1% and 31.3%). Thus, significantly, the group makes up the same proportion of the avifauna from the tropics to the cold temperate regions, and is adapted to all latitudes (and altitudes). In this it contrasts with tropical groups like the tanagers (Thraupidae), and groups that are obviously recent arrivals in South America, such as the Sylviinae (Table 1).

The striking feature of the tyrannids is the range of body forms and ecological types developed, a diversity second only to that in the Neotropical Furnariidae. Thus, apart from generalized "flycatchers" of all sizes there is a range of warbler- and tit-like types (Figs. 1 and 2); relatively large-bodied drongo-like aerial predators; a ground-feeding shrike-like genus (Fig. 3), terrestrial thrush, chat, and wagtail (motacillid) counterparts, and various large and small, broad-billed, genera—see illustrations in Johnson (1967), Haverschmidt (1968), and elsewhere. A few species are largely berry feeders (Fig. 4). A second feature of the family is the broad range of habitats occupied, and the fine degree to which these may be divided. Thus, amongst the "medium-sized aerial feeders," for example, are species the are largely confined to the deep forest, forest glades, the forest edge, treetops, creek margins, rocks in the middle of swift-flowing streams, reedbeds, marsh edges, and a variety of different open country situations—see de Schauensee (1964), and others. This level of habitat specialization clearly exceeds that found in the muscicapid flycatchers of both Africa and Australia.

The family, as noted, varies widely in body size. There are genera and species with body lengths as small as 3 inches, and as large as 14 inches (Table 2). Most species fall within the size range of 3½ to 9½ inches. The greatest numbers of species are in the 4½ to 6½ inch body-length range (Table 2), although there is a fair-sized segment in the 7½ to 8 inch length range. Fifteen of the 150 species occurring in Colombia fall into this category. There is a continuum, it will be noted, from small-bodied to large-bodied species.

This situation contrasts with that in Africa, as will be discussed later. Here a series of families and subfamilies of insectivorous birds occupy segments of this size range. At the one end of the scale are the penduline tits (Remizidae) with body lengths of 3 to 3¾ inches, followed by the warblers (Sylvinnae) and tits (Paridae), most of which range from 4 to 6 inches in length, then the Old World flycatchers (Muscicapinae) which, with the exception of the long-tailed *Terpsiphone*, range from about 5 to 8½ inches. Still larger are the cuckoo-shrikes (Campephagidae), with body lengths to 8 to 10½ inches, the drongos (Dicruridae), ranging from 7½ to 10½ inches, and the shrikes (Laniidae), which mostly fall into the size range of 7½ to 10½ inches. For full details of body length measurements see McLachlan and Liversidge (1958).

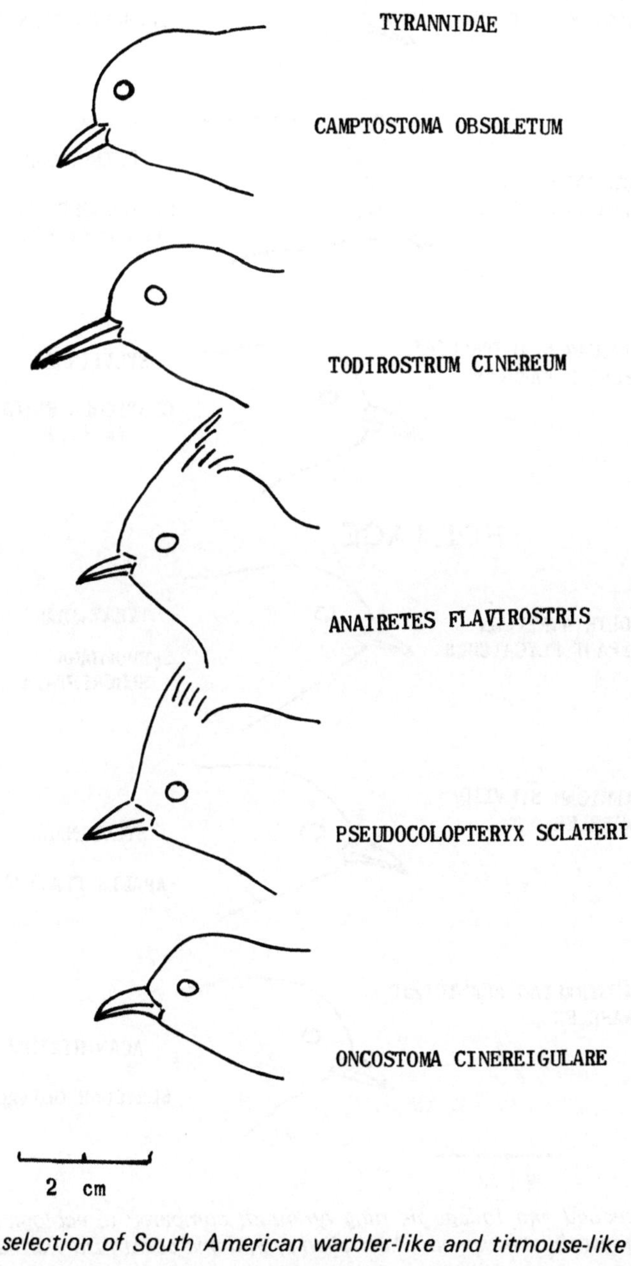

Fig. 1. A selection of South American warbler-like and titmouse-like tyrannids.

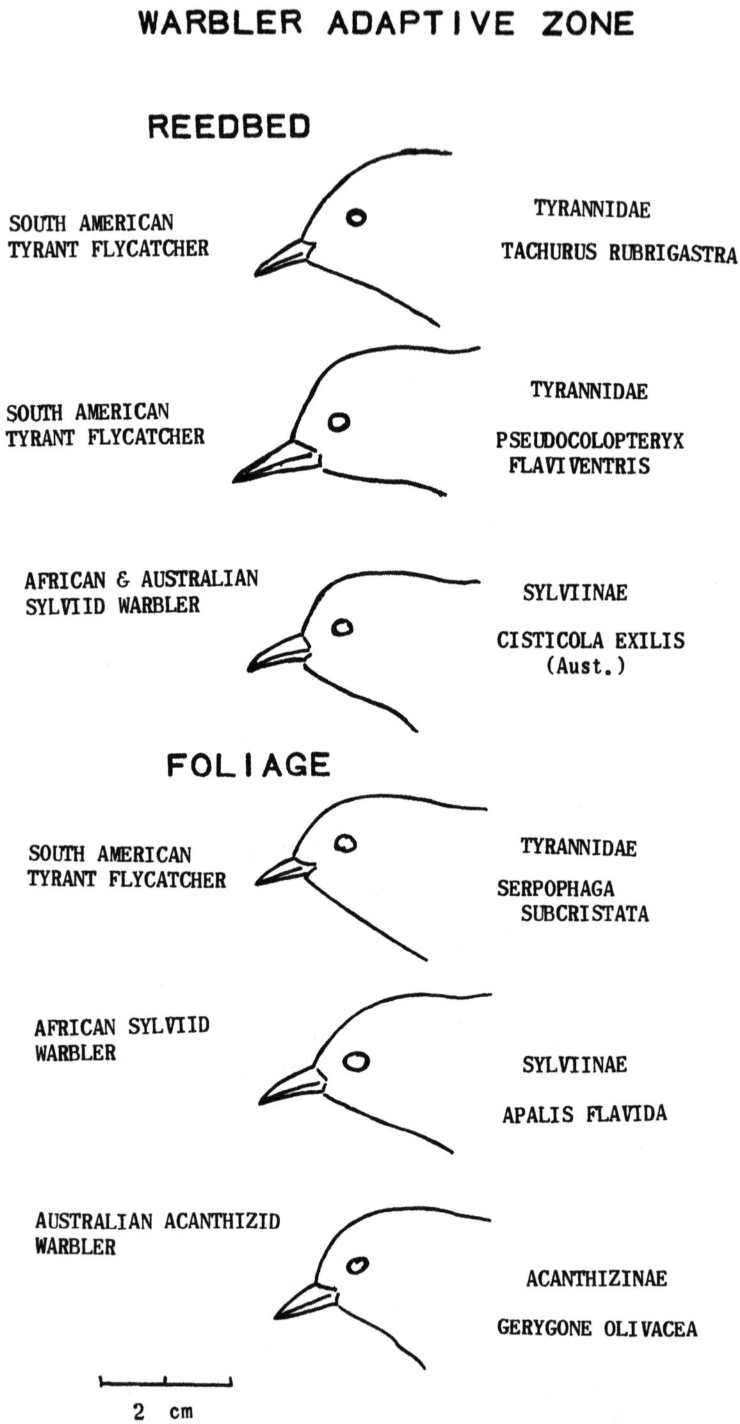

Fig. 2. Reedbed and foliage-gleaning tyrannids compared to ecological counterparts amongst the African and Australian sylviid and acanthizid warblers.

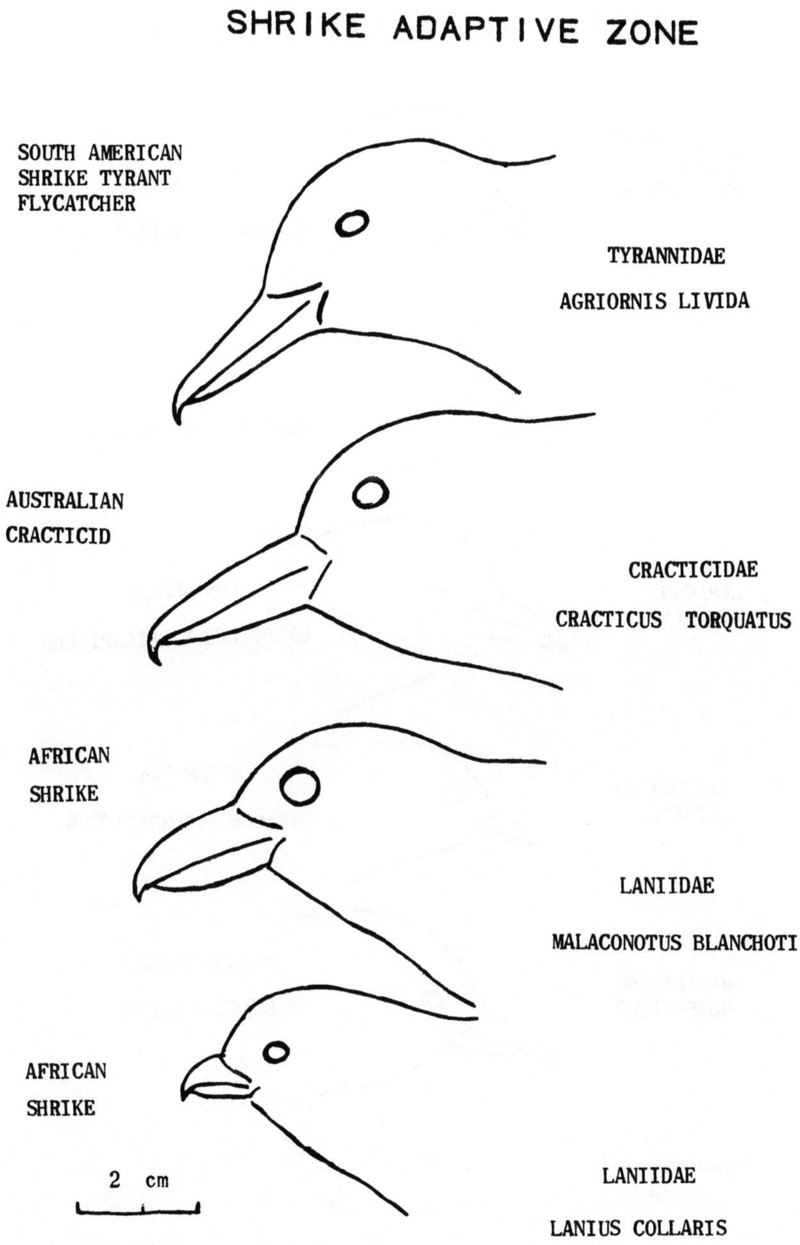

Fig. 3. The head of the South American shrike tyrant, Agriornis livida, *relative to ecological counterparts in Australia (*Cracticus torquatus, *Cracticidae), and Africa (various shrikes, Laniidae).*

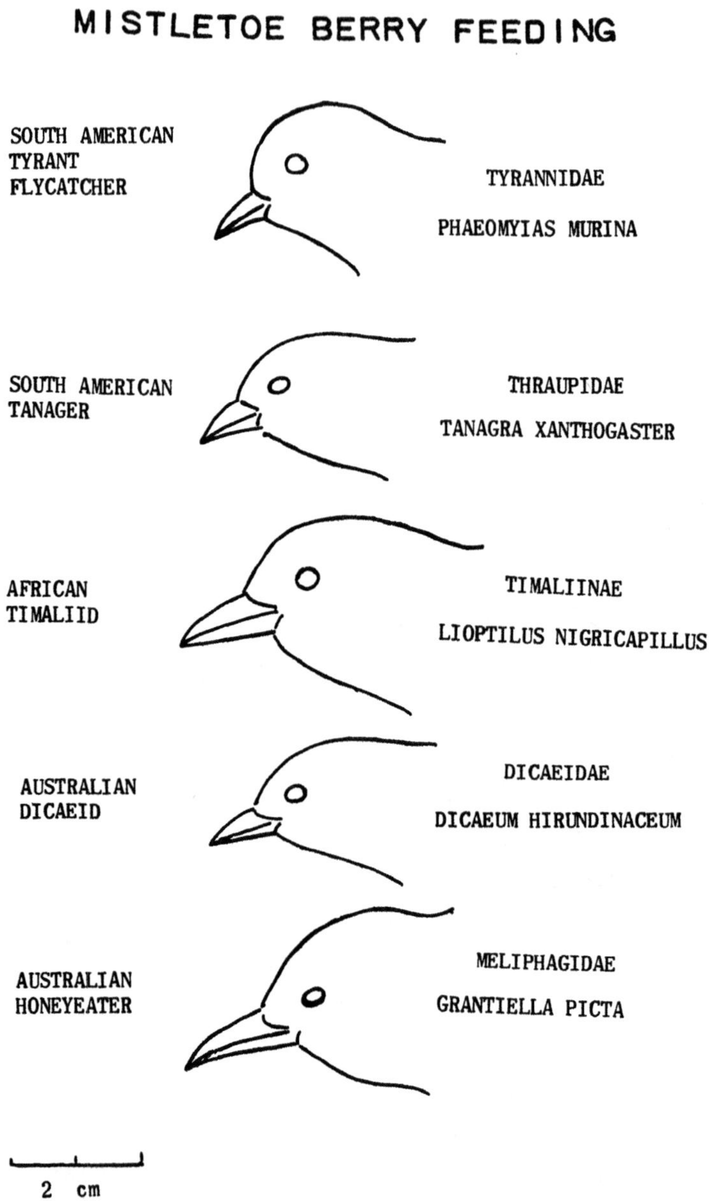

Fig. 4. The mistletoe-berry feeding tyrannid Phaeomyias murina, *relative to ecological counterparts, the South American tanager,* Tanagra xanthogaster, *African timaliid,* Lioptilus nigricapillus, *Australian flower-pecker,* Dicaeum hirundinaceum, *and Australian honeyeater,* Grantiella picta.

Table 2. Tyrannidae of Costa Rica, Colombia, and Chile Frequency of Different Body Lengths[a]

	Costa Rica		Colombia		Chile	
Total number of breeding species	69		150		35	
Number on which data are available	57		142		35	
Body Length (inches)[b] (means)	No.	%	No.	%	No.	%
3			1	1		
3½	3	5	3	2		
4	5	9	12	8		
4½	4	7	15	11	5	14
5	9	16	24	17	3	9
5½	11	19	21	15	2	6
6	7	12	16	11	3	9
6½	3	5	12	8	5	14
7	5	9	7	5	2	6
7½	2	4	8	6	4	11
8	2	4	10	7	1	3
8½	1	2	3	2	3	9
9	3	5	5	4		
9½	2	4	3	2	3	9
10			2	1		
10½						
11					1	3
11½					2	6
12						
12½						
13					1	3

[a]Data from Skutch, 1960; Slud, 1968; de Schauensee, 1964; Johnson, 1967.

[b]1 inch ~ 2.54 cm.

ECOLOGICAL SEPARATION IN REGIONAL FAUNAS, MECHANISMS PERMITTING A DIVERSITY OF SPECIES TO OCCUR TOGETHER

The only part of Neotropica from which sufficiently comprehensive ecological data are available to permit an analysis of ecological separation *within* regional faunas is Costa Rica. This is on the basis of two large monographs, those of Skutch (1960) and Slud (1964). Costa Rica "in relation to its small size (20,000 square miles) has possibly the richest bird life of any country of comparable size in the world" (Slud, 1964, p. 22). This country, moreover, has considerable physical relief and a distinct series of life-zones comparable to those of the Andes, vide:

(1) a Tropical Lowland belt, containing dry, moist, and wet forest divisions, which rises from sea level to a height of between 500 and 700 meters;
(2) a Subtropical belt, with a vertical range of 1000 meters, extending from 700 meters up to an altitude of 1400 or 1500 meters;
(3) a Lower Montaine belt, rising from 1500 to 2500 or 2600 meters; and
(4) a Montane Belt, extending from 2500 meters to about 3500 meters where, on the highest peaks, the subalpine paramo begins (Slud, 1964). Costa Rica has a rich tyrannid fauna of 40 genera and 69 breeding species. Its ornithological diversity and small size, therefore, make it an ideal region for an analysis of spatial, habitat, and feeding specializations.

Comparative ecological data for the 60 breeding species of tyrannids, for which information is available, is briefly summarized in Table 3. This includes measurement of body length for the various species, details of habitat occupied, the life zone or zones where each occurs, and a summary of the feeding habits. Included in the last named are: (1) details of the height at which the bulk of the feeding is carried out; (2) whether or not the bird catches insects by aerial sallies, or feeds within the confines of the tree (i.e., in the branches or foliage); and (3) if it takes its prey in the air, whether it feeds in exposed situations (e.g., fluttering out from the tops of trees or bare branches), or in sheltered ones (i.e., within or below the canopy). If a species is migratory or is, to a significant extent, a fruit eater, this is stated. Some of the information contained in Table 3 is summarized in Table 4 (number and percentage of species confined to each of the different vertical life zones); and Table 5 (numbers of species feeding at each of nine vertical levels, and numbers of species hunting in each of four different ways, both relative to body size).

Table 3. Body Size, Habitat Utilization and Feeding Habits in the Tyrant Flycatchers (Tyrannidae) of Costa Rica[a]

Species	Body Length (inches)	Habitat	Feeding Habits
Sayornis nigricans	7¼	Along mountain streams ST[b]	Makes short sallies after *Odonata*, etc. from boulders or low branches, also feeds on ground at waters edge
Colonia colonus	5½ (plus 5 inch tail)	Forest edges, clearings, in humid lowlands. T[b]	Sallies (1-23 yards) from tops of dead trees, leafless branches

Muscivora tyrannus	6 (plus 10) inch tail)	Open savannahs (southwestern). ST[c]	Aerial sallying from low bushes; also insects from ground
Tyrannus melancholicus	9	Open situations in woodland or scattered trees. T, ST	Aerial sallying from bush or medium tree. Berries eaten
Legatus leucophaius	6	Open country or fringes of trees along rivers; in tops of leafy trees. Migratory. T, lower ST	Some short sallying flights. Much fruit eaten
Conopias parva	7¼	Lower Caribbean slopes Forest borders and semi-open areas. T	Hunts for insects in foliage at medium height; rarely sallies
Megarhynchus pitangua	9½	Woodland, forest edges, scattered trees. T	Rarely sallies; large insects (e.g., cicadas) from foliage and branches at medium heights
Myiodynastes luteiventris	8½	Clearings with scattered trees. Humid. ST. Migratory	Short sallying flights high in trees. Berries eaten
M. maculatus	9	Dryer lowlands, mainly forest edge and scattered trees in clearings. Migratory. T, ST,	Sallies less than previous sp. Keeps high
M. hemichrysus	9	Forest edges. St, LM[d]	Makes forays at or through the foliage, shrub height to high in trees
Myiozetetes similis	7	Clearings and semi-open country up to 6000 ft. Most of feeding lower down. T, ST, LM	Forages for insects in foliage and branches medium to low heights. Sometimes sorties or drops to ground
M. cayanensis	7	Scattered trees in open swampy areas and lake and lagoon edges. T	Not stated but presumably like *M. similis*
M. granadensis	7	Scattered trees in open places; river edges in forest. Extreme southeast. T, lower ST	Aerial sallying often well above treetops. Berries eaten
Pitangus sulphuratus	9¼	Semi-open terrain. Medium heights down to tops of shrubs. T	Aerial sallying insects and fruits from within foliage; even eats lizards, fish, etc.

Myiarchus nuttingi	8	Dry forest; open woodland and scrubby country. Northern Pacific coast. T, lower ST	Feeds low in trees; insects from branches, by hovering, and some sallying
M. ferox	8	Mangroves, Pacific coast, T	Feeds in low vegetation and amongst mangrove branches
M. tyrannulus	8¾	Dry forest and semi-open country, Pacific coast. T	Feeds amongst twigs and branches; and by short aerial sallies
M. tuberculifer	6¾	Woodland edges, second growth, clearings with scattered trees; low to medium heights. T, ST	Forages inside foliage, hopping along branches and flitting at leaves; some short aerial sallies from exposed branches
Contopus cinereus	5¼	Semi-open terrain with trees and tall bushes, low to medium heights. T	Strictly a sallying flycatcher; but does not have fixed lookout perch
C. lugubris	5½	Semi-open woodland and forest openings in highlands. Feeds from topmost branches of big trees. ST	Strictly a sallying flycatcher; "is the only abundant, high-perching, longsallying flycatcher in middle altitudes" (Slud)
C. ochraceus	5½	High altitude forests, Irazu and Turrialba volcanos. Very rare. M	unknown
Empidonax flavescens	5½	Undergrowth and shrubbery within forests and at forest edges. ST, LM	Insects from undersides of leaves and ground; also short aerials allies
E. albigularis	5½	Caribbean side of central plateau. ST, LM. Very local and rare	Unknown
E. atriceps	5½	Enclosed openings in forest. M	Makes short sallies from bushes and small trees
Mitrephanes phaeocercus	5	Clearings in high oak forests. ST, M	True sallying flycatcher, usually from outer twigs and branches
Terenotriccus erythrurus	4	Lower levels (5-25 feet) forest. T	Captures small insects on short darts from twig to twig
Aphanotriccus capitalis	5¼	Forest second growth, 5-15 ft. ST	Darts at foliage for prey

Myiobius barbatus	5	Forest understory. T	Secures prey acrobatically in the air, the fan-shaped tail permitted rapid twists and turns (like *Setophaga*)
M. atricaudus	5½	Riverside, coastal thickets. T	Similar to *M. barbatus*
M. fasciatus	5	Low bush growth; limited to a few restricted localities. Upper T; lower ST	Catches small insects by flitting through close-set branches
Onychorhynchus mexicanus	6¾	Dry forest understory. T	Catches insects in flight as they dash from branch to branch
Platyrinchus cancrominus	3¾	Dry to subhumid forest. Northern. T, lower ST	Flits through undergrowth and lower story for insect prey
P. mystaceus	3¾	Southern forests. Undergrowth and lower story. ST	Similar to *P. cancrominus*
P. coronatus	3¾	Heavy woodland, from near ground to lower under story. T	Flits up to pick up insects from leaves
Tolmomyias sulphurescens	6¼	Light woodland, forest edges. Low to medium heights. T, ST	Hunts insects in foliage crowns of lower trees and flits at insects on leaves
T. assimilis	5½	Humid, heavily forested foothills, semi-open areas. T	As above, except feeds in medium to very high situations
Rhynchocyclus brevirostris	6¼	Shady situations in humid forests, up to 30 feet. T, ST	Feeds mainly by fluttering up to pick insects off leaves
Todirostrum nigriceps	3½	Woodland edges; canopy of small to very tall trees. T, part ST	Feeds by flitting amongst canopy
T. cinereum	4	Thickets and shrubbery. Semi-open country. Feeds within a few yards of the ground. T, ST	Moves through shrubbery and leaps at insects on leaves
T. sylvia	4¼	Dry to semi-humid lowlands; in undergrowth thicker and more tangled than *T. cinereum*. T	Feeds by hopping through twigs and vines

Oncostoma cinereigulare	4¼	Bushes and herbs near ground. T	Plucks insects from undersides of leaves
Lophotriccus pileatus	4¼	Lower understory of forest. ST	Plucks insect prey by hovering at leaves
Myiornis atricapillus	3½	Heavy forest; forages from upper understory to the middle levels. Also low shrubbery growth. T	Flits at leaves
Phylloscartes superciliaris	5	Wet forests. ST	Feeds, gnatcatcherlike, mostly within foliage
Capsiempis flaveola	4¾	Tangled thickets in scrubby pastures, up to 20 feet. T	Darts at leaves, rarely flutters into air
Serpophaga cinerea	5	Boulder-strewn courses of swift mountain rivers. ST	Gleans prey from mossy stones and banks, or by fluttering over water
Elaenia flavogaster	6½	Trees and shrubs of open country. T, ST, LM	Short aerial sallies, plus foraging amongst branches for insects, small fruits
E. chiriquensis	6	Savannahs and scrubby pastures. Southwest Pacific coast. T, lower ST	Aerial feeding and foraging within branches. Eats fruits
E. frantzii	6	Forested highlands, above 6000 feet; semiopen habitat or inside woods. M	Like above two species but engages in longer and more dashing flights
Myiopagis viridicata	5¾	Semi-open country and scrubby woods. Pacific mountain slopes. T	Feeds by moving through foliage or fluttering at leaves
Sublegatus arenarum	6	Mangroves, within a few yards of the ground. T	Makes short aerial sallies
Camptostoma imberbe	4¾	Scrubby vegetation, woodland borders in northern dry forests. T	Feeds warbler-like, fluttering at, and moving through, foliage, mainly at heights of 5-15 feet
C. obsoletum	4¾	Southern humid Pacific slope forests. T	Similar to foregoing but feeds more commonly higher up
Tyranniscus vilissimus	5¼	Edges and open parts of forests, crowns of small to medium trees. T, ST, LM	Chiefly mistletoe a berry feeder. Gathers insects by flitting, vireo-like, amongst foliage
Acrochordopus zeledoni	4¼	Forest borders. ST, LM	Groups move warbler-like, through lower, or medium levels of the trees

Ornithion semiflavum	3½	Borders of woodlands on Pacific coast. Rare. T	Feeds in foliage of lower trees
O. brunneicapillum	3¾	Forests and second growth of Caribbean coasts. Medium heights to high in trees. T, lower ST	Feeds in folage
Leptopogon superciliaris	5¾	Forests, keeping low in trees. ST	Short aerial sallies from bare branches
Mionectes olivaceus	5½	Forests, feeding at 5-10 feet. ST, LM	Forages by hovering or clinging at leaves
Pipromorpha oleaginea	5¼	Deep forest and woodland borders. Replaces previous species altitudinally. T, lower ST	Snatches insects from foliage as it flits from twig to twig, lower levels of trees. Small fruits

[a]Data from Skutch (1960), Slud (1964).

[b]T–Inhabitant of lowland tropical life zone.

[c]ST–Subtropical life zone (1500 to 2500 m).

[d]M–Montane life zone (2500 to 3000 m). LM–Lower montane life zone.

The survey brings out that the various species of Costa Rica Tyrannidae separate out ecologically in a wide variety of ways. These include:

(1) *Differing in body size (and hence in bill size, etc).* The species range in body length from 3 to 9½ inches, with the greatest number of species, as noted, in the 5 to 6 inch range. There is a broad correlation between body size and way of life. The larger aerial feeding tyrannids tend to live in more exposed situations, and range much more widely in their feeding than the smaller ones, most of which keep to the shrubbery and trees. Amongst the aerial feeders small-bodied ones like those of the genus *Elaenia*, with a body length of 6 inches, make much shorter aerial feeding sallies than larger species.

(2) *Differences in structure.* Structural differences, particularly in the bill, predispose species for different ecological roles and foods. The wide range of bill lengths found in species of equivalent size will be seen in Fig. 5. Thus, for example, small tyrannids with a wing-length of 50 mm vary in their bill-lengths from about 7.0 mm to 10.5 mm. Species with wing lengths of 70 mm have bills ranging in size from about 9.0 mm to 17 mm. Figure 6 provides equivalent data for the range of widths to be found in bills of these different lengths, Fig. 7 for bill depth relative to width, and Fig. 8 for the ratio bill width/length plotted against wing length. Figure 9 provides data on tarsal length relative to wing length; changes in the leg, however are less important in forest species than in those with semiterrestrial habits–see the large number of species that fit into this category on the Argentinan pampas, etc., in Wetmore(1926).

Table 4. Numbers and Percentages of Species Inhabiting the Various Vertical Life Zones. Tyrannidae of Costa Rica and Colombia[a]

	Costa Rica		Colombia	
Total No. breeding species	69		150	
No. for which data are available	61		141	
Distribution in Life Zones	No.	%	No.	%
Tropical Zone	24	39	68	48
Tropical and Lower Subtropical	8	13	16	11
Tropical and Subtropical	5	8	9	7
Tropical, Subtropical, and Lower Temperate (Montane)	4	7	4	3
Subtropical	13	21	17	12
Subtropical and Temperate (Montane)	4	7	15	11
Temperate (Montane)	3	5	7	5
Paramas			3	2
Subtropical, Temperate and Paramas			2	1

[a]Data from Slud (1964) and de Schauensee (1964).

Certain flycatchers (and pipits) have evolved the tailwagging habit, apparently as a device for flushing prey. This occurs in *Serpophaga* and *Todirostrum cinereum* (Skutch, 1960) amongst the tyrannids, and has been evolved independently in muscicapids like the Australian *Rhipidura leucophrys*. The tail, also, may be modified into a "fan," which increases maneuverability in confined spaces, permitting rapid twisting and turning, and producing a tumbling kind of flight. This modification has been achieved independently in *Myiobius barbatus* amongst the tyrannids, amongst the Australian muscicapid flycatchers, and by *Trophocercus* in Africa (writer's observations).

(3) *Separation on the basis of vertical life zone inhabited.* The 61 Costa Rican tyrannids for which data are available proved to separate out into seven groups, inhabiting, respectively: the tropical zone; tropical plus lower subtropical; tropical and entire subtropical; subtropical only; subtropical and montane; tropical, subtropical, and montane; and montane region only (Table 3). When these data are summarized (Table 4) 24 species prove to be confined to the tropical zone, 13 to the subtropical, and 3 to the montane. Others occur in two

Table 5. Feeding Levels, and Hunting Methods, Relative to Body Size in the Tyrannidae of Costa Rica[a]

	Mean Body Length (inches)							
	3-3¾	4-4¾	5-5¾	6-6¾	7-7¾	8-8¾	9-9¾	Total
Vertical Level at which most feeding carried out								
Ground			1		1			2
Low Shrubs and Bushy Growth		3	2	2				7
Shrubs and Lower Levels of Trees	3	2	1					6
Trees, Lower Levels	2	2	5	1		1		11
Trees, Lower to Middle Heights		2	3	4	1		2	12
Trees, Middle Heights					1		1	2
Trees, Middle to Higher Levels	1		3	1				5
Trees, Higher Levels			2	1	1	1	1	6
Trees, all Levels	1						1	2
Total	7	9	17	9	3	2	5	52
Methods of Obtaining Insect Prey								
Aerial sallying in exposed situations			6	3	1	1½	1½	13
Aerial sallying in sheltered situations (e.g., within branches)		1	2	4				7
Obtaining prey by darting at foliage	2	3	3	1				9
By hopping or fluttering through branches or foliage	5	5	5	2	3	2½	3½	26
Total	7	9	16	10	4	4	5	55

[a]Data from Skutch, 1960; Slud, 1964. Only species on which data is available are included.

of these zones. There is thus a significant level of spatial separation or isolation as a result of altitudinal distributional differences.

Table 3 gives comparative data for Colombia, based on the species accounts in de Schauensee (1964). Here 68 species are confined to the tropical zone, and a total of 39 to the subtropical and montane zones. There are also 3 exclusively paramas (high grassland) species. This confirms that separation into life zones is an important mechanism for the spatial separation of species.

(4) *Geographic separation in the nonaltitudinal Sense.* Costa Rica is the contact, or overlap, zone for several pairs of southern and northern species (Table 3). *Platyrinchus mystaceus* and *P. cancrominus*, which are sometimes regarded as conspecific, are an example. In another case counterparts are essentially confined to the Pacific and Caribbean coasts of Costa Rica, respectively; e.g., *Ornithion semiflavum* and *O. brunneicapillum*.

(5) *Habitat separation within a life zone.* Birds in general show a marked preference for one or another vegetation type, or "habitat," and rarely, or never, occur outside of it. Habitat separation is one of the commonest ways in which closely related species avoid interspecific competition. There are many examples of this in the Costa Rican tyrannids. For example, in the genus *Myiarchus*, *M. nuttingi* and *M. tyrannulus* occur in dry forest in the tropical zone, and *M. ferox* in mangroves. A similar kind of separation occurs in species of the genera *Myiobius*, *Tolmomyias*, and *Todirostrum* (Table 3). *Todirostrum sylvium* is an example of a species whose habitat varies relative to the presence, or absence, of a potentially competitive close relative (Slud, 1964).

South American Tyrannidae

1. *Agriornis livida*
2. *Xolmis irupero*
3. *Muscisaxicola rufivertex*
4. *Muscigralla brevicauda*
5. *Lessonia rufa*
6. *Ochthoeca ornanthoides*
7. *Sayornis nigricans*
8. *Colonia colonus*
9. *Gubernetes yetapa*
10. *Alectrurus tricolor*
11. *Hymenops perspicillata*
12. *Muscipipra vetula*
13. *Arundicola leucocephala*
14. *Pyrocephalus rubinus*
15. *Ochthornis littoralis*
16. *Muscivora tyrannus*
17. *Tyrannus melancholicus*
18. *Tyrranopsis sulphurea*
19. *Megarhynchus pitangua*
20. *Myiodynastes luteiventris*
21. *Pitangus sulphuratus*
22. *Myiarchus tyrannulua*
23. *Empidonax euleri*
24. *Onychorhynchus coronatus*
25. *Platyrinchus mystaceus*
26. *Tolmomyias sulphurescens*
27. *Rhynchocyclus olivaceus*
28. *Ramphotrigon ruficauda*
29. *Todirostrum cinereum*
30. *Onchostoma olivaceum*
31. *Colopteryx galeatus*
32. *Myiornis auricularis*
33. *Pseudocolopteryx flaviventris*
34. *Tachurus rubrigastra*
35. *Anairetes flavirostris*
36. *Serpophaga subcristata*
37. *Elaenia parvirostris*
38. *Phaeomyias murina*
39. *Camptostoma obsoletum*
40. *Tyranniscus vilissimus*
41. *Ornithion inerme*
42. *Mionectes striaticollis*
43. *Pipromorpha oleaginea*

Africa

Muscicapinae

1. *Muscicapa adusta*
2. *Parisoma subcaeruleum*
3. *Namibornis herero* (transferred to the Turdinae by White, 1962, p. 149).
4. *Bradornis pallidus*
5. *Sigelus silens*
6. *Pedilorhynchus comitatus*
7. *Chloropeta natalensis* (placed in the Sylviinae by White, 1960, p. 415).
8. *Hyliota australis*
9. *Batis molitor*
10. *Stenostira scita*
11. *Trochocercus cyanomelas*
12. *Terpsiphone viridis*
14. *Megabyas flammulata*
15. *Bias musicus changamwensis*
Myioparus griseogularis (placed in the Muscicapinae by Hall and Moreau, 1970, p. 211).

Turdinae

T1. *Cercomela familiaris*
T2. *Oenanthe monticola*
T3. *Oenanthe pileata*
T4. *Saxicola torquata*
T5. *Myrmecocichla formicivora*
T6. *Monticola brevipes*

Sylviinae

S1. *Eremomila ictopygialis*
S2. *Apalis flavida*
S3. *Camaroptera toroensis*
S4. *Prinia subflava*
S5. *Sylvietta whytii*
S6. *Acrocephalus boeticatus*
S7. *Cisticola galactotes*

Timaliinae

X1. *Lioptilus nigricapillus*
X2. *Turdoides jardinei*

Paridae and Remizidae

P1. *Parus niger*
P2. *Anthoscopus caroli*

Dicruridae

D1. *Dicrurus adsimilis*
D2. *Dicrurus lugwigii*
D3. *Dicrurus atripennis*

Eurylaimidae

E1. *Smithornis capensis*
D2. *Pseudocalyptomena graueri*

Motacillidae

M1. *Motacilla capensis*
M2. *Macronyx croceus*
M3. *Anthus leucophrys*
M4. *Anthus caffer*

Laniidae

L1. *Lanius collaris*
L2. *Malaconotus blanchoti*
L3. *Dryoscopus cubla affinis*
L4. *Prionops plumata*
L5. *Corvinella corvina*

Meropidae

B1. *Merops pusillus*
B2. *Bombylonax breweri*

Australia

Muscicapinae

1. *Rhipidura fuliginosa*
2. *R. leucophrys*
3. *Myiagra rubecula*
4. *Seisura inquieta*
5. *Arses telescophthalmus*
6. *Monarcha alecto*
7. *M. melanopsis*
8. *Peltops blainvillii*
9. *Machaerirhynchus nigripectus*
10. *Microeca leucophaea*
11. *Petroica multicolor*
12. *Poecilodryas pulverulenta*
13. *Eopsaltria australis*

Malurinae

*1. *Malurus cyaneus*
*2. *Clytomyias insignis*
*3. *Chenorhamphus grayi*

Campephagidae

C1. *Coracina novaehollandiae*
C2. *Edolisoma tenuirostre*
C3. *Lalage sueurii tricolor*

Grallinidae

G1. *Grallina cyanoleuca*

Cracticidae

CR. *Cracticus torquatus*

Artamidae

A1. *Artamus cyanopterus*
A2. *A. superciliosus*

Pachycephalinae

P1. *Pachycephala pectoralis*
P2. *Myiolestes megarhynchus*
P3. *Colluricincla harmonica*

Falcunculinae

F1. *Eulacestoma nigropectus*
F2. *Falcunculus frontatus*
F3. *Oreoica gutturalis*

Sylviinae

S1. *Cisticola exilis*

Acanthizinae

A1. *Sericornis magnirostris*
A2. *Acanthiza lineata*
A3. *Gerygone olivacea*

Epthianurinae

W1. *Epthianura albifrons*
W2. *Ashbyia lovenis*

Dicaeidae

Y1. *Dicaeum hirundinaceus*
Y2. *Pardalotus striatus*

Dicruridae

D1. *Dicrurus hottentotus*

Motacillidae

M1. *Anthus novaeseelandiae*

Meropidae

B1. *Merops ornatus*

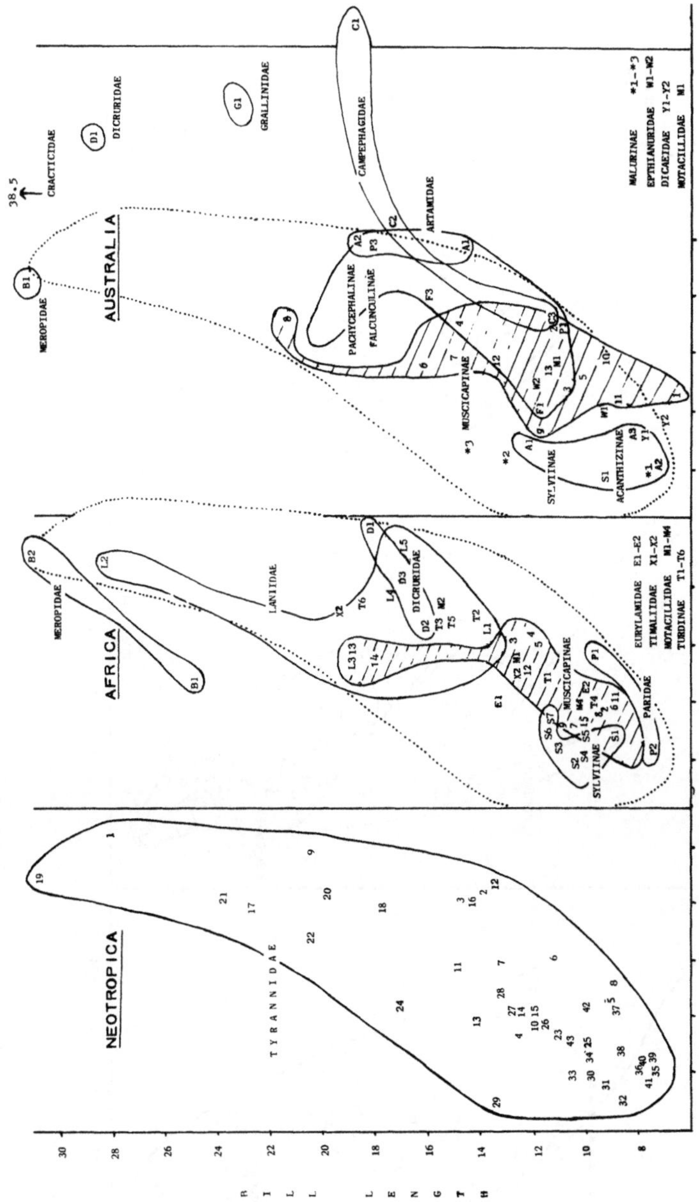

Fig. 5. Neotropical Tyrannidae and potential ecological counterparts in the avifaunas of Africa and Australia, comparisons on the basis of wing-length (which may be used as an indication of overall size) and bill-length (which varies with the size and kinds of food consumed). On the basis of these two morphological characters it will be seen that the tyrannids are equivalent to about 9 families and subfamilies of birds in Africa and Australia. The contour line of the Tyranniadae in the first section is supermiposed over the African and Australian segments as a dotted line. See key for species identifications.

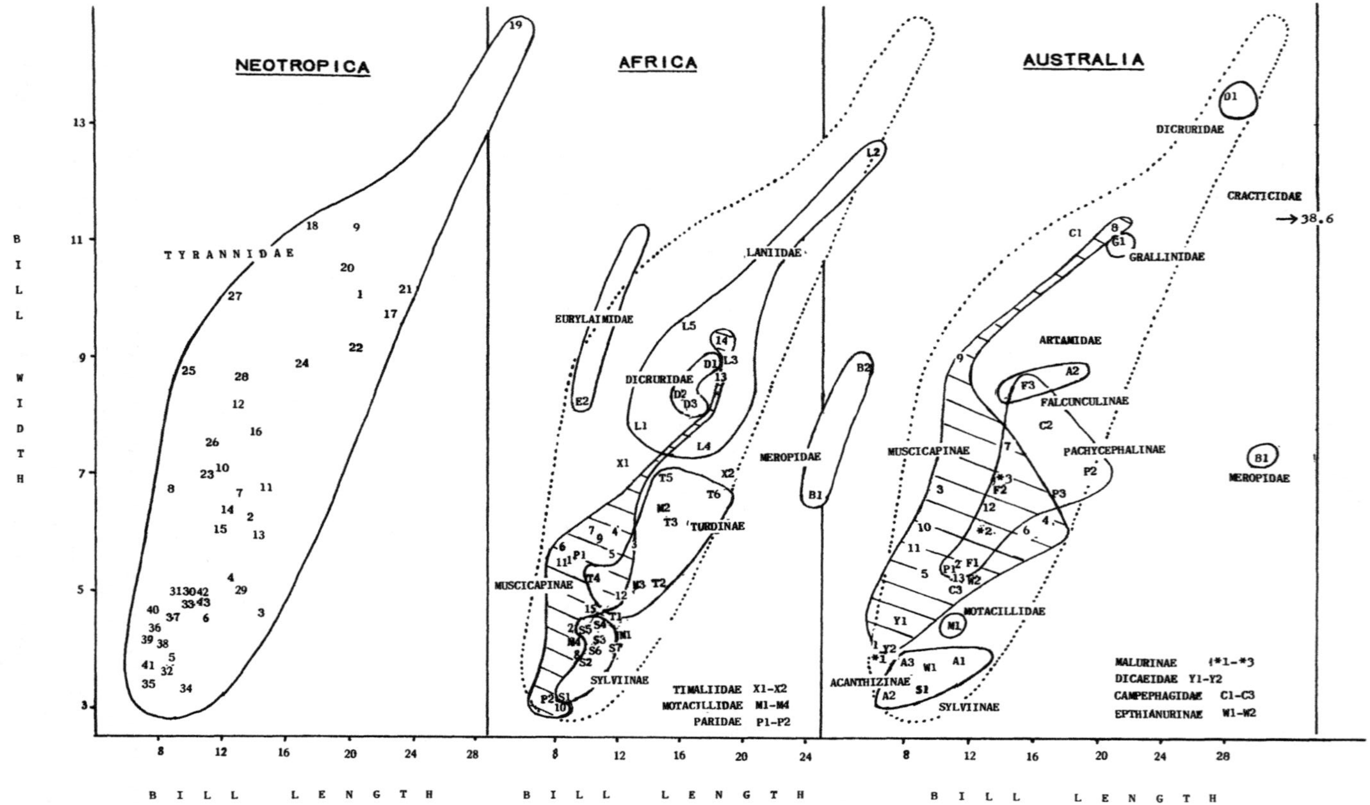

Fig. 6. A comparison of the three continental faunas in terms of bill width and bill length. (See text.) See key for species identifications.

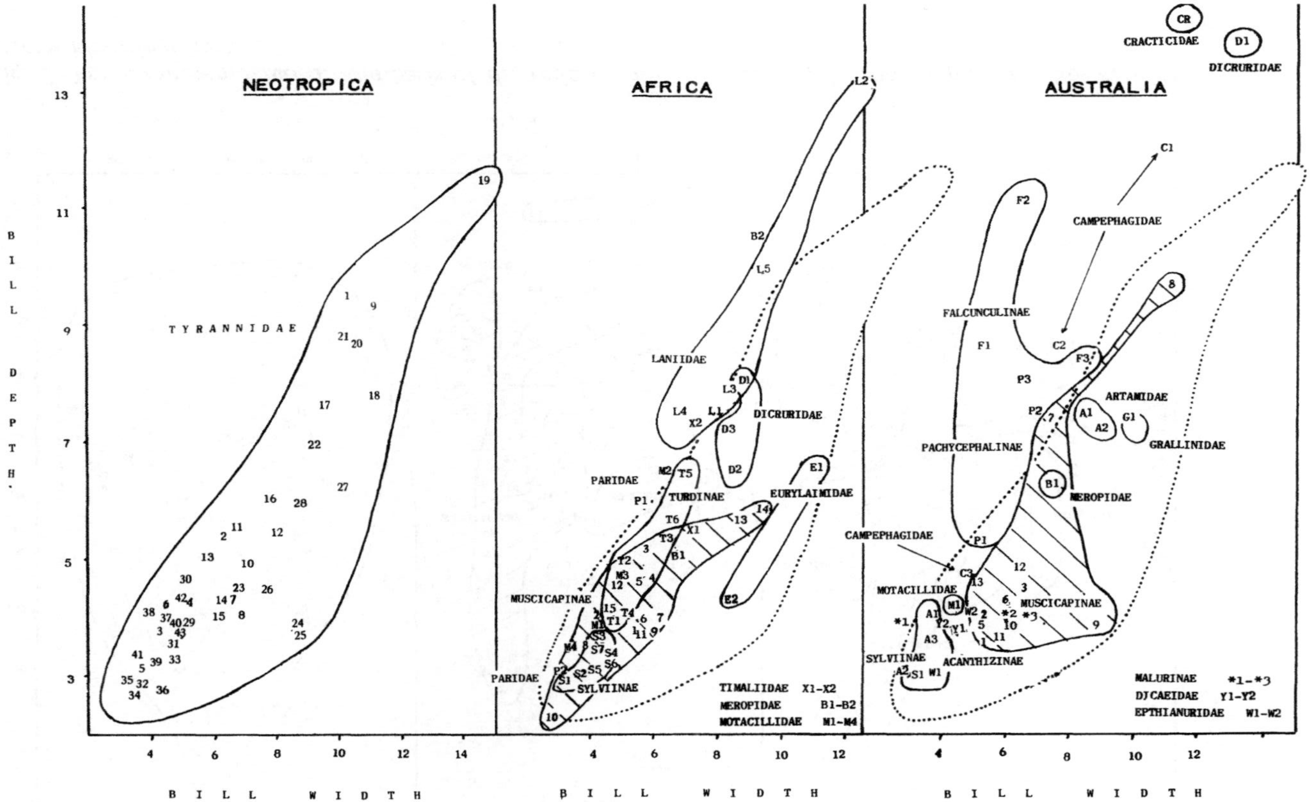

Fig. 7. The tyrannids and African and Australian counterparts compared in terms of bill depth and bill width. See key for species identifications.

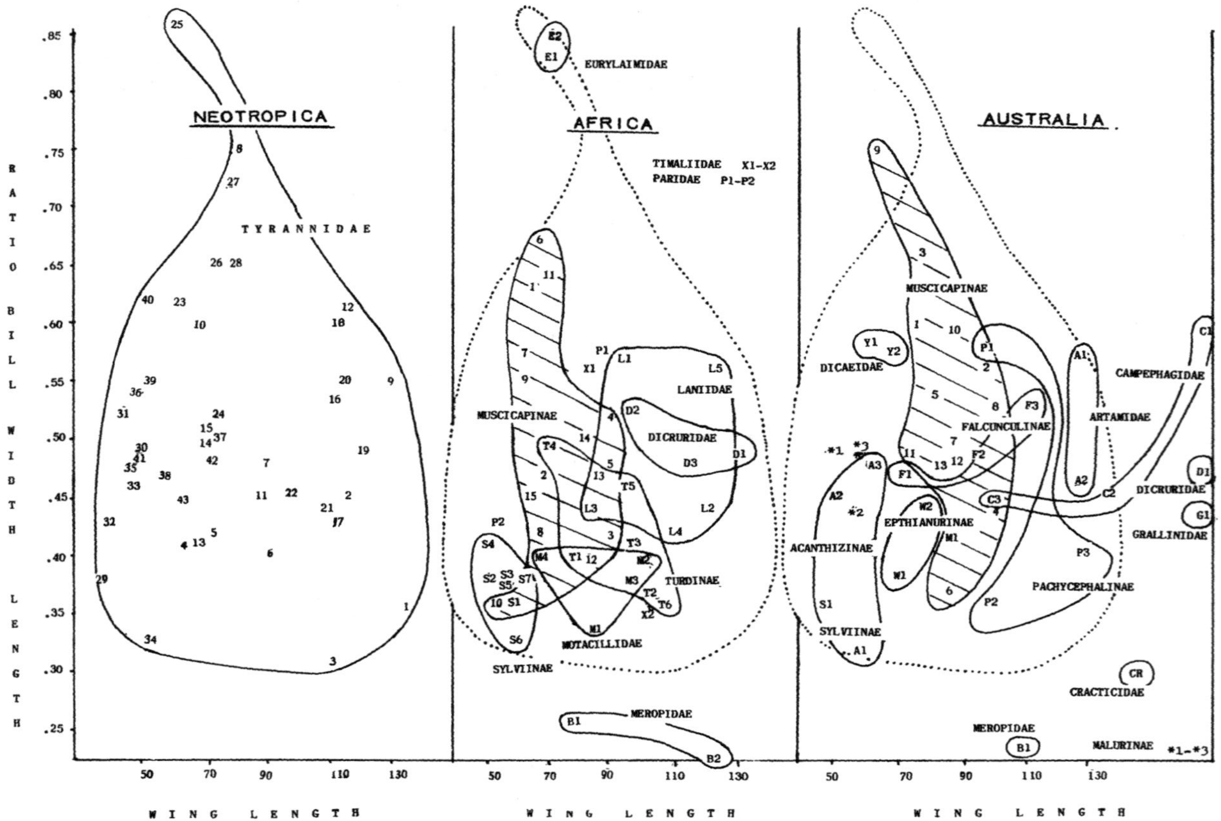

Fig. 8. The same compared on the basis of the ratio of bill width to length, and wing length. See key for species identifications.

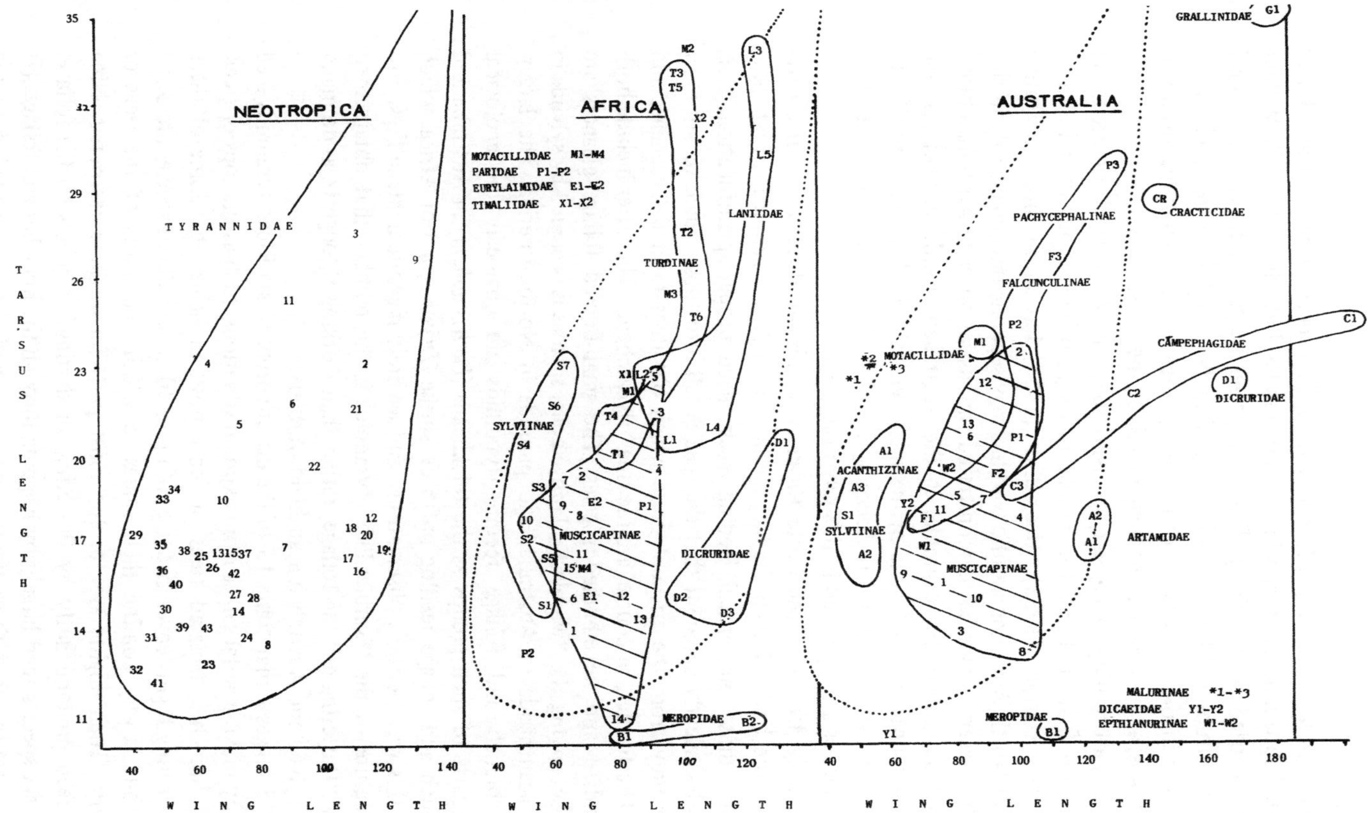

Fig. 9. Comparisons on the basis of tarsus length (which is correlated with perching habits, and increases with greater emphasis on terrestrial habits), and wing length (see text). See key for species identifications.

(6) *Separation on the basis of feeding level.* The species listed in Table 3 may be divided into nine categories in terms of the level at which the bulk of their feeding is carried out (Table 5). Thus two species are ground feeders; seven feed in the low shrubs and bush growth; six in shrubbery and the lower levels of the trees; eleven in the lower levels of the trees; twelve range through the lower and middle levels; and six feed in the higher levels of the trees.

Vertical partitioning of habitat, or living space, is of universal occurrence in birds. It was first discussed for Paridae in England by Colquhoun and Morley (1943), and Hartley (1953). MacArthur (1958) has shown how cohabiting species of North American parulid warblers may separate out ecologically in this way. The present writer has demonstrated various instances in Australian birds. It is well developed, for example, in the Australian acanthizid warblers, the ecological counterparts of some of the small tyrannids in that continent (Keast, 1968, 1970). Within the Costa Rican tyrannid genera *Tolmomyias* and *Todirostrum* species are separated in this way (Table 3).

(7) *Feeding Methods.* The Costa Rican tyrannids are divided into five categories on the basis of their feeding methods in Table 5. Admitting that these data, like those on vertical feeding levels, need proper quantification, it is nevertheless clearly apparent that the species fall into those which obtain most of their prey from the air, and those which obtain it from the branches and leaves. The former, of course, are the true "flycatchers." The latter include both small-bodied and large-bodied species. The small-bodied foliage gleaners are clearly occupying the "warbler" adaptive zone, such as is occupied by sylviids in Africa, acanthizids in Australia, and parulids in North America. The larger-bodied branch and foliage feeding tyrannids are obviously the ecological counterparts of such groups as the thickheads (*Pachycephala*) in Australia and the branch and foliage feeding puffback shrike (*Dryoscopus*) of Africa, which feed in this way (writer's observations, and see body measurements in Figs. 5 to 9). A feature of the "warbler-like" tyrannids is the number that obtain their food by "fluttering at" the foliage, rather than working through it, as do many African sylviid and Australian acanthizid warblers.

Table 5 suggests that some 13 of the insectivorous Costa Rican tyrannids feed mainly by making aerial sallies from exposed situations, such as the tops of trees, exposed branches, isolated trees or fence posts in fields, etc. Some of these insect-pursuing flights cover distances of up to 30 meters. Seven species are aerial feeders but largely confine their sallies to within the limits of the trees or shrubbery. Their flights are for the most part, much more restricted. Nine species feed predominantly by fluttering, or darting, at insects on the foliage. Twenty-six species feed largely by hopping through the branches and foliage. (It must be stressed that these figures must be regarded as provisional since the authors who carried out the field studies did not do so with the method of analysis used here in mind. Further quantitative studies would, accordingly, be worth while. Nevertheless, the figures bring out segregation patterns very nicely.)

It is obvious that differences in feeding behavior clearly serve to provide a measure of ecological isolation.

(8) *Consumption of different foods.* There is little information on the actual foods eaten by Costa Rican tyrannids, though Haverschmidt (1968) provides some comparative data for Surinam. Some tyrannids are partly berry eaters, and in this they differ from the purely insectivorous species. (It might be noted that typically African and Australian muscicapids and warblers do not eat fruit.) Amongst the Costa Rican species that eat berries are: *Tyrannus melancholicus, Lagutus leucophaius, Myiodynastes luteiventris, M. granadensis, Elaenia flavogaster, E. chiriquensis, Pitangus sulphuratus,* and *Pipromorpha oleaginea*. These species are in the 5 to 9 inch body-length range.

(9) *Seasonai changes in spatial distribution.* A few Costa Rican breeding tyrannids are migratory, spending part of the year in continental South America. Included here are *Legatus leucophaius, Myiodynastes luteiventris, M. maculatus*. The seasonal withdrawal of a species should theoretically reduce interspecific competition. Some 10 species of North American tyrannids, however, overwinter in Costa Rica, potentially increasing the intensity of interspecific competition within the family.

"TROPICAL DIVERSITY" RELATIVE TO THE NEOTROPICAL TYRANNIDS

The diversity of life in certain parts of the world's tropics is legendary, and famous evolutionists from Darwin and Wallace to Dobzhansky, Huxley, and Mayr, have commented upon it. Recent writers who have discussed the subject with reference to birds include Pianka (1966), MacArthur (1969), and Mayr (1969). It is generally agreed that avian diversity is to a considerable agree the result of a very fine division of the available habitats and niches, this being made possible by the uniform climate and productivity of the tropics. MacArthur (1969) also suggests that "patchiness" in distribution is important in permitting the simultaneous occurrence of many species.

A discussion of this subject lies beyond the scope of the present paper. It is worth noting, however, the extent to which the tyrannids as a group, and the Costa Rican fauna in particular, supports these concepts. The following points may be made: (1) Part of the ecological species separation is altitudinal, 24 species being confined to the tropical zone, 13 to the subtropical, and 3 to the montane zone. The list could be further extended by bringing in species that occupy two, or part of two, zones. An equally great vertical isolation of species is characteristic of Colombia and Equador as Chapman (1917, 1926) has shown. De Schauensee (1964) gives the following figures for Colombia out of a total fauna of 120 species: confined to tropical zone, 46; tropical and subtropical, 7; upper tropical and subtropical, 32; subtropical, 16; subtropical to temperate, 13; temperate, 5; temperate and lower paramos, 1. This means, in effect, that the fauna is divided in two halves, one confined to the lowlands, and the other to heights exceeding 5,000 ft. A similar division occurs in the tanagers (Thraupidae) where, out of a fauna of 116 species, 53 are confined to the tropical (including lower subtropical zone), 48 to the upper tropical and subtropical, 12 to the subtropical and temperate, and 5 to the temperate, zones.

In a recent analysis of altitudinal species replacement in the Andes of Peru, Terborgh (1971) concludes that ecotones account for less than 20% of the distributional limits, competitive exclusion for about one third of the limits, and gradually changing conditions along the gradient for about one half of the limits.

Since the presence of life zones, or habitats in the vertical sense, is simply an "alternative" to their being distributed latitudinally, or longitudinally, as in nonmontane areas, duplication of faunas in this way can hardly be an example of tropical richness. It would appear, however, that writers on tropical diversity have not given adequate weight to the altitudinal separation of species.

(2) The second most important way in which the cohabiting tyrannids of Costa Rica are separated is on the basis of body size and morphology. A wide spectrum of different structural types occur. Slight differences in bill length may be an important mechanism for minimizing interspecific competition in birds–see, for example, the discussion in Schoener (1965) concerning differences between closely related species pairs. There is, however, no evidence that the fineness of morphological differences between cohabiting species is necessarily any greater in tropical than in non-tropical areas–note, for example, the degree of species diversity in the South African *Cisticola* (Sylviidae), ploceid finches, etc. (McLachlan and Liversidge, 1958).

(3) Whilst it is undeniable that niches are finely divided up in the Costa Rican (and Colombian) tyrannids it is significant that the ways in which they separate out are no different from those found in birds occurring elsewhere. These include occupying slightly different habitats, allopatry, separating vertically into feeding zones within the plant association, and having different feeding habits. The fineness of this division is apparently no greater than that found in some of the more diversifed temperate faunas. The subject warrants further study.

(4) The present survey does not provide any data on whether or not a greater "patchiness" in distribution characterizes tropical, as against nontropical, avifaunas. The subject should receive detailed analysis in the field. Certainly, of course, species are isolated into different habitats, or subhabitats, as elsewhere, and physiographic barriers lead to duplication of species on either side of them. Thus, amongst the tanagers occurring in the lowland tropical zone of Colombia, 18 species are confined to the rain forests to the east of the Andes (Amazonian forests), 8 to those of the Pacific coast, and 5 to the forests and dry scrubs of the Caribbean sector of the country.

The final important question is whether or not the diversity of tyrannids in the tropics of Neotropica is substantially greater than in temperate parts of South America. Colombia, with an area of 455,000 square miles, has 78 genera and 150 species, Costa Rica (2,000 square miles) 40 genera and 69 species, northern Argentina (78,000 square miles), 55 genera and 72 species, and southern Argentina (322,000 square miles) has 23 genera and 66 species (see Table 1).

It is apparent from these figures that the tyrannids, in contrast with predominantly tropical families like the Thraupidae, are relatively rich in temperate regions. Thus, Northern Argentina, with an area only about one fifth that of Colombia, has half as many species as the latter. It is, of course, spurious to arbitrarily compare two areas so dissimilar in size as these two segments of the continent. It might be noted, however, that the figures for Colombia are almost halved if those species confined to the highlands (i.e., areas above 5,000 ft.) be eliminated. When this is done the numbers of species of tyrannids occurring in tropical Colombia is very little different from those occurring in temperate Northern Argentina.

Proper treatment of this subject is obviously beyond the scope of the present analysis. It should be noted, however, that individual families of birds may be remarkably diversified in temperate regions. Southern Africa, for example, is remarkably rich in species of resident sylviid warblers (16 genera and 43 species), the genus *Cisticola* having 18 species. It has 45 genera and 84 species of breeding finch-like birds (Ploceidae, Fringillidae, Estrildidae), and 14 genera and 25 species of larks (Alaudidae)–figures from McLachlan and Liversidge (1958). The subcontinent has of course, a land area of about 1¼ million square miles. Nevertheless, the diversity of savannah-adapted groups of birds and mammals is great. It can be linked to the dominance of this vegetation type on the continent (Keast, 1971), and suggests a long and continuous history of savannah conditions there.

THE NEOTROPICAL TYRANNIDAE RELATIVE TO AFRICAN AND AUSTRALIAN COUNTERPARTS, SOME COMPARISONS OF GENERAL MORPHOLOGICAL TYPES

The South American tyrant flycatchers can be broadly (and arbitrarily) grouped into a series of categories as a basis for a comparison with ecological equivalents in Africa and Australia:

(1) *Generalized small to medium-sized aerial-feeding flycatchers*. There is a large number of genera of these, e.g., *Elaenia, Empidonax, Pyrocephalus, Sayornis,* and *Phylloscartes*. They range in body length from 5 to 7½ inches. Members of this group are the obvious size and ecological counterparts of the generalized muscicapid flycatchers of Africa and Australia–see such African genera as *Muscicapa, Pedilorhynchus, Trochoceros,* and *Sigalis,* and Australian genera such as *Microeca, Myiagra,* and *Arses*. Frequently bill types and sizes can be matched in the members of these different faunas; e.g., in *Arundicola leucocephala* the bill closely approximates to that of the Australian *Monarcha* (*Piezorhynchus*) *alecto*, a species that also feeds on flying insects taken near water.

(2) *The tody flycatchers* (*Todirostrum*, etc.). This diminutive group, which have short tails, and very long parallel-sided bills (Fig. 1), lacks counterparts in the other faunas. The bill shape is duplicated in the much larger Australian *Seisura*. The feeding is partly warbler-like. Much of the insect food is obtained by fluttering up to the foliage to remove objects from the leaves, or catching flying insects in the outer branches (Skutch, 1960; Slud, 1964).

(3) *The tiny olivaceous tyrannids*. This diversified group of small-bodied warbler counterparts have bodies 3 to 5 inches in length. The genera include *Myrornis* (very similar in appearance to the Australian warbler *Acanthiza lineata*), *Tyranniscus, Ornithion, Camptostoma, Colopteryx, Onchostoma* (which has a massive curved bill that it uses for getting insects from the underside of leaves—see Fig. 1). This group also obtains its prey by fluttering about the foliage and outer branches: picking up the items in situ, or when the prey is frightened into taking flight. See Table 3 for a summary of the data of Skutch (1960) and Slud (1964) on the feeding behavior of this group. See also Fig. 2 for a comparison of *Serphophaga subcristata*, with African and Australian warbler counterparts (*Apalis flavida*, Sylviinae; *Gerygone olivacea*, Acanthizinae), which also feed in this fashion and occupy this adaptive zone (writer's observations). Inhabiting the dry shrubby hillsides of Argentina and Chile are genera that have the size, appearance, and hunting behavior of titmice, gnatcatchers, and kinglets; e.g., *Anaeretes* (see Fig. 1) and *Spizitornis* (Wetmore, 1926, p. 332; Johnson, 1967, p. 283). These are presumably also the ecological equivalents of these North American groups.

Whilst it is obvious from the absence of other birds in this size category that the 3 to 5 inch-long tyrannids are the counterparts of the sylviid warblers in Africa, and the acanthizid and sylviid warblers in Australia, it should be noted that Africa has a few muscicapid flycatchers (e.g., *Stenostira*, *Hyliota*) that fall within this size range. Taxonomists are now, however, tending to reclassify such genera as warblers (e.g., Traylor, 1970). Since in so doing they are greatly influenced by their warbler-like feeding habits this course may ultimately turn out not to be valid.

(4) *The small tyrannids of the reedbeds*. Inhabiting the reedbeds of southern South America are a couple of genera of small tyrannids that have largely dispsensed with aerial hawking in favor of obtaining their insect prey from the reedstems and aquatic growth. Wetmore (1926, p. 318) has written of *Pseudocolopteryx sclateri*, "These odd little birds were found among sedges and other low growth at the borders of lagoons, often above shallow water covered with floating vegetation. As they worked about through such growth they were so well concealed that it was difficult to locate them." *Pseudocolopteryx flaviventris* (Fig. 5) is also essentially a rush-dwelling bird (Johnson, 1967). Wetmore (1926) writes of it: "these little birds worked about so quietly among weeds and low bushes over the water that there was a distinct surprise to find them more alert and active in willow thickets. . ." Best adapted of all for reedbed dwelling is the

"many-colored rush-tyrant," *Tachurus rubrigastra* (Fig. 2), which Wetmore describes as clambering about among the rush stalks "where their long legs fitted them for progress," and being very shy. Johnson refers to them "flitting endlessly from rush to rush in search of insects."

It is obvious that these two tyrannid genera occupy the adaptive zones of such specialized marsh-dwelling sylviid warblers (see comparisons of bills in Fig. 2) as *Acrocephalus* (reed warblers) in Africa and Australia and *Megalurus* in Australia. Dr. K. C. Parkes informs me that in its behavior *Tachurus* is also reminiscent of the reedbed-frequenting American parulid warbler, *Geothylpis trichas*.

(5) *The broad-billed tyrannids.* Flattening, broadening and, commonly, shortening of the bill, like the development of pronounced rictal bristles, are widespread structural adaptations in aerial feeding birds. In a few cases the broadening reaches an extreme degree; e.g., in the Australian muscicapid *Machaerirhynchus* (feeding habits described in Rand and Gilliard, 1968), and in the Asian and African family Eurylaimidae (broadbills). The tyrannids have an exceptional diversity of small- to large-bodied, broad-billed species. These range in size from the 3 to 4 inch-long *Platyrinchus mystaceus* to the 5-inch-long *Tolomyias flavoolivaceus* and *Rhynchocyclus olivaceus* (the latter with a bill similar to that of the African broadbill *Smithornis capensis*), 6 to 7 inch-long *Ramphotrigon ruficauda*, and 9½ inch-long *Megarhynchus pitangua.* The smaller species feed by catching insects on sallying flights or by flitting at the leaves (Slud, 1964; Skutch, 1960). Heavy-bodied insects like Hemiptera and Coleoptera are, however, prominent in the diet (Haverschmidt, 1968). *Megarhynchus* feeds extensively on cicadas (Skutch, 1960) but also takes a range of other invertebrates and small vertebrates. It commonly scoops small minnows from pools, a habit also shared by *Myiozetetes granadensis* and *Pitangus sulphuratus* (Slud, 1964). Possibly the somewhat kingfisher-like bill of *Megarhynchus* partially preadapts it for this role.

(6) *The larger aerial-feeding tyrannids.* Many large-bodied genera (*Pitangus, Tyrannus, Myiarchus, Myiodynastes,* etc.) and the long-tailed tyrannids are aerial feeders but some take their food from the foliage or ground. The great kiskadee, *Pitangus sulphuratus*, which does some of its feeding above the treetops, takes lizards, snakes, and small fish, in addition to a wide range of invertebrates.

This group of flycatchers is obviously convergent with, and presumably are the ecological counterparts of, the aerial-feeding African drongos: *Tyrannus melanocholicus* has a bill almost identical with that of the African *Dicrurus adsimilis*. The long-tailed *Gubernetes yetapa* by contrast, is remarkably similar, in bill shape, body form, and in its long tail, to the African roller, *Coracias spatulata* (Coraciidae). It does not fly high in the air like the roller, however, feeding near the ground; shrubs and fence-posts are used as lookout points (L. Short, pers. commun.).

(7) *The giant shrike tyrants.* The genus *Agriornis* of the higher Andes includes a group of species that are structurally convergent with, and occupy the same

adpative zones as, the Australian butcher-birds (*Cracticus*) and larger shrike-thrushes (*Colluricincla*), and certain African shrikes (e.g., *Lanius collaris*). The largest of them, *Agriornis livida fortis*, has a body 14 inches in length. It feeds on mice, lizards, frogs, nestlings, and insects (Johnson, 1967), dropping down on to them from a lookout perch. The bill is strongly hooked in *Agriornis* and quite similar to that of *Cracticus* and *Lanius* (Fig. 3).

(8) *The terrestrial and semi-terrestrial thrush-like and chat-like tyrannids.* In southern South America are a number of genera that do most of their feeding on the ground and resemble generalized thrushes in body form and general ecology. *Lichenops* is an example. More strongly terrestrial, convergent with, and filling the role of the terrestrial chats and wagtails (motocillids) of Africa are the ground-tyrants (*Muscisaxicola*). All are conspicuously long-legged. Johnson (1967, p. 240) describes *Muscisaxicola* as "a very active bird, continually running hither and thither over the ground and taking short flights to catch the flies and other insects which form its stable food." Perhaps the most aberrant of this group is *Muscigralla*, which walks with an unusual up-down jerking of the tail (Short and Morony, 1969).

(9) *The berry-feeding tyrannids.* The Neotropical Tyrannidae differ from the Old World Muscicapidae in that berry-feeding is a pronounced habit. In this regard they partly enter the adaptive zones of bulbuls in Africa and meliphagids and *Dicaeum* (Dicaeidae) in Australia. Many of the smaller, and larger, tyrannids are part berry feeders—see food data in Skutch (1960), Slud (1964), and Haverschmidt (1968). Even *Muscisaxicola* is partly a berry feeder (Johnson, 1967). A couple of the smaller genera are almost entirely mistletoe-berry feeders and thus are ecologically equivalent to small tanagers, the dicaeid *Dicaeum hirundinaceum* and the meliphagid *Grantiella picta* in Australia, and bulbuls and small timaliids in Africa. A typical one of these is *Phaeomyias murina* (see Haverschmidt, 1968)—see Fig. 4 for a comparison of *Phaeomyias* with ecological counterparts in Africa and Australia, and note how similar it is in bill form to the Australian *Dicaeum*.

REASONS UNDERLYING THE EXCEPTIONAL MORPHOLOGICAL DIVERSITY OF THE TYRANNIDS: THE LATE PENETRATION OF SOUTH AMERICA BY POTENTIAL COMPETITORS FROM THE NORTH

Table 1 shows the relative numerical importance of the Tyrannidae, and various other families of passerines from north to south in Neotropica. The families fall into three main categories: (1) endemic groups that are equally well developed from north to south (Tyrannidae, Furnariidae); (2) endemic groups that are essentially tropical in their occurrence (Thraupidae); and (3) northern groups that penetrate the continent to various degrees from the North (Parulidae, Sylviinae, Motacillidae, Turdinae).

The tyrannids and the furnariids make up a significant proportion of the avifauna throughout the continent. They are, relatively speaking, even better developed in the far south than the north. The tyrannids account for 31% of the genera and species in southern Argentina, and the furnariids 21.6% and 24%, respectively. They are obviously well entrenched.

How much penetration of Neotropica, both in terms of numbers of species and distance, have the Nearctic and Palearctic insectivorous birds been able to achieve against the endmic tyrannids and furnariids? The Parulidae (North American Warblers) have 7 breeding genera and 12 species in Costa Rica (making up 3.7% of the total genera, and 3.6% of the species), 4 genera and 11 species (1.1% and 1.9%) in central Peru, and 3 genera and 5 species in northern Argentina (1.6% and 1.8%)–see Table 4. None extend to southern Argentina. This overall representation is very small compared to the remarkable Nearctic diversity of this family in North America. The Nearctic Vireonidae, also warbler-like, follow a similar pattern but have fewer genera and species, e.g., 2 and 7 in Costa Rica, 3 and 6 in central Peru, 2 and 2 in northern Argentina, and nil in southern Argentina (Table 1). The Sylviinae, a Palearctic subfamily of warblers with a few North American representatives, is quite insignificant in South America. There are 3 genera and 4 species in Costa Rica, 3 and 3 in central Peru, 1 and 2 in northern Argentina, and none in southern Argentina (Table 1).

The typical thrushes (Turdinae), well established in North America, achieve a modest diversity in Neotropica. There are 3 genera and 11 species in Costa Rica, and 5 and 17 in central Peru. These are largely forest inhabitants, but some are high montane species. One genus and three species occur in southern Argentina. All the Neotropical thrushes are relatively generalized types and they have not developed long-legged chat-like forms. Only one genus of the terrestrial, insect-eating motacillids (*Anthus*) occurs in South America, and the group inhabits high grasslands. Three species are widespread. The Paridae (tits) and Laniidae (shrikes), Palearctic families and poorly represented in North America, fail to reach Neotropica. Other African-Oriental-Australian groups like the large-bodied, aerial feeding Dicruridae (drongos) fail to reach the Americas.

There would appear to be three explanations for the poor penetration of South America by these northern families: competition from the tyrannids and furnariids, being poorly adapted to the tropics through which they would have to pass in order to reach the temperate parts of the continent, and, in the case of some, relatively late arrival in the Hemisphere. Surely, however, these latter two explanations are of no significance in the case of the parulids, two genera of which have radiated in South America and overwinter there in large numbers (28 species).[1] Presumably this family has been in direct contact with South America

[1]Mayr (1964) mentions that there is a possibility that the two genera in question, *Myioborus* and *Basileuterus*, are not actually parulids.

since at least the formation of the Panamanian land bridge in the Late Pliocene. The conclusion would appear inescapable, hence, that it is the well-adapted warbler-like tyrannids which have limited their establishment and diversification.

Competition from the endemic tyrannids, in this case the semiterrestrial ones (e.g., *Muscisaxicola, Muscigralla*) may well also explain the failure of the turdids to evolve chat-like types in South America.

THE ADAPTIVE ZONES OF THE NEOTROPICAL TYRANNIDAE, A DETERMINATION OF ECOLOGICALLY EQUIVALENT GROUPS IN THE AVIFAUNAS OF AFRICA AND BY A STUDY OF STRUCTURE

More precise comparisons between the species comprising the South American Tyrannidae and their apparent, or presumed, ecological counterparts in the avifaunas of Africa and Australia can be obtained from measurements.

The data are plotted for pairs of morphological features as follows: (1) bill length and wing length; (2) bill width and bill length; (3) bill depth and bill width; (4) ratio of bill width to length, relative to the wing length; and (5) tarsus length to wing length. (Treatment of these data by multifactorial analysis would permit fewer figures but would lack the clarity of this presentation.)

So that visual comparisons can readily be made, where space permits, the limits of the more important families are marked by a contour line on each of the figures. The "demarkation line" of the South American Tyrannidae is superimposed over the African and Australian segments in order to more clearly mark out what families and subfamilies it duplicates in terms of the pair of structures being considered.

(1) *Bill Length/Wing Length* (Fig. 5). It will be seen that the South American Tyrannidae extend from small-bodied species with a wing length of about 40 mm to large-bodied ones with a wing length of about 130 mm, and from short-billed (length about 7 mm) to long-billed (length, 31 mm) species. By contrast, the typical flycatchers of Africa and Australia, the Muscicapinae, have a much more restricted size range (see cross hatched areas in Fig. 5). The wing length range is 50 to 100 mm and the bill length 8 to 19 mm in the former, and 60 to 105 mm, and 6.5 to 22 mm, respectively, in the latter. (See Vaurie, 1953, for an account of variation in the tribe Muscicapini of the Old World flycatchers.) This accords with the rather more restricted range of adaptive zones and ecological niches occupied by the African and Australian flycatchers–see descriptions of ecology in such standard works as McLachlan and Liversidge (1958), Chapin (1953-1954), and Bannerman (1936).

At the same time the Australian muscicapids are slightly more diversified structurally than the African ones (Fig. 5). This may be correlated with three distinct subgroups occurring there (Muscicapini, Monarchini, Rhipidurini). In Africa only the first two occur, although Traylor (1970) has recently suggested that the platysteirines really constitute a distinct group. The size range in both the African and Australian muscicapids would be still smaller, it might be noted, if it were not for the occurrence of single massive-billed forms in each, e.g., *Megabyas* in Africa (see no. 13, African segment of Fig. 5), and *Peltops* (no. 8 in the Australian segment), in Australia. The following other families and subfamilies fall within the size range of the South American Tyrannidae, so far as these two morphological characteristics are concerned: *Africa*–Sylviinae, Turdinae, Timaliinae, Paridae, Motacillidae, Eurylaimidae, Dicruridae, and Laniidae; *Australia*–Acanthizinae, Sylviinae, Pachycephalinae, Falcunculinae, Epthianurinae, Dicaeidae (including *Pardalotus*), Dicruridae, Motacillidae (only one species on the continent) Of the two African Meropidae one falls within the size range of the tyrannids, the other well outside of it. In Australia one of the Artamidae falls within the range, the other is outside of it; whilst two of the three genera of Campephagidae fall outside of it. The single Australian bee-eater, drongo, cracticid (*Cracticus torquatus*), and the ground-feeding grallinid, *Grallina cyanoleuca*, lie well outside of the tyrannid size range.

Many of the African and Australian genera, of course, occupy adaptive (or feeding) zones that the tyrannids do not utilize; e.g., the bush shrikes and aerial-feeding bee-eaters. The objective here is simply to detail the total diversity of the families in terms of these two morphological variables.

(2) *Bill Width/Bill Length Ratio* (Fig. 6). As noted, aerial-feeding birds tend towards broad, flattened bills. Bill width in the South American tyrannids ranges from 3.5 mm to about 14 mm, compared to 7.0 to 18.5 mm in the African muscicapids, and 7.0 to 22.0 mm in the Australian ones. In Australia there is a very broad-billed flycatcher, *Machaerirhynchus* (no. 8 in the Australian segment of Fig. 6), but in Africa this adaptive zone is occupied by a different family, the broadbills (Eurylaimidae)–see E1 and E2 in the African segment of the diagram. In addition, the following families and subfamilies fall into the range of the tyrannids with respect to these two characters: *Africa*–Turdinae, Sylviinae, Paridae, Motacillidae, Dicruridae, Laniidae; *Australia*–Acanthizinae, Sylviinae, Epthianurinae, Pachycephalinae, Falcunculinae, Motacillidae, Dicaeidae, Artamidae, Campephagidae, Dicruridae, and Grallinidae (*Grallina cyanoleuca*). In both continents the Meropidae (whose long, narrow bills are possibly a special adaptation for consuming stinging hymenopterous insects [Fry, 1969, 1970]) lie well outside of it, as does the larger Australian *Cracticus torquatus* (Cracticidae).

(3) *Bill Depth/Bill Width Ratio* (Fig. 7). The bills of the tyrannids range from a depth of 2.0 to 11.5 mm, those of the African muscicapids from 3.0 to 5.5 mm and Australian muscicapids from 3.5 to about 10.0 mm (e.g., *Peltops*, no. 8). Again the great range of the tyrannids is obvious. In bill depth (massiveness), however, the tyrannids cannot match the shrikes of Africa, and falcunculids, pachycephalids, and campephagids, of Australia.

(4) *Bill Width/Length Ratio relative to Wing Length* (Fig. 8). All the groups here fall within the size range of the tyrannids, except for the bee-eaters in Africa and Australia, and the Australian campephagids *Coracina novaehollandiae* (C1) and *Edoliisoma tenuirostre* (C2), and *Dicrurus, Grallina,* and *Cracticus*.

(5) *Tarsus Length/Wing Length Ratio* (Fig. 9). Increased terrestriality is associated with an increase in the length of the tarsus. Accordingly, it is seen that some of the tyrannids are distinctly long-legged. In the course of this development certain forms have become convergent with the African chats and pipits. It will be seen that the tyrannid *Muscisaxicola rufivertex* (no. 3 in Fig. 9) is a good counterpart to the African chat *Oenanthe monticola* (T2 in the African segment of the figure) as is *Hymenops perspicallata* (11) with the African Chat, *Monticola brevipes* (T6), and motacillid *Anthus leucophrys* (M3). The chat-tyrant *Ochthoeca ornanthoides* (6) matches up well with the African chat *Saxicola torquata* (T4). The contour line for these characters in the tyrannids, it will be seen, includes four of the six turdid chats, and two of the three motacillids. *Oenanthe pileata* (T3) and *Myrmecocichla formicivora* (T5) of the chats, and the motacillid *Macronyx croceus* (M2), fall outside of it. These are longer legged, and hence could be said to be more specialized for terrestrial life, than any of the tyrannids.

In contrast to the above it is interesting to note that the so-called Australian "chats," *Epthianura* (W1 and W2 on the Australian segment of the map), an endemic terrestrial group (Epthianurinae) given the name because of their chat-like appearance, habits, and ecology, do not show any marked lengthening of the leg.

Compared to all these long-legged birds the truly aerial bee-eaters, which as a group, are distinctly short-legged, stand apart. Amongst the Australian birds *Dicrurus, Cracticus, Grallina*, the two artamids, and two of the three campephagids, whilst falling within the range of the tyrannids in tarsal length, are larger birds.

In summary, the above review demonstrates the following: (1) The South American Tyrannidae are much more diversified structurally than the Muscicapinae and this is true even if the Sylviinae, Turdinae, Acanthizinae be grouped with them to constitute the larger family Muscicapidae.

(2) In ecology, size and structure, and range of adaptive zones utilized, it is obvious that the tyrannids are equivalent to a whole range of families and subfamilies on these other two continents (Table 6). In Africa these include the Muscicapinae, Sylviinae, Turdinae, Eurylaimidae, Paridae, Dicruridae and, to a partial extent, the Campephagidae, Motacillidae, Laniidae, and Oriolidae. The analogy might, of course, be still further extended to include an individual genus or species in other families, e.g., the bulbuls (Pycnonotidae).

In Australia (Fig. 6) these counterparts are: Muscicapinae, Sylviinae, Acanthizinae, Epthianurinae, Falcunculinae, and Pachycephalinae (partly), Campephagidae (partly), Dicruridae (one species in Australia, two in New Guinea).

Table 6. Ecological Counterparts, and Partial Counterparts, of the Neotropical Tyrannidae in the Avifaunas of Africa and Australia. Breeding species only[a]

Family and Subfamily	Africa Genera	Africa Species	Australia Genera	Australia Species	Probable Origin (Region)	Apparent Degree to Which Tyrannids Penetrate Adaptive Zone
Meropidae	1	17	1	2	Oriental (see Fry, 1969)	Marginally
Eurylaimidae	2	4	absent		Oriental	Totally
Alaudidae	9	50	1	1	Palearctic Eremian	Very Marginally
Pycnonitidae	12	45	absent		African or Oriental	Very Marginally
Campephagidae	2	6	4	17	Oriental	To considerable degree
Muscicapinae	19	60	13	59	Old World	Totally in Africa most in Australia
Pachycephalinae	absent		6	32	Australian	To considerable degree
Falcunculinae	absent		2	4	Australian	To moderate degree
Timaliinae	6	29	9	23	Sino-Hima-layan	Very Marginally
Sylviinae	29	135	5	8	Old World	Totally
Malurinae	absent		7	32	Australian	To minor extent
Acanthizinae	absent		17	54	Australian	To considerable degree
Epithianurinae	absent		1	5	Australian	To considerable degree
Turdinae	20	91	4	5	Palearctic or African	Partly
Motacillidae	4	23	1	1	Palearctic	To modest extent
Laniidae	10	55	1	1	African	Partly
Artamidae	absent		1	7	Australian	Marginally

Paridae and Remizidae	2	16	absent		Eurasian	To large extent
Dicaeidae	absent		6	19	Oriental	To moderate extent
Meliphagidae	absent		30	117	Australian	To modest extent
Oriolidae	2	5	1	7	Old World	To moderate degree
Dicruridae	1	3	2	2	Oriental	Totally
Cracticidae	absent		3	11	Australian	Very Marginally
Grallinidae	absent		3	4	Australian	Very Marginally

[a]Classification largely as per Mayr and Amadon, 1951. Figures for numbers of genera and species obtained as follows: *Africa*–White (1960, 1961, 1962a, 1962b, 1963, 1965). *Australia*–Mayr (1941), Rand and Gilliard (1968); plus writer's calculations for Australia proper.

Dicaeidae (partly), Oriolidae (partly), Motacillidae (only a single species in Australia) and, to some extent, the Artamidae, *Grallina* in the Grallinidae, *Cracticus* in the Cracticidae, and *Merops* (Meropidae). Certain of the insectivorous Meliphagidae, not included in Figs. 5 to 9, apparently also find counterparts in the tyrannids.

THE AVIFAUNAL HISTORIES OF SOUTH AMERICA, AFRICA, AND AUSTRALIA RELATIVE TO THE TYRANNIDS AND THEIR ECOLOGICAL COUNTERPARTS

The striking range of adaptive zones occupied by the Neotropical Tyrannidae can be attributed to their being a versatile and aggressive group, and, in larger measure, to the protracted geographic isolation of South America that minimized the entry of competitors from outside. Endemic groups thus had the opportunity for a prolonged period of evolution and diversification without interference. They were thus not only able to become established in all the major adaptive zones on the continent but thoroughly consolidate their hold on these. Competition came mainly from within the continent but since Neotropica is a large land mass there was doubtless competition between groups and subgroups that evolved in regions of differentiation within the continent. The prolonged isolation of South America explains the marked endemism at the family level there. It amounts to some 30 families (30% of the total). This compares to only 7 families in the case of Africa, two of which were known from France in the Miocene, and 13 families for Australia (Keast, 1971). South America was isolated throughout the 60,000,000 years of the Tertiary and became joined to North America only with the formation of the Panamanian land bridge at the end of the Pliocene (Woodring, 1954; Harrington, 1962). The Tertiary mammalian fossil record, which is particularly rich in endemics, provides ample confirmatory record of this prolonged isolation (Patterson and

Pascual, 1968). Birds undoubtedly had a similar history to the mammals but the fossil record is skimpy. The tyrannids and most of the other endemic families undoubtedly originated and achieved a considerable radiation during the later Tertiary. At the same time, it must be noted, the *tropical* Neotropical avifauna as we know it today is a compound of two former tropical avifaunas, one in southern North America, which was then tropical, and the other in northern South America (Mayr, 1946, 1964). As this author notes, some of the seeming South American endemics undoubtedly arose in central America, which was formerly a promontory of North America.

Irrespective of this the faunistic history of South America is strikingly different from that of Africa, as the present writer has recently traced out (Keast, 1971). Nearly all African avian families, as has been stressed by Moreau (1966), are shared with one or more of the other zoogeographic regions. It is difficult today to decide the exact origins of most of these. However, if the "rule" be applied that a group tends to retain unspecialized types, and/or be most numerous and diversified in its area of origin, we can come up with certain deductions (Table 6). On this basis the bee-eaters are probably Oriental (see discussion in Fry, 1969), the eurylaimids Oriental, the larks "Palearctic African," the motacillids Palearctic (with the longclaws and endemic African radiation), the timaliids Sino-Himalayan, the dicrurids Oriental, and so on (see Keast, 1971).

Again there is no avian fossil record to go on but the mammalian record indicates that faunistic interchange occurred between African and Southern Eurasia at the end of the Oligocene, end of the Miocene, and end of the Pliocene, these presumably being times when the intervening Tethys Sea became narrowed (Cooke, 1968). Birds, with their greater mobility are, however, much more able than mammals to cross water gaps. Hence, the avian history of Africa must have been one of periodic, and frequent, interchanges with the northern continents.

The reasons for the very different subdivision of the avian insectivorous adaptive zones in Africa and South America now become apparent. In contrast to the continued evolution in isolation enjoyed by the tyrannids, and most of the other South American groups, Africa has received continued waves of colonists so that the avifauna has been built up as a series of strata superimposed one upon the other. As each new family became established there must have been a period of intense competition and a resorting and redivision of adaptive zones and niches, until stability was again reached. Since the new colonists had themselves evolved in a large land mass they were, presumably, highly efficient and competitive. The result would have been that each of the individual families of birds tended to become restricted, or confined, to the narrow zone where it was optimally efficient. The result was a series of families occurring together, each relatively well-delineated morphologically, and fairly well separated ecologically. Muscicapid flycatchers were held in the main to the aerial flycatching role, and a body size intermediate between the small-bodied sylviid warblers and parid and remezid tits, which were specialized for catching insects from the leaves and undergrowth, on the one hand, and the larger branch- and ground-feeding shrikes, and aerial-feeding drongos, on the other.

The long-standing isolation of South America accords well, of course, with the great differences in the diversity and abundance of the sylviids, turdids, and motaciilids here, as compared to Africa. Each of these groups, strikingly diversified in Africa, is represented here only by one or a few generalized genera, and they make only a marginal impact on the rest of the avifauna. They are obviously all late arrivals in South America.

The Australian avifauna has had a history of development in isolation similar to the South American one. Extensive areas of sea, and the Indonesian Islands have provided a "faunal filter" along which only a few Eurasian groups have been able to pass. The distances between Asia and Australia have been so much greater than between the two Americas during the Tertiary. It is worth noting, however, that colonization has been along the same lines of latitude, and hence climatic belts, rather than from north to south. The new colonizers, hence, have not had to pass through a series of "hostile" climatic regions. As both Mayr (1944) and Simpson (1961) have suggested, the later arriving Australian birds, and the rodents amongst the mammals, both provide evidence of having colonized Australia in several "waves." This can be seen from their showing several "levels of difference" from their Oriental counterparts today. Australia has no fewer than a dozen families of Eurasian birds represented by only one or two species, it might be noted (Keast, 1971).

THE SIGNIFICANCE OF THE ASSOCIATED FAUNA IN INFLUENCING, OR LIMITING, EVOLUTIONARY DIVERSIFICATION IN A GROUP

The influence of the associated fauna on the evolution of component groups has received little attention from evolutionists. What discussion there is has mostly been at the species level. Thus Lack (1944, 1946, 1971) has explained certain distributional facts, such as the altitudinal separation of the two chaffinch species on the Canary Islands, in terms of interspecific competition. Some other biologists, and particularly Andrewartha and Birch (1954), have voiced reservations about accepting interspecific competition as a possible determinant of species ecology, pointing out that some of the arguments for it involve circular reasoning, and a failure to distinguish cause and effect. The basic difficulty, of course, is that here, as in most evolutionary processes, we have only the end results with which to work; the time span involved is too great for us to actually see them in action. This has been pointed out by Mayr (1964).

Most later biologists and paleonotologists draw on competition and species interactions as the simplest explanation for a variety of observed phenomena, including the extinction of evolutionary lines. Island faunas, in particular, provide ample proofs of the effectiveness of interspecific competition as an evolutionary force.

The present study demonstrates very clearly that ecological interactions must be sought not only at the species level but in terms of the whole associated fauna. It provides, moreover, ample demonstration of how the associated fauna may influence the direction taken by an evolutionary line. The evidence may be summarized as follows:

(1) The avian insectivorous adaptive zones are very differently divided up on the three continents. This, in part, can be directly linked to the basically different taxonomic compositions of the three faunas.

(2) There are both major geographic and historic components to these differences. Because of their long-standing isolation, interaction with non-endemic families has been minimal in the case of South America and Australia, maximal in the case of Africa. The African fauna, as noted, has had a long history of successive "superimpositions" of new families from outside on pre-existing ones. There must, accordingly, have been a whole series of major "competitive interactions" and resorting, or redivision, of ecological roles at different times in order to produce the contempory picture. This end result can be seen not only in the fact that some eight or more African families and subfamilies occupy the passerine insect-eating adaptive zones on that continent, but in the fact that these groups tend to be well delineated morphologically and ecologically. There is no proof, of course, that the restriction of these groups to the postulated "zones of optimum efficiency," in the middle of their potential ecological spectrum, occurred only after arrival in Africa. Some of it must have occurred elsewhere as, for example, when an ecologically aggressive African family colonized the Oriental region. Presumably, also, in some cases groups evolved on the same continent right from the beginning, so that the division of the available adaptive zones occurred very early. But the overall concept that it is past "interactions" which have restricted African families to the particular range of adaptive zones that they occupy today, seems valid. The African muscicapids, which are much less morphologically and ecologically variable than the Australian ones, provide a good example.

(3) To say that the Neotropical Tyrannidae are ecologically equivalent to, say, eight African families and subfamilies is true, of course, in broad terms only. In the course of the competitive sorting out of ecological roles with other groups, in this case with other endemic South American familes like the Furnariidae and even the Thraupidae, they have progressed or penetrated further in some directions than others. The same thing has occurred during the evolution of each of the other continental bird families. One would not, therefore, except to match individual, or combinations of, families as "exact equivalents." Whilst it is justifiable to say that the Neotropical tyrannids "embrace" the range of adaptive zones of the African muscicapids and eurylaimids, for example, this is not completely true of the Australian muscicapids. In Australia there is a series of

ground-feeding robin-like, or thrush-like "robin flycatchers," which are the equivalents (or partial equivalents) of some of the forest-dwelling thrushes of Africa, and thrushes and other groups, in South America. The tyrannids have not produced such forms. Again, whilst it is true to say that the warbler-like tyrannids are the ecological and morphological equivalents of sylviid warblers in Africa this is really true only of the arboreal and thicket-dwelling assemblages. In undergrowth situations in South America furnariids, probably with some help, small formicariids occupy a range of "warbler-like" niches. Again, whilst the tyrannids occupy most of the warbler niches in reedbeds and swamp-edge undergrowth, a reed-bed dwelling furnariid (*Phleocryptes*) has still evolved.

(4) Thus, just as each continent has entrenched endemic groups filling particular or specialized roles, which must be competed with by newcomers, so widespread families do not necessarily fill exactly the same range of ecological niches on each continent. The differences between the African and Australian muscicapids find parallels in the campephagids, which are also much more diversified in Australia. In the case of groups like the dicrurids that are probably late colonizers of both Africa and Australia, the morphological and ecological types present, of course, reflect the characteristics of the particular genera that colonized each continent. As each family diversifies, however, it would seem inevitable that it should come into competitive contact with other elements in the fauna.

(5) To sum up, it is fairly obvious that within each fauna evolution takes place so as to utilize all available ecological opportunities and potential ways of life. Supplied with different evolutionary raw material in the first place, with very different histories of colonization from outside and, doubtless, different potential for the development of new forms within, it is to be expected that the ecological opportunities will be differently divided up on each continent. Something must determine how far, and in what direction, any particular group will evolve. The most potential individual force is undoubtedly the associated fauna inhabiting that continent. This is not losing sight of the fact, of course, that different continents provide different ranges of ecological opportunities. Note the tremendous range of savannah-adapted animal groups in Africa where that vegetation type is particularly prominent.

(6) Another important facet of faunistic evolution is the production of convergent body types in different continents. The most famous case of this is, of course, the Australian marsupials relative to the placental mammals of other continents. The production of similar body forms by unrelated groups infers that there is an optimum body form for a particular role. The more specialized the role the more necessary are specialized body structures. Arboreal mammals can remain rather generalized in their body structures; burrowing ones like moles, and running carnivores like canids and thylacinids, cannot.

Because, in small passerines, we are dealing with generalized types (and, at any event, are considering only parts of the body) it is not to be expected that extreme convergent types will occur. This, of course, is so. But there are many interesting morphological parallels in the three continental avifaunas. Note the independent development of long legs in African turdid chats, and South American chat-tyrants, for example, and the independent development of an excessively broadened, flattened bill in such distinct groups as the Neotropical tyrannids (e.g., *Platyrhynchus*), Australian muscicapid flycatchers (*Machaerirhynchus*), and African Eurylaimidae (*Smithornis*). Again, there is the shrike-like bill of the tyrannid *Agriornis* relative to that of the Australian *Cracticus*, and various African laniids (Fig. 3). The "berry-eating" bills of the species shown in Fig. 4 could be said to be another example of convergent counterparts. Likewise, it is possible to match individual bills in some of the insect-eating species on each of the three continents.

Notwithstanding the above, however, it is apparent when Figs. 5 to 9 are looked at as a group, that exact counterparts or "matches" at the species level in three faunas are relatively few indeed. Varying bill lengths, widths, depths, leg lengths, and body sizes are seldom drawn together in exactly the same combination. Rather, each family on each continent embraces a general range of structural types in which these variables are combined in slightly different ways. This is presumably because there are no vigorous selective forces operating in particular directions to produce extreme body forms. Rather, the emphasis is simply in exploiting a generalized range of ecological opportunities.

To sum up further, the present comparative analysis of continental avifaunas provides a good demonstration of how associated faunas influence and limit the direction of evolution and radiation in individual groups. The adaptive zones and ecological niches occupied by each continental family is obviously the result of long periods of evolution in association with a range of other faunal elements. An ecological vacuum does not last for long. If an ecological opportunity evolves, something moves in to fill it. The completeness of the "ecological coverage," the variety of groups involved in the subdivision of the opportunities, and the different kinds and patterns of exploitation involved, are proof of this.

The present analysis represents a further demonstration that faunas tend to evolve towards an overall state of integration and balance, in which all the major ways of life and ecological opportunities are filled. As such it represents an interesting parallel to the writer's generalized discussion of the contemporary mammal faunas of Africa, South America, and Australia (Keast, 1969), and his more detailed analysis of the resorting of ecological roles in the impoverished insular avifauna of Tasmania (Keast 1968, 1970).

CONCLUSION

The Neotropical Tyrannidae, with 103 genera and 302 species, represents a striking example of a dominant continental family. Its numerical richness stems from its having an unusually wide range of different morphological types which in additional to "typical" flycatchers, include warbler-, tit-, thrush-, and chat-like types. A review of patterns of space and habitat utilization within a regional avifauna (that of Costa Rica) shows the species to be separated out ecologically on the basis of altitudinal life zone occupied, on habitat, feeding habits, vertical levels at which feeding is carried out, and on food. These patterns of separation are the same as those separating closely related species in temperate avifaunas. They amount to a relatively fine pattern of niche subdivision.

When the Neotropical tyrannids are compared with ecological counterparts, and potential counterparts, in the avifaunas of Africa and Australia, it is found that they occupy the adaptive zones, and ecological niches of some 8 to 10 families, and subfamilies, on those continents. This is demonstrated on the basis of body structure as well as general ecology.

This "success" can be explained by the past histories of isolation of the continents. South America was spatially isolated throughout the Tertiary, and came into direct contact with the rest of the world only with the formation of the Panamanian land bridge at the end of the Pliocene. Hence, its fauna has evolved in isolation, and with a minimum of interference from outside groups. Endemism is particularly high in South America, 30 families being endemic. During this prolonged isolation the tyrannids, as well as other endemic groups like the furnariids, were able to penetrate and establish themselves in a wide diversity of adaptive zones. In contrast to this, the African avifauna has been subject to repeated colonizations from, and interchanges with, southern Eurasia. That is why Africa has only 7 endemic families of birds. The African avifauna could be said to have been built up by the progressive superimposition of new families from outside on the older members. Each new arrival doubtless initiated a series of vigorous new competitive interactions, and a resorting of ecological roles before a state of balance within the avifauna was once again achieved. The result is a large number of cohabiting insectivorous families, each more or less clearly differentiated on size, body morphology, and ecology.

Analysis further shows that the various insectivorous adaptive zones and ecological niches may be quite differently divided up on each continent. Each, of course, has its range of entrenched endemic families with their ecological specializations. Where groups (e.g., Muscicapinae, Campephagidae) occur on two or more continents they are not equally developed on each. They by no means precisely occupy the same range of niches.

It is apparent that all available ecological opportunities, and ways of life, are taken up on each continent. The faunas are integrated and reasonably balanced units. This could be achieved of course, only by repeated interactions within the fauna. Accordingly, there can be no doubt that the associated fauna is a very important factor, and in many cases is obviously the dominant factor, in determining the direction that evolution will take in individual groups. The present review provides many demonstrations of this.

Acknowledgments

This research was carried out whilst the writer was holding a grant from the National Research Council of Canada and the writer would like to thank that body for its support. All the measurements were made on the collections of the American Museum of Natural History, New York, and I should like to thank the Chairman of that Museum's Department of Birds, Dr. Dean Amadon, for the many facilities afforded me.

REFERENCES

Amadon, D. 1943. Bird weights as an aid in taxonomy. Wilson Bull., 55:164-177.

Andrewartha, H. G., and L. C. Birch. 1954. The Distribution and Abundance of Animals. Chicago, University of Chicago Press.

Bannerman, D. A. 1936. The Birds of Tropical West Africa. IV. London, Publ. Secretary of State for the Colonies.

Chapin, J. P. 1953-1954. The birds of the Belgian Congo. Bull. Amer. Mus. Nat. Hist., 75A:1-820, and 75B:1-846.

Chapman, F. M. 1917. Distribution of bird life in Colombia. Bull. Amer. Mus. Nat. Hist., 36:1-729.

——1926. The distribution of bird life in Ecuador. Part I. Bull. Amer. Mus. Nat. Hist., 55:133-144.

Colquhoun, M. K., and A. Morley. 1943. Vertical zonation in woodland bird communities. J. Anim. Ecol., 12:75-81.

Cooke, H. B. S. 1968. The fossil mammal fauna of Africa. Quart. Rev. Biol., 43:234-264.

de Schauensee, R. M. 1964. The Birds of Colombia and Adjacent Areas of South and Central America. Narbith, Pennsylvania, Livingston Publishing Co.

——1966. The Species of Birds of South America and their Distribution. Philadelphia, Academy of Natural Sciences of Philadelphia.

Dilger, W. C. 1956. Adaptive modification and ecological isolation mechanisms in the thrush genera *Catharus* and *Hylocichla*. Wilson Bull. 68:171-199.

Fry, C. H. 1969. Evolution and systematics of bee-eaters. Ibis, 111:525-557.

——1970. Convergence between jacamars and bee-eaters. Ibis, 112:257-258.

Harrington, H. J. 1962. Paleogeographic development of South America. Bull. Amer. Assoc. Petrol. Geol., 46:1773-1814.

Hartley, P. H. T. 1953. An ecological study of the feeding habits of the English titmice. J. Anim. Ecol., 22:261-288.

Haverschmidt, F. 1968. Birds of Surinam. London, Oliver, & Boyd, 445 pp.

Johnson, A. W. 1967. The Birds of Chile and Adjacent Regions of Argentina, Bolivia and Peru, Vol. II. Buenos Aires, Platt Establecimientos Graficos, 446 pp.

Keast, A. 1968. Competitive interactions and the evolution of ecological niches as illustrated by the Australian honeyeater genus *Melithreptus* (Meliphagidae). Evolution, 22:762-784.

——1969. Comparisons of the contemporary mammalian faunas of the southern continents. Quart. Rev. Biol., 44:121-167.

——1970. Adaptive evolution and shifts in niche occupation in island birds. Biotropica, 2:61-75.

——1971. Faunal elements and evolutionary patterns; some comparisons between the continental avifaunas of Africa, South America, and Australia. Proc. 15 Int. Ornith. Cong. (In Press.)

Lack, D. 1944. Ecological aspects of species-formation in passerine birds. Ibis, 1944:260-286.

——1946. Competition for food by birds of prey. J. Anim. Ecol., 15:123-129.

——1971. Ecological Isolation in Birds. Oxford, Blackwells.

MacArthur, R. H. 1958. Population ecology of some warblers of northeastern coniferous forests. Ecology, 39:599-619.

——1969. Patterns of communities in the tropics. Biol. J. Linn. Soc., 1:19-30.

McLachlan, G. R., and R. Liversidge. 1958. Roberts' Birds of South Africa. Capetown, Cape Times Ltd., 504 pp.

Mayr, E. 1941. List of New Guinea Birds. New York, American Museum of Natural History.

——1943. Animal Species and Evolution. Cambridge, Mass., Harvard University Press.

——1944. The Birds of Timor and Sumba. Bull. Amer. Mus. Nat. Hist., 83(2):123-194.

——1946. History of the North American bird fauna. Wilson Bull., 58:3-41.

——1964. Inferences concerning the Tertiary American bird faunas. Proc. Nat. Acad. Sci. U. S. A., 51:280-288.

——1969. Bird speciation in the tropics. Biol. J. Linn. Soc., 1:1-17.

——and D. Amadon. 1951. A classification of recent birds. Amer. Mus. Novitates, No. 1496:1-42.

Moreau, R, E. 1966. The Bird Faunas of Africa and its Islands. London and New York, Academic Press, 424 pp.

Osterhaus, M. H. 1962. Adaptive modifications in the leg structure of some North American warblers. Amer. Midl. Natural, 68:474-486.

Patterson, B., and R. Pascual. 1968. The fossil mammal fauna of South America. Quart. Rev. Biol., 43:409-451.

Pianka, E. 1966. Latitudinal gradients in species diversity: a review of concepts. Amer. Natural, 100:33-46.

Rand, A. L., and E. T. Gilliard. 1968. Handbook of New Guinea Birds. New York, American Museum of Natural History.

Schoener, T. W. 1965. The evolution of bill size differences among sympatric congeneric species of birds. Evolution, 19:189-213.

Short, L. L., and J. J. Morony, Jr. 1969. Notes on some birds of central Peru. Bull. Brit. Ornith. Club, 89:112-115.

Simpson, G. G. 1944. Tempo and Mode in Evolution. New York, Colombia University Press.

——1961. Historical zoogeography of Australian mammals. Evolution, 15:431-446.

Skutch, A. F. 1960. Life histories of Central America birds II. Cooper Ornithological Society Pacific Coast Avifauna, No. 34, 593 pp.

Slud, P. 1964. The birds of Costa Rica, distribution and ecology. Bull. Amer. Mus. Nat. Hist., 126:1-430.

Smith, W. J., and F. Vuilleumier. 1971. Evolutionary relationships of some South American ground tyrants. Bull. Mus. Comp. Zool., 141:181-268.

Terborgh, J. 1971. Distribution of environmental gradients: theory and a preliminary interpretation of distributional patterns in the avifauna of the Cordillera Vilcabamba, Peru. Ecology, 52:23-40.

Traylor, M. A. 1970. Notes on African Muscicapidae. Ibis, 112:395-397.

Vaurie, C. 1953. A generic revision of flycatchers of the tribe Muscicapini. Bull. Amer. Mus. Nat. Hist., 100:457-538.

Wetmore, A. 1926. Observations on the birds of Argentina, Paraguay, Uruguay, and Chile. Bull. U. S. Nat. Mus., 133:488 pp.

White, C. M. N. 1960. A check list of the Ethiopian Muscicapidae (Sylviinae), Part 1. Occas. Papers Nat. Mus. Southern Rhodesia, No. 24B:399-430.

——1961. A Revised Check List of African Broadbills, Pittas, Larks, Swallows, Wagtails and Pipits. Lusaka, Government Printer, 84 pp.

——1962a. A Revised Check List of African Shrikes, Orioles, Drongos, Starlings, Crows, Waxwings, Cuckoo-shrikes, Bulbuls, Accentors, Thrushes, and Babblers. Lusaka, Government Printer, 176 pp.

——1962b. A Check List of the Ethiopian Muscicapidae (Sylviinae). Parts II and III. Occas. Papers Nat. Mus. Southern Rhodesia, No. 26B:653-738.

——1963. A Revised Check List of African Flycatchers, Tits, Tree Creepers, Sunbirds, White-eyes, Honey Eaters, Buntings, Finches, Weavers, and Waxbills. Lusaka, Government Printer, 218 pp.

——1965. A Revised Check List of African Non-Passerine Birds. Lusaka, Government Printer, 299 pp.

Woodring, W. P. 1954. Caribbean land and sea through the ages. Bull. Geol. Soc. Amer., 65:719-732.

7

Behavioral Genetics and the Fine Structure of Mouse Populations: A Study in Microevolution

J. C. DEFRIES AND G. E. McCLEARN
Institute for Behavioral Genetics, University of Colorado
Boulder, Colorado 80302

INTRODUCTION

Although research in behavioral genetics began over a century ago, its status as a distinct interdiscipline has been recognized for little more than a decade (McClearn, 1970; Vandenberg and DeFries, 1970). Since 1960, this intersection between genetics and the behavioral sciences has given every evidence of being a fertile hybrid. Research in behavioral genetics has been undertaken at an accelerating rate, but has largely been preoccupied with demonstrating that individual differences in behavior are due at least in part to gene differences. However, the study of the impact of behavioral differences on the genetic composition of populations is also clearly within the purview of behavioral genetics (King, 1967).

The importance of behavior as a force in evolution has become particularly evident from recent studies of the "population fine structure" of the house mouse (*Mus musculus*). Investigations of biochemical polymorphisms in wild populations, field and laboratory studies of their ecology and behavior, and behavioral genetic analyses of the social behavior of laboratory mice all indicate that stable populations of house mice are comprised of small breeding units, or demes, with little genetic interchange among them. The effective breeding size within such populations is sufficiently small that chance factors must play an important role in determining differences in gene frequencies among demes in the same geographic area. In addition, this breeding structure may facilitate evolutionary trial and error among local populations (Selander, 1970). Since the social behavior of house mice is apparently largely responsible for their population structure, behavior must be considered as an important factor in the evolutionary biology of this species.

BIOCHEMICAL POLYMORPHISMS

The elegant work of Selander (1970) on the biochemical genetics of wild house mice clearly demonstrates the feasibility of population fine structure research. Selander has demonstrated allelic variation at various genetic loci which control the structure of hemoglobin and several esterases in wild house mice and has estimated allelic frequencies from samples trapped in various regions of North America, various barns within a region, and at various locations within a barn. Geographic patterns of frequencies among regions were found to be clinal, indicating that selection was probably the principal factor responsible for differences in allelic frequency at the regional level. However, a marked heterogeneity of allelic frequencies was also found when samples of wild mice were collected from different barns within the same region. These differences were found to persist, even when the barns were located only a few yards apart on the same farm; thus, there is little effective genetic interchange among mice from barn to barn. This interbarn variation in allelic frequency is apparently random, indicating that the frequencies within a barn population are largely due to the genetic composition of a relatively few founder individuals.

Of even greater interest, however, was the variation in allelic frequencies found among samples trapped at different locations within the same barn. A detailed analysis of the spatial variation in allele frequencies within a barn was accomplished by live-trapping mice in a 1.6 foot[1] grid-pattern throughout one of the large (192 x 24 foot) chicken barns in Texas. These barns house laying hens, and dense populations of mice (5.9 mice per square yard[1]) occur throughout. The mice burrow into the ground and mounds of dried chicken feces which accumulate beneath the cages of laying hens. The population size is relatively constant throughout the year; reproduction occurs at a low rate in these populations and seasonal variation in allelic frequencies has not been found. Within such a barn a total of 870 traps was set in the grid-pattern on the west side and then a similar trapping was accomplished on the east side five days later, yielding a total collection of 698 adult and 128 juvenile mice. A heterogeneity between samples caught in the east and west sides of the barn was found, but the grid sampling indicated that this variation was not clinal. Instead, a complex mosaic pattern was found for each locus; small regions of high and low frequencies of alleles were found on both sides of the barn. This spatial

[1] foot ~ 30 cm.; 1 yard = 3 feet.

clustering of similar genotypes is regarded as direct evidence of the tribal subdivision of wild mice (Selander, 1970).

ECOLOGICAL AND BEHAVIORAL FIELD STUDIES

Selander (1970) fully realized that his results were highly consistent with expectations based upon the findings from earlier ecological and behavioral studies. For example, several investigators have shown that the home range of wild mice is relatively small. Young et al. (1950) found that the average distance between successive captures of wild mice in two large buildings in Madison, Wisconsin, was only 12 feet and that 90% of all recoveries were within 30 feet of the previous capture. Average distances moved did not increase as a function of the interval between captures, thus indicating stability of home ranges. Although openings for ingress and egress were numerous, little or no evidence for migration was found. In a study of house mice occupying corn ricks in England, Southern and Laurie (1946) concluded that the home range must be something of the order of about 50 square feet and that vertical movement within the rick was less than lateral.

Similar evidence for small home range has been found by Anderson (1965) in his extensive investigations of the breeding structure of wild populations of house mice which inhabit small grain buildings in the prairie regions of Alberta, Canada. In these studies, the average population size within a building was found to consist of about four adult females and three adult males. In addition, it was found that individual mice seem to restrict their activities to a single building, even though the buildings might be separated by less than a meter. Reproductive isolation is also indicated by the observation that neighboring buildings housed populations which were persistently different genetically at the *T* locus. Experimental introduction of mice into already inhabited buildings resulted in no recaptures in these granaries, although two mice were recaptured in a previously unoccupied area nearby. These results clearly indicate that mice that inhabit a granary building are a distinct breeding unit.

Although most environments occupied by wild mice lack the environmental discontinuities of these small granary buildings, the evidence for a mosaic of small breeding units was also found in field work on feral mice which inhabit a grassland of Great Gull Island, off the coast of New York (Anderson, 1965). Although an attempt to introduce a mutant allele at the *T* locus in this population was successful, it failed to spread over the whole inhabited area of the small island. Ecological and behavioral studies of mice on Great Gull Island indicate that the area is subdivided into male territories and that the territory is passed from the resident male to one of its own offspring.

These field studies indicate that in stable situations, wild mice live in relatively small breeding units with little genetic interchange among units. According to Anderson (1965), emigration to previously unoccupied areas apparently dissipates reproductive surplus and thus accounts for the stable population sizes within the demes. However, such a process may or may not lead to outbreeding. In Anderon's view, this system avoids overexploitation of local resources and simultaneously assures a continuous search for sites where new colonies may be founded. Considerable intrademe homogeneity could result, leading to chance changes in gene frequency among demes. Such random drift coupled with the differential success of established demes in the founding of new breeding units through emigration could provide an effective evolutionary force in populations of wild mice.

ECOLOGICAL AND BEHAVIORAL LABORATORY STUDIES

The findings from field investigations discussed above are in general agreement with those from laboratory studies. Crowcroft (1955) observed that fighting tended to disperse populations of wild mice released in a large enclosure (250 square feet) into spatially distinct breeding areas. This finding was later substantiated by Crowcroft and Rowe (1963). In this latter paper, the authors emphasized the effectiveness of breaking up the area of the enclosure into a number of incompletely separated smaller units, thus providing the mice with physical landmarks for orientation and thereby facilitating the establishment of discrete territories.

More recently, Rowe and Redfern (1969) have studied aggressive behavior in small family groups of confined wild house mice. They observed that aggressive behavior was rare within a family group (presumably where social hierarchies were already established), but that marked aggressive behavior occurred when unrelated mice were introduced into the family group. In the presence of strangers, intrafamily aggressive behavior was also found to be increased. When two family units were allowed to meet, much fighting took place and no permanent interchange of mice between the groups was observed. When males or females within a group were isolated, residents were found to be aggressive towards them when they were returned after a two- or three-week period; thus, it appears that emigration results in a more or less permanent separation from the family group.

Reimer and Petras (1967) have utilized both wild and laboratory stocks in their excellent study of the breeding structure of house mice in a population cage. Mice were released into a cage which consisted of a series of nest boxes connected by runways. In each of four experiments, mice formed small breeding units, each generally composed of one dominant male, several females, and several subordinate males. In longer established demes, dominant males were replaced by their offspring. It was concluded that the formation of these breeding units was primarily due to male territoriality. Interdemic migration was found to be rare and appeared possible only through females. When intruders were experimentally introduced, one at a time, 80% of the intruders were dead within 12 hours of introduction. Eventually all male and most female intruders were killed. However, as the number of intruders was increased, the percent killed was reduced. Resident females, as well as males, were found to drive off intruders.

The family unit appeared to be stable over a period of several generations. In one instance, seven mice were found to successfully defend an area against 168 mice in an adjacent territory. Several cases of female movement between established male territories were observed, suggesting a possible mechanism for gene exchange between breeding units. However, no males were found to enter successfully an established deme.

The study of Reimer and Petras (1967) is especially important because it was found that this deme structure resulted regardless of the source and genotype of the mice involved. Thus, the finding of male territoriality followed by the formation of small breeding units seems to be general across field and laboratory investigations and to occur with both wild and laboratory mice. Crowcroft, in his delightful little book, *Mice All Over* (1966), describes an amusing incident which illustrated to him that the fighting ability of laboratory mice had not been modified by generations of domestication. He found a wild mouse which had managed to climb into an enclosure containing laboratory mice, but could not get out. Presumably because of the disadvantage of being a stranger in someone else's territory coupled with a weight handicap, the wild mouse was found to be at the bottom of the hierarchy in the enclosure. Crowcroft relates that it seemed very odd to find a "genuine" mouse being chased about by an "artificial" one. Although anecdotal, this observation plus that of Reimer and Petras (1967) indicate that studies of the social behavior of laboratory mice may have some generality to wild populations.

SOCIAL DOMINANCE AND DARWINIAN FITNESS

Although the field and laboratory studies discussed above indicate that stable populations of house mice are subdivided into small breeding units (demes, family groups, tribes) with little effective genetic interchange among demes, a valid estimate of the effective population size is not available from these studies since information about the differential contribution of individuals to the gene pool was not obtained. Until quite recently, the contribution of the subordinate male was not clear: Selander (1970) states that subordinate males presumably do not contribute to the gene pool; on the other hand, Brown (1953) observed little relationship between copulation order among males and their relative position in the social hierarchy. Crowcroft (1966) has even suggested that the dominant male may be so busy defending his territory that the subordinate males may actually do the majority of the mating.

Knowledge of the reproductive success of subordinate males is quite important because it can greatly influence the effective population size (N_e), where

$$N_e = \frac{4N_mN_f}{N_m + N_f}.$$

From the above expression, it may be seen that when the number of males that contribute to the gene pool (N_m) = 1, $N_e < 4$, regardless of number of females (N_f).

Recently, some definitive evidence concerning the reproductive success of subordinate males has been obtained (DeFries and McClearn, 1970). The purpose of this study was to assess strain differences in social behavior of laboratory male mice and to test for possible relationships with Darwinian fitness, as measured by relative success in siring litters. For this purpose, standardized social living units were constructed, utilizing three standard mouse cages (9 x 4 x 4 inches[2] high) for each unit. An opening was cut into one end of each of the cages, which were then connected with a Y-shaped plastic manifold constructed of 1¾-inch clear Plexiglas tubing. Thus, animals placed within each "triad" unit (see Fig. 1) could move freely from one cage to another. Adequate food, water, and bedding were placed in each cage.

In the first experiment, three males, each from a different inbred strain, and three inbred females were placed in each of 22 triads. Choice of male combinations was dictated by the possibility of ascertaining paternity of subsequent litters from the coat colors of the pups and by availability. Males were examined daily for number of tail wounds and then weighed. Spatial location of the animals within the triad was recorded three times per day. After two weeks of such observation, males were discarded and females retained until litters were born and old enough to permit determination of paternity.

In general, males began to fight within a few minutes of introduction into the triad, and within one or two days one male was usually established as the clearly dominant animal. Wounds were usually found on the tails of subordinate males, although some wounding was also found on the hind quarters. A comparison of number of tail wounds with evaluations of dominance based upon behavioral observations indicated that a criterion of least number of tail wounds was an almost infallible index of social dominance.

The two subordinate males were usually found together in one cage (84% of observations). Females, however, were apparently free to come and go from one cage to another, including that of the subordinate males. Highly consistent strain differences in social dominance were found. A/Ibg and BALB/cIbg males were usually dominant, whereas DBA/2Ibg males were usually most subordinate (greatest number of tail wounds). C57BL/Ibg males tended to be intermediate. Of greater interest, however, was the relationship between social dominance and

[2] 1 inch ~ 2.4 cm.

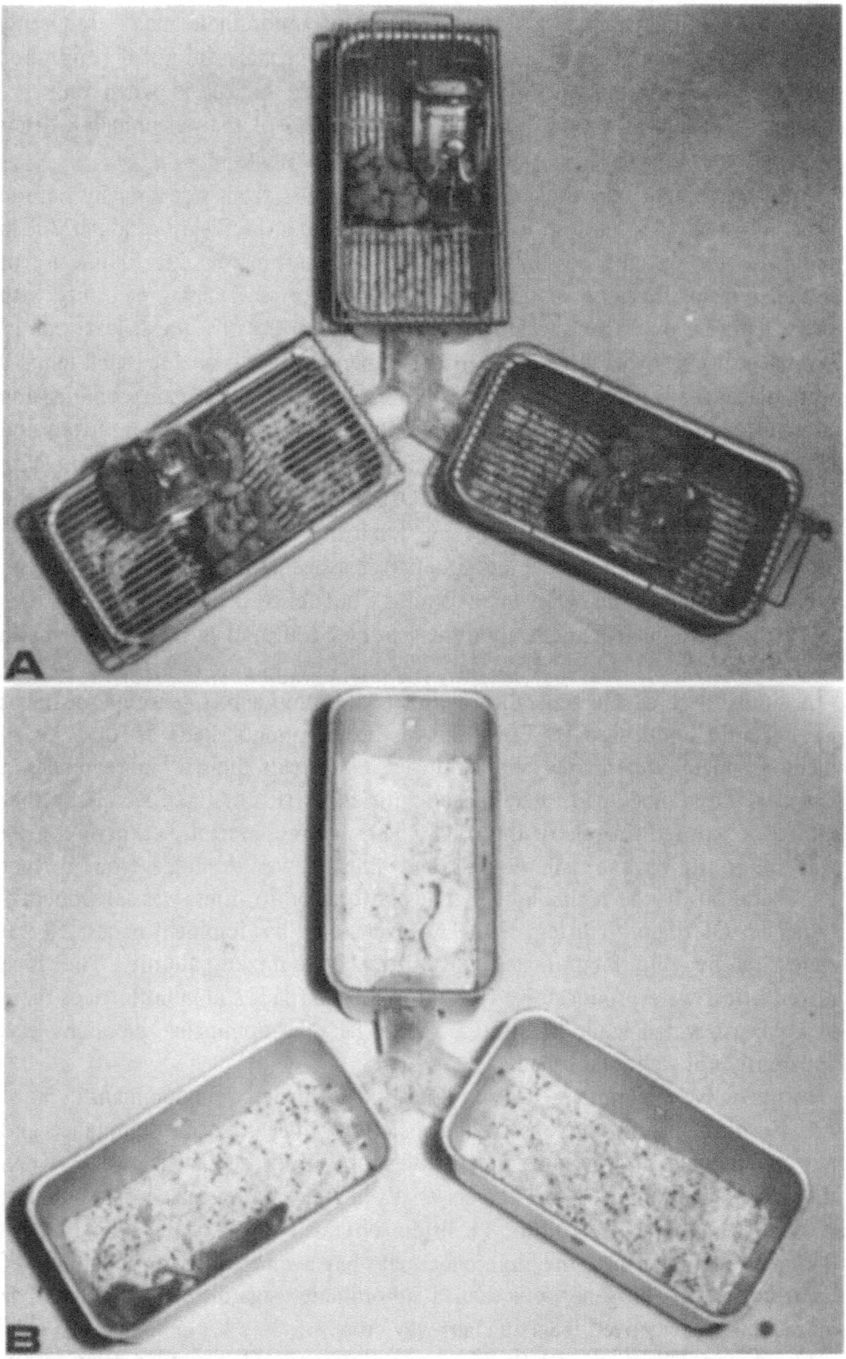

Fig. 1. Triad unit used in studies of the social behavior of laboratory mice: A. Assembled triad; B. Interior view.

Darwinian fitness: of the 61 litters obtained in this experiment, 56 (92%) were sired by dominant males. Although death of subordinate males frequently occurred, the reproductive success of the dominant male did not depend upon death of his subordinates. When litters that were conceived when only the dominant male was alive were excluded, 53 (91%) of the remaining 58 litters were still found to have been sired by the dominant males.

This experiment was then repeated using females from a genetically heterogeneous stock (HS) derived from an eight-way cross (McClearn et al., 1970) in order to rule out the possibility that this relationship was due to the use of inbred females. Because of the greater difficulty of ascertaining paternity with these females, only those which were homozygous recessive for at least one of several coat color loci were used. In this experiment, two or three such females were placed with pairs of inbred males in each of 20 triads. Again, combination of males was dictated by ability to determine paternity of subsequent litters and by availability. Strain differences were again found which were consistent with those observed in the first experiment. Of the 42 litters obtained in this second experiment, 40 (95%) were sired by dominant males. This result not only indicates the reliability of the test situation in assessing the relationship between social dominance and Darwinian fitness, but also demonstrates that this relationship is not dependent upon the genotype (outbred vs. inbred) of females employed.

A similar test of the generality of this relationship across genotype of males has recently conducted (DeFries and McClearn, unpublished). In this experiment, F^1 males were paired with HS males in 36 triads and with inbred males in 5 triads. These male pairs were placed with either two or three females, either inbred or outbred. In one of the triads, there was no contest, i.e., neither male was found to have a tail wound, and thus it was excluded from further consideration. In the remaining 40 triads, a total of 76 litters was subsequently obtained. Of these 76 litters, 70 (92%) were sired by dominant males, 3 (4%) were sired by subordinate males, and 3 litters were mixed paternity. The three mixed litters were produced by females from two triads and in both triads there were very few tail wounds on either male, i.e., the dominance hierarchy was apparently not well established.

In the above experiment, it was found that F^1 males were dominant in 34 of the 40 triads. Thus, the observed association between social dominance and reproductive success was confounded with genotype of the male. However, when the data were examined from the six triads in which an HS male was dominant, it was found that 10 of the 11 litters obtained (91%) were still sired by dominant males. Thus, the dominant male has a much higher probability of contributing to the gene pool than a subordinate, regardless of whether it is inbred, an F^1 or derived from an eight-way cross.

Lloyd and Christian (1969) have recently reported a similar relationship between social rank and reproductive success of female house mice in freely growing laboratory populations. The criteria used to establish female rank included distance moved, size of area in which moved, and percent of observation periods in which the females were active. High-ranking females were found most frequently in association with the dominant male and were reported to be responsible for almost all of the litters produced.

If the association between social dominance and Darwinian fitness is as high in wild populations as observed in our laboratory stock, the effective population size within a deme is indeed small. In fact, if the dominant male sires more than 90% of the offspring in a deme, the effective population size will be less than 4, even if the number of reproducing females is large. If the observation of Lloyd and Christian (1969) has generality to wild populations, the number of females that reproduce within a deme may also be quite small. Thus, it is quite conceivable that the effective population size in natural populations of house mice is less than 4. Such a finding has important implications for the evolutionary biology of house mice. With such a small effective population size, random drift would be an important factor in determining allelic frequencies in local populations.

This finding supports an hypothesis of Lewontin and Dunn (1960) made more than 10 years ago to account for the discrepancy between observed and expected frequencies of the *t* allele in natural populations of house mice. Although *t* alleles are recessive lethal or male sterile, most natural populations are polymorphic for alleles at the *T* locus. A transmission ratio of about 0.95 in heterozygous males is believed to be responsible for the maintenance of this polymorphism. However, the observed frequency of the *t* allele in natural populations is about half that calculated on the basis of deterministic models that assume infinite population size and random mating. Lewontin and Dunn (1960) suggested that this discrepancy might be due to random genetic drift. Assuming that mice live in small family units, a large fraction of these demes would become fixed for the normal allele. When a geographical population consisting of such demes is sampled, the observed frequency of the *t* allele would be less than expected on the basis of deterministic models.

Levin et al. (1969) have recently employed a Monte Carlo simulation of a geographical population of house mice and examined the range of deme sizes and interdemic migration rates that were compatible with the random-drift hypothesis. For interdemic migration rates of 0%, 1%, and 3%, it was found that random drift would have an important effect on *t*-allele frequencies only when the effective population size was less than 12, 8, and 4, respectively. Since the effective breeding-unit size may be less than 4, the original hypothesis of Lewontin and Dunn (1960) remains tenable.

The above findings concerning the relationship between social dominance and Darwinian fitness provide clear-cut evidence that behavior can influence the genetic composition of populations. Because of the importance of this point, it was deemed advisable to verify this relationship by alternative means. Characters that are closely (and linearly) associated with fitness should manifest relatively little additive genetic variance, i.e., the genetic variance in fitness characters should be largely nonadditive (Roberts, 1967). Since heterosis or hybrid vigor is due to nonadditive gene effects, such characters should display considerable heterosis. When two inbred lines are crossed to produce an F_1, and an F_2 generation is subsequently obtained, the F_1 should have the highest mean phenotypic value for fitness characters, and the F_2 mean should be intermediate between the F_1 mean and the mean of the inbred lines. Thus, if individual differences in social dominance are due largely to nonadditive gene effects, i.e., if social dominance is closely related to Darwinian fitness, the predicted rank order of males in a triad would be: F_1, F_2, and inbred.

In a study still in progress in our laboratory (Kuse and DeFries, unpublished), significant nonadditive gene effects have been found for this character. In ten triads, four males were introduced, one each of the following: DBA/2Ibg, C57BL/Ibg, and their F_1 and F_2. Maximum number of tail wounds during the first three days in the triad was used to assess dominance and the following results were obtained: F_1, 0.2; F_2, 0.4; C57BL, 1.6; and DBA/2, 11.8. Thus, the predicted rank order is obtained; however, the only comparison which reached levels of statistical significance was that of inbred vs. outbred ($t = 4.4$, $df = 36$, $p < .01$). Outbred males were also found to have fewer tail wounds than inbreds in eight other triads, four of which utilized DBA/2, BALB/c, and their F_1 and F_2 cross males, and four of which used C57BL, BALB/c, and their F_1 and F_2 crosses. However, only in the latter four triads were F_1 males found to have fewer tail wounds than F_2 males. Clearly these samples are too small to permit generalizations, but it does appear to be the case that individual differences in social dominance among males are due at least in part to nonadditive gene effects.

One very marked effect found in this latter study was that due to the presence of females. In half of these triads, females were not placed with males. In these all-male triads, relatively less aggression was observed. Thus, it seems that presence of females in the triad situation leads to increased aggression and thereby hastens the establishment of the hierarchy.

The triad is a highly artifical apparatus in that emigration and escape are not possible. Thus, it might be argued that the intensity of aggression and subsequent wounding observed in the triad would never be observed in natural populations. However, a relatively high incidence of wounding was found in English corn ricks by Southwick (1958); the higher the density, the higher the incidence of wounds among males. At very high densities, 45.2% of the males over 17.5 grams were found with conspicuous wounds or scars. The incidence of wounded females was small (less than 2%) and seemed to be independent of density. DeLong (1967) has also observed a high incidence of tail wounds among feral mice in California. In a densely populated area, 42% of the males and 12% of the females were freshly wounded during one month in which immigration was high. No more than 5% of the individuals in the other less densely populated areas were found to be wounded at any time. The number of newly tagged animals in the more densely populated area was less than that observed in the other areas, indicating that attempts to immigrate were less successful. Thus, high levels of aggression apparently do occur in natural populations and the level seems to be especially high during periods when immigrants are attempting to colonize occupied territories.

SPECULATIONS

The previous comments relating to the small deme size of house mice with little effective gene exchange among demes all refer to stable populations. However, dispersal phases exist which may provide an opportunity for gene exchange. Natural populations of mice are not homozygous at all loci; thus, some mechanism(s) must exist to maintain variability within demes. It would be most interesting to examine offspring found in newly established demes in order to determine whether the level of heterozygosity was higher than that found in older populations. If so, this would suggest that gene exchange does take place during periods of migration.

In stable populations, however, inbreeding must occur. With an effective population size of 4, the increase in the coefficient of inbreeding within a deme should be about 12.5 per generation. Thus, populations of house mice in their natural habitat are at least to some extent inbred. This might account for the success of establishing highly inbred strains of mice for laboratory research. House mice may already have been selected for their ability to withstand relatively high levels of inbreeding. If so, some of the objections occasionally raised regarding the use of highly inbred strains of mice for research may not be valid. In fact, a heterogeneous stock produced by a four-way or eight-way cross may be more "artificial" than an inbred strain. When mice from four-way (Kessler, 1966) and eight-way (McClearn and DeFries, unpublished) crosses were used to establish freely growing laboratory populations, aberrantly high population densities were achieved. Although direct evidence is not available, usual processes that regulate population growth in house mice may somehow be altered in such "artificial" mice.

Some of the recorded mouse "outbreaks" (Hall, 1927; Crowcroft, 1966) might also be attributed to such a phenomenon. Due to decreased predation or increased food supply, outmigration might sharply increase. Eventually a point may be reached where the number of potential immigrants is too great for a successful defense of the home territory. At this point, gene exchange could occur between demes on a massive scale, resulting in an immediate increase in the mean level of fitness characters and an apparent population explosion.

Christian (1970) has recently hypothesized that intraspecific competition is a major force in mammalian evolution and that genetic changes on which selection operates are found in socially subordinate animals. In stable populations, dominant animals will remain within their territory as long as the habitat remains favorable; thus, animals that emigrate to less favorable habitats will usually have been subordinate. Although the vast majority of the dispersing animals would not survive in these marginal habitats, rare genetic recombinants may occasionally occur that make survival and reproduction possible. In a territorial species, spatial separation restricting gene flow between the new colony and the old population could occur in the absence of geographic isolation. Recombination, followed by survival in a previously unfavorable habitat, spatial separation restricting gene flow, and natural selection for adaptation to the new environment would seem to include the necessary ingredients for speciation.

CONCLUDING REMARKS

The data obtained from investigations of biochemical polymorphisms in wild populations of house mice, from ecological and behavioral field studies, and from laboratory research with both wild and laboratory stocks, all indicate that stable populations of house mice are comprised of a mosaic pattern of small breeding units or demes with little genetic interchange among them. Recent evidence concerning the relationship between social dominance and reproductive success in laboratory mice indicates that the effective population size within such demes may be less than 4. This population fine structure clearly has implications for the evolutionary biology of the species. The effective population size is sufficiently small that chance factors will play an important role in determining gene frequencies at the local level. In addition, the differential success of established demes in starting new breeding units could be an important force in the evolution of house mice. Since the population fine structure of house mice is largely due to their social behavior, it would seem that the validity of the point that behavior can have important effects on the genetic composition of populations is clearly demonstrated.

REFERENCES

Anderson, P. K. 1965. The role of breeding structure in evolutionary processes of *Mus musculus* populations. *In* R. Honcariv, ed., Mutation in Populations, pp. 17-21. Prague, Academia.

Brown, R. Z. 1953. Social behavior, reproduction, and population changes in the house mouse (*Mus musculus* L.). Ecol. Monogr., 23:217-240.

Christian, J. J. 1970. Social subordination, population density, and mammalian evolution. Science, 168:84-90.

Crowcroft, P. 1955. Territoriality in wild house mice, *Mus musculus* L. Mammal., 36:299-301.

——1966. Mice All Over. Chester Springs, Pennsylvania, Dufour Editions, Inc.

——and F. P. Rowe. 1963. Social organization and territorial behaviour in the wild house mouse (*Mus musculus* L.). Proc. Zool. Soc. London, 140:517-531.

DeFries, J. C., and G. E. McClearn. 1970. Social dominance and Darwinian fitness in the laboratory mouse. Amer. Natural., 104:408-411.

DeLong, K. T. 1967. Population ecology of feral house mice. Ecology, 48:611-634.

Hall, E. R. 1927. An outbreak of house mice in Kern County, California. Univ. California Publ. Zool., 30:189-203.

Kessler, A. 1966. Interplay Between Social Ecology and Physiology, Genetics and Population Dynamics of Mice. Ph.D. Thesis, The Rockefeller University, New York.

King, J. A. 1967. Behavioral modification of the gene pool. *In* Hirsch, J., ed., Behavior-Genetic Analysis, pp. 22-43. New York, McGraw-Hill.

Levin, B. R., M. L. Petras, and D. I. Rasmussen. 1969. The effect of migration on the maintenance of a lethal polymorphism in the house mouse. Amer. Natural., 103:647-661.

Lewontin, R. C., and L. C. Dunn. 1960. The evolutionary dynamics of a polymorphism in the house mouse. Genetics, 45:705-722.

Lloyd, J. A., and J. J. Christian. 1969. Reproductive activity of individual females in three experimental freely growing populations of house mice (*Mus musculus*). J. Mammal., 50:49-59.

McClearn, G. E. 1970. Behavioral genetics. Ann. Rev. Genet., 4:437-468.

—— J. R. Wilson, and W. Meredith. 1970. The use of isogenic and heterogenic mouse stocks in behavioral research. *In* Lindzey, G., and D. D. Thiessen. eds., Contribution to Behavior-Genetic Analysis: The Mouse as a Prototype, pp. 3-22. New York, Appleton-Century-Crofts.

Reimer, J. D., and M. L. Petras. 1967. Breeding structure of the house mouse, *Mus musculus*, in a population cage. J. Mammal., 48:88-89.

Roberts, R. C. 1967. Some evolutionary implications of behavior. Canad. J. Genet. Cytol., 9:419-435.

Rowe, F. P., and R. Redfern. 1969. Aggressive behaviour in related and unrelated wild house mice (*Mus musculus* L.) Ann. Appl. Biol., 64:425-431.

Selander, R. K. 1970. Behavior and genetic variation in natural populations. Amer. Zool., 10:53-66.

Southern, H. N., and E. M. Laurie. 1946. The house-mouse (*Mus musculus*) in corn ricks. J. Anim. Ecol., 15:134-149.

Southwick, C. H. 1958. Population characteristics of house mice living in English corn ricks: Density relationships. Proc. Zool. Soc. London, 131:163-175.

Vandenberg, S. G., and J. C. DeFries. 1970. Our hopes for *Behavior Genetics*. Behav. Genet., 1:1-2.

Young, H., R. L. Strecker, and J. T. Emlen, Jr. 1950. Localization of activity in two indoor populations of house mice, *Mus musculus*. J. Mammal., 31:403-410.

8

Evolutionary Consequences of Differential Fertility and Assortative Mating in Man

BRUCE K. ECKLAND
Department of Sociology, The University of North Carolina
Chapel Hill, N. C. 27514

INTRODUCTION

It is tempting to argue that the organic development of man's intellect has been more fundamental than any other human trait in setting him apart from other forms of life. His ascent in the animal kingdom, for example, obviously would not have been so spectacular if we were left only to point to his grasping hands and erect posture. Yet, both of these conditions contributed substantially to his progress since, as LaBarre some years ago noted, they enabled him literally to "monkey" with his environment (LaBarre, 1955). Moreover, man's acquisition of language, his becoming a culture-bearing animal, and his ability to reflect upon himself as an object of concern (and to contemplate his future) are all probably more the consequence than the cause of his intellectual development.

Although we now rely almost exclusively upon environmental means (e.g., formal education) to improve them, man's cognitive skills surely are as important for his adaptedness today as at any point in his evolutionary history. Yet, man may soon be taxing his genetically conditioned capacities to effectively use and control the very technology that his skills have enabled him to create. Most of man's "social problems," of course, are largely social and not biological in origin, including his run-away technology. Too, he has hardly exhausted his own resources, as evidenced by the wastage and underutilization of human talent in even our most technologically advanced societies. Nevertheless, it is quite probable that the exponential growth of science and technology will eventually require a new breed of man if such knowledge is to be effectively used by any but a select few.

We are suggesting then that the continuing evolution of man–in particular his mental capacities–may be as important to his future adaptedness as it was to his past. In this chapter we will trace the genetic reconstitution of relatively recent publications in regard to man's intelligence and outline some probable trends for the not too distant future. Our discussion focuses upon the genetic consequences of two basic factors in the bisexual reproduction of human societies, i.e., differential fertility and assortative mating. Moreoever, we will argue that these factors do not operate entirely independent of each other. The effects of differential fertility upon the reconstruction of man's mental abilities may depend, especially in the future, upon mate selection or, in other words, who mates with whom.

Briefly, assortative mating increases the average homozygosity of polygenic traits like intelligence, which, in turn, increases the opportunity for natural selection even when rates of fertility and mortality are declining. Thus, while modernization has led to smaller family size and lower death rates, the same general set of social forces is leading to positive assortative mating for intelligence, which may intensify changes in gene frequencies. Assortative mating redistributes the genes in the population in such a manner as to increase the rate of change in those portions of the population where departures from average family size are greatest. Since this set of propositions depends in part upon at least moderate differentials in fertility, despite declining birth rates, we will begin with a review of the relationship between reproduction and natural selection.

THE OPPORTUNITY FOR NATURAL SELECTION

When the frequency that a particular gene appears in a population exceeds its expected rate of mutation, its frequency usually is due to some form of natural selection–migration, mortality, or differential fertility.

The addition or loss of members of a breeding population through migration can substantially alter the character of the gene pool over time and largely accounts for the complexity of most racial taxonomies. Although exceedingly important in understanding race differences, genetic migration will not be dealt with in this chapter–mainly because we know so little about how the polygenes related to intelligence in the past might have been correlated with any particular pattern of migration.

Aside from migration, mortality and differential fertility are the only other two mechanisms by which the genetic structure of a given population can change through natural selection, and both no doubt have influenced the evolution of man's intellectual abilities. Differential fertility simply refers to the amount of variance in the number of children people have, each of whom contributes a different set of genes to its progeny. However, since the progeny also must live and reproduce in order to maintain the genetic pool, mortality also enters into the picture.

Figure 1 describes the historical relationship betwen mortality and fertility. The death rate, especially infant mortality, rapidly declines in most countries as medical technology is introduced, followed by a much slower decline in birth rates, which together produce the population growth curve shown in Fig. 2. Unlike strong sociocultural supports for lower death rates, limiting family size usually runs counter to various religious and moral sanctions in most countries. The result almost invariably is an unprecedented growth in population.

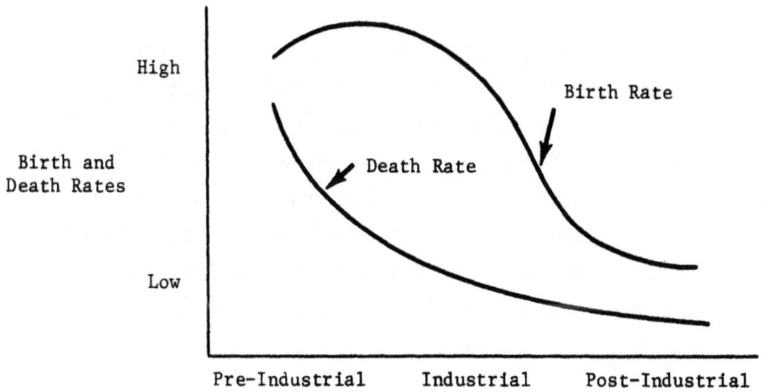

Fig. 1. Trends in Birth and Death Rates.

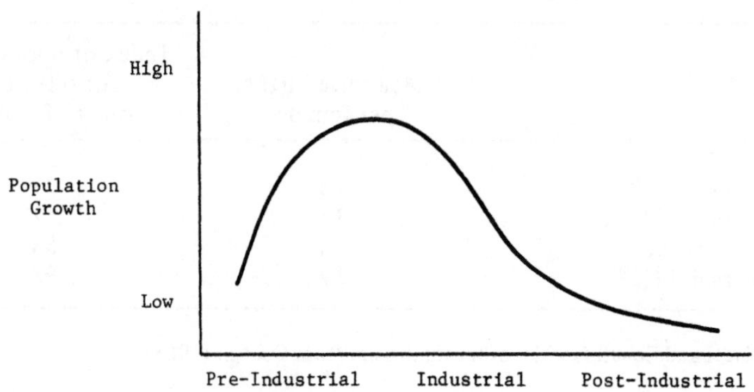

Fig. 2. Normal Population Growth Curve.

In 1840, just prior to the industrial revolution in America, about two out of five white males born that year did not survive to age 20 (Jacobson, 1964). In many undeveloped nations today the death rate is even higher. Thus, as the figures of Neel and Schull (1972) on agricultural societies indicated, mortality at one time was probably the major force for genetic selection in human societies.

However, mortality now plays only a minor role in the opportunity for genetic selection in technologically advanced societies. Almost no one dies before reaching the reproductive period in America today, or at least so few that the effect on the genetics of intelligence must certainly be of little consequence.

If all other things had remained equal, the opportunity for natural selection might have nearly vanished in any of the contiguous populations exhibiting such a dramatic reduction in death rates. However, everything usually does not remain the same. Most importantly, when the *birth* rates eventually declined in developing countries, the number of departures from average family size actually appears to have increased. When the average size of families is small, the variation in fertility between families tends to be greater than when the birth rates are high.

According to Osborn (1968), for instance, the opportunity for potential selection attributed to differential fertility is higher in the general population of the United States today, where the average family has less than three children, than among the Hutterites (U. S.) or in rural Quebec where, the average family has more than nine children. The relationship between birth rates and the opportunity for selection during the period of modernization in the U. S. is shown in Table 1. Generally, as the number of live births per female has gone down, the index of opportunity for genetic selection due to fertility has gone up.

Table 1. The Relationship Between Fertility Rates and the Opportunity for Selection in the United States

	Mean live births per female	Index of opportunity for selection due to fertility
Women born in 1839	5.5	.23
Women born in 1871-75	3.5	.71
Women born in 1909	2.1	.88
Women born in 1922[a]	2.2	.92

[a]From Neel and Schull (1972); other figures are from Osborn (1968).

All of the figures in this table for women born after 1939 in the U. S. show a higher opportunity for potential selection due to fertility than any of the indices computed by Neel and Schull (1972) for either early or agricultural man.

The relative contributions of mortality and fertility to the opportunity for selection are projected over time in Fig. 3. In summary, losses in the opportunity for natural selection due to declining mortality have largely been offset by the increasing variance in fertility, despite declining birth rates. Thus, the index as a whole quite probably has remained fairly stable during the course of modernization.

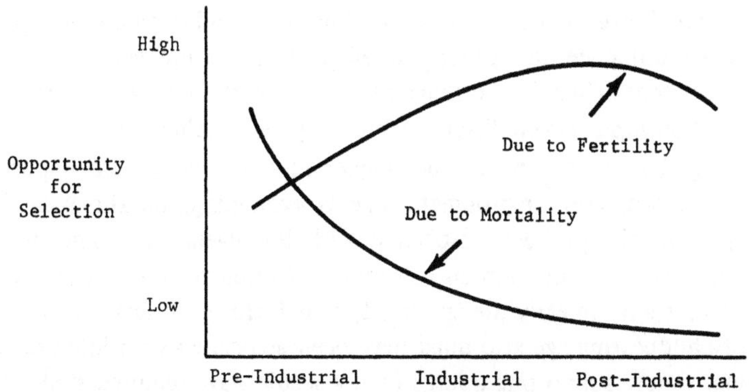

Fig. 3. Indices of Opportunity for Selection, by Fertility and Mortality.

INTELLIGENCE AND NATURAL SELECTION

Thus far, we have spoken only in terms of whether or not individual differences in reproductive habits and mortality patterns have become so small in modern societies as to preclude the opportunity for selection. Obviously, if every child that was born lived to reproduce the same number of progeny, the gene pool would remain virtually unchanged. Each member of the population would pass proportionately the same number of genes on to the next generation. However, we have concluded that this is not the case. The available evidence in fact suggests that the opportunity for selection probably is at least as high today in technologically advanced countries as it was before the onset of the industrial revolution.

Whether *actual* selection has occurred, however, depends on the degree to which differential fertility or mortality is correlated in some manner with the distribution of genes in the breeding population. If the "overproducers" carry different genes than the "underproducers," the population frequency of some genes will change. And, if these genes are related to a polygenic trait like intelligence, a quantitative change in the population mean of that trait should be observed (all other things being equal).

During most of man's evolution, mortality undoubtedly has favored the natural selection of genes for higher mental ability, which generally confers an advantage in the competition for survival. However, there is little evidence that mortality has much influence on the gene pool for intelligence in most contemporary societies, with perhaps one or two exceptions. The first is the somewhat higher death rate usually found for seriously retarded individuals. The second possibly is the unknown effect of war and famine. Although the genetic consequences of wartime fatalities due directly to military incursions have never been investigated, there is some evidence that severe malnutrition during wartime favors the survival of the more intelligent members of the group. As noted earlier, however, death rates under normal conditions are now probably too low to influence the genetic intelligence of future generations to any significant extent. Thus, in the remainder of this section we will focus primarily on the relationship between differential fertility and intelligence.

Assuming that, as a general rule, higher mental abilities are important to "success" in whatever environment one is examining, variations in fertility probably have always been associated with intelligence to some degree. In preindustrial societies the burden of genetic selection for intelligence no doubt fell upon mortality in early life, as already noted. However, fertility rates among adults of childbearing age also must have been associated with intelligence. For one thing, simply supporting a large family would have required skills that not all men possessed. While we are not claiming that intelligence was the only or always the necessary prerequisite, it certainly must have aided early and preindustrial man in his adaptedness to a physically harsh environment—one in which childbearing was almost always a precarious event.

Also contributing to this relationship was polygynous mating, a practice that not only prevailed but probably predominated throughout most of human history. The most successful, and we presume intelligent, males had more mates and as a consequence sired a disproportionately large share of the next generation. This, moreover, suggests that natural selection, to the extent that reproduction was involved, at one time may have depended far more upon variation in the sexual activity of males than females, most of whom probably reproduced at a level closely approximating their biological potential.[1]

In any case, before industrialization began to change the basic social structure of human societies, including plural mating, intelligence and fertility were positively related, as shown in Fig. 4a. With the beginning of industrialization, the relationship probably became curvilinear and perhaps even slightly negative, as shown in Fig. 4b.

Although those at the lowest end of the IQ distribution were now more likely to live to childbearing age, it is doubtful that they were producing many more children than before modernization began. Mentally retarded persons were often sterilized, institutionalized, or in other ways insulated from normal sexual unions.

However, the overproducers, i.e., those at the opposite end of the scale, now became underproducers. This situation largely reflects the rise of professional and managerial classes and the changing functions of the family. As economic production shifted away from the home, children became liabilities, particularly those born to a second wife or concubine. This led to the more frequent use of

[1] The point also suggests that the index of opportunity for selection used earlier in this chapter may seriously underestimate the potential influence of differential fertility, since the index usually is based upon the reproductive performance of females when in fact it often was the males who were most variant in their contributions to the gene pool.

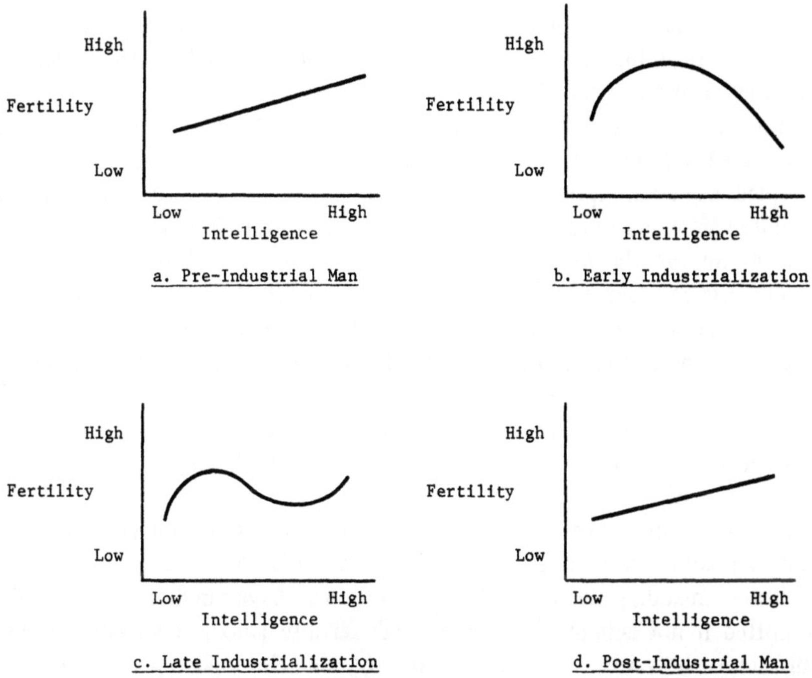

Fig. 4. The Relationship between Differential Fertility and Intelligence at Four Historical Periods.

birth control among the most ambitious and successful adults. Class differences in fertility were particularly notable in most industrializing countries around the turn of the century. The highest socioeconomic groups, in fact, were producing below their level of "replacement."

With improved control of infant mortality, on the other hand, the lower and lower-middle class (with the exception of the mentally retarded) generally were producing an unprecedented number of children. Still largely rural and uneducated, the reproductive habits of this group were least affected by the technological changes that elsewhere in the society were beginning to undermine the traditional relationships between the extended family and the rest of the social order.

Evenutally with the growth of the middle classes and the rise of mass education, the relationship between fertility and intelligence becomes slightly more complicated–mainly as a result of the diffusion of birth control practice into the expanding middle class and an upswing in family size among the most successful adults. Figure 4c is intended to illustrate where we are today and Fig. 4d where we will likely be in the near future.

The bimodal distribution in Figure 4c is based upon the combined results of several recent studies in the U. S., which indicate that birth rates are now higher among the more educated and intelligent members, and remain depressed at the very bottom of the distribution as well as among the middle class (Reed and Reed, 1965; and Bajema, 1963, 1966, and 1971).

Once below the level of replacement, professional families are now producing above it–if only because, as Davis (1965) notes, "children have become items of conspicuous consumption which the better-off can afford" and because "children are easier to care for if the family can afford to live in spatious suburbs and to withdraw the wife completely from the labor force." On the other hand, fertility has been depressed among those in the middle "who are too educated to have children carelessly and too poor to have them abundantly" (Davis, 1965). In general, the relationship between intelligence and fertility comes out on the positive side, but only slightly.

In the not too distant future we anticipate that the relationship will become more positive and linear. All that is required, incidentally, is a drop in the lower mode of the IQ distribution.

Of prime importance in accomplishing this shift is the diffusion of birth control practice into the lower working-class. Early welfare programs in many countries, including the U. S., heavily subsidized large families–programs that permitted if not actually encouraged high fertility among low-income groups. This trend, however, is not likely to continue. Not only are the structural supports for carelessly having children beginning to loosen (as evidenced by the general shift in federal aid from rehabilitative to preventative programs), but married working-class women are continuing to enter the labor force in increasing numbers, making large families more difficult to sustain emotionally. Moreover, sterilization, legal abortions, and improved contraceptive devices are becoming more available to persons in all walks of life, often at modest costs. Add to this our nation's (and especially our young people's) concern over ecology and population problems, our increasingly liberalized sex mores, including those sanctioned by ecclesiastical institutions (which serve to legitimize sex without childbearing), and increasing opportunities for social and geographic mobility (due both to rising educational levels and higher standards of living), I reach the conclusion that, whether self-limiting or imposed by law, the high fertility in the lower mode of the SES and IQ distributions will drop.[2] This shift, moreover, does not require a radical change in our reproductive institutions, only a more rational and humane vision of the future into which the majority of people will bring their children.

In summary, it appears that the correlation between intelligence and fertility originally was positive, which is consistent with evolutionary theory regarding the early development of man–although during periods of transition from food

[2] We have referred occasionally to the relationship between intelligence and social status (whether measured by income, occupation, or education is somewhat incidental). This relationship, so far as we know, always has existed and in all probability is becoming stronger as our educational institutions act as a giant sorting and selecting mechanism in the allocation of human resources.

gathering to horticultural and from horticultural to agrarian societies, temporary reversals in the selection for intelligence may have occurred. We have no way of knowing. Nevertheless, an actual reversal probably did accompany early industrialization when children suddenly became economic liabilities and birth control was routinely practiced only among the educated classes. As the technology advanced, persons who were better off began again to have more children than before and the negative correlation between fertility and intelligence disappeared. This is where we are today. Eventually, however, as birth control measures are more widely adopted in the future and those near the bottom of the distribution begin to limit family size, the cycle will be complete and it is likely that the correlation will once again be positive.

ASSORTATIVE MATING AND INTELLIGENCE

Unlike differential fertility and mortality, assortative mating alone does not change gene frequencies in the population. It does, however, change the distribution of genes by increasing the average homozygosity and increasing the total population variance (Crow and Felsenstein, 1968). What this essentially means is that for any highly heritable trait, like intelligence, the tendency for "likes to beget likes" may be increased markedly under nonrandom (assortative) mating, limited possibly only by conditions of dominance and epistasis. Assortative mating for intelligence reduces the within-family variance while increasing the between-family variance, thus reducing the tendency for children to regress to the adult population mean while increasing the proportion of individuals in the tails of the IQ distribution. It has been estimated, for instance, that positive assortative mating for intelligence may directly account for about 80% of the present frequency of individuals with IQs above 145 (Jensen, 1969).

This kind of reassortment of many interacting genes occurs quite independent of selection pressures. If all persons lived to contribute an identical number of progeny to the next generation, positive assortative mating for intelligence would still increase the population variance as well as the genetic parent-child correlation for this trait.

Of course, all persons do not contribute the same number of genes to the next generation, as we have been arguing up to this point. Now, it also seems quite probable that when the opportunity for selection due to differential fertility or mortality is significant, the rate at which certain gene frequencies actually change will be influenced by the average homozygosity of the trait, especially if assortative mating itself occurs at different rates among different people carrying different genotypes, or if the differentials in assortative mating for a particular trait correspond to differentials in fertility for the same trait. In other words, we are suggesting that the population variance indirectly is an extremely important factor in natural selection. If the range of individual differences is very small in the parent population, then even strong selection pressures may not radically change the genetic make-up of the species. On the other hand, if the average homozygosity of the trait and individual differences are large, relatively minor selection pressures conceivably may have significant and rapid effects on the gene pool. Before expanding upon these ideas, however, let us look more closely at the relationship between assortative mating and intelligence, particularly in the U. S. today.

Historically, it appears that assortative mating for intelligence, i.e., specifically the IQ correlation between marriage partners, has been increasing in somewhat the manner shown in Fig. 5. To prove the point, we would need IQ data on several cohorts of married couples at different points in time, or similar data on several countries at different stages of development. Lacking such information, we nevertheless can surmise what has happened.

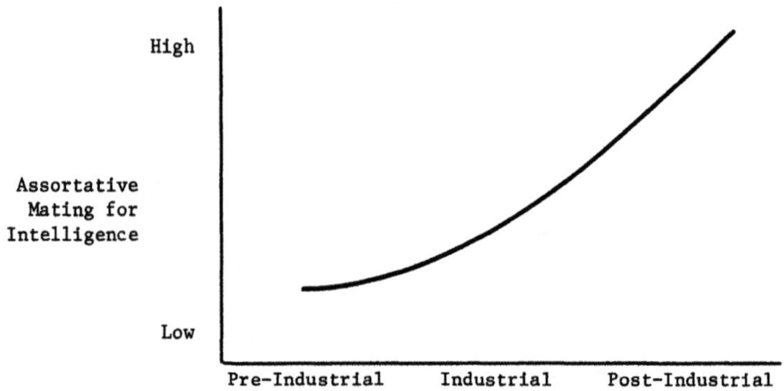

Fig. 5. Hypothesized Trends in Assortative Mating for Intelligence.

Before the rise of public education, a moderate degree of assortative mating for intelligence probably existed, if only as an indirect result of social endogamy and inbreeding. However, it must have increased quite markedly when large proportions of the population began to receive a formal education and when the extent to which most persons progressed in the school system became increasingly dependent upon individual ability (Eckland, 1967, 1970). Indeed, schools are not entirely meritocratic. Yet, there can be little doubt that the two major transitions in secondary and higher education (the first between 1910 and 1940, the second between 1950 and 1970) led to a system of mass education in the U. S. which now rather effectively redistributes the nation's human resources in a way in which even the poor but clever man has an exceedingly good chance to succeed (Blau and Duncan, 1967).

Moreover, our educational institutions have always been well known for creating mating pools. That is, schools and colleges are ecological units, each setting more or less well-defined boundaries within which courtship and marriage tends to occur. But even when propinquity alone is not a factor, a person's academic credentials are often exceedingly important in determining who is eligible to marry whom. Thus, to the extent that educational attainment depends on intellectual skills (including *where* one goes to school or college) and educational selection operates in substantially the same manner for both men and women (and for the most part it does), then assortative mating for intelligence should be expected to increase.

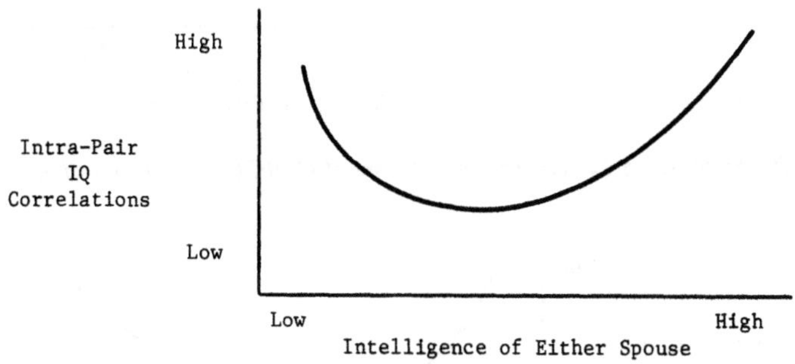

Fig. 6. Differential Assortative Mating for Intelligence.

One other point. Assortative mating may be more intense at the high and low ends of the IQ distribution, as illustrated in Fig. 6. Apparently persons who progress exceptionally far in the educational system almost invariably are well above average in intelligence, whereas those who do not make it beyond the eighth grade today, with few exceptions, are far below average. At each of these extremes, the tendency for "likes to marry likes" is greater than at the intermediate levels, which are more heterogeneous. United States Census data tends to support this line of reasoning. Whereas assortative mating is found to occur at all levels of education, the differences by which observed exceed expected frequencies of men who choose wives with the same amount of education are almost twice as great among both college graduates and men with less than an eighth grade education than among those in any of the intermediate categories (Kiser, 1968). We infer from these findings that the same pattern probably holds true for assortative mating by intelligence.

DIFFERENTIAL FERTILITY AND ASSORTATIVE MATING

Assortative mating for intelligence increases the genetic parent-child correlation for this (as any other) trait, as noted earlier and now shown in Fig. 7. Since assortative mating probably is higher at the upper and lower levels of intelligence, it follows that the parent-child correlations vary accordingly, as shown in Fig. 8.

Now, if the tendency of children to resemble their parents in IQ is highest in the tails of the distribution (as Fig. 8 suggests) and if fertility is associated with

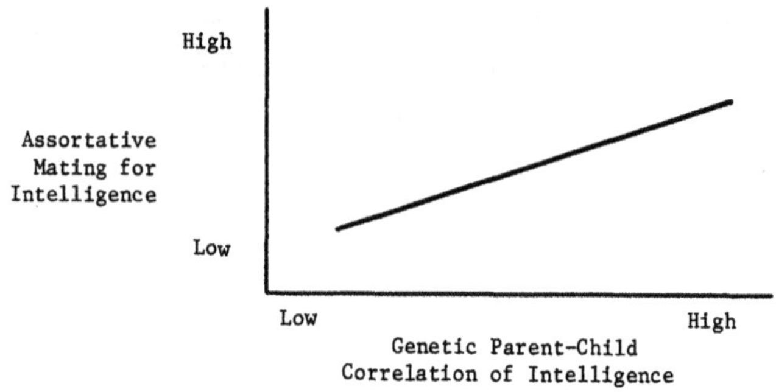

Fig. 7. Effects of Assortative Mating on Parent-Child Correlations of Intelligence.

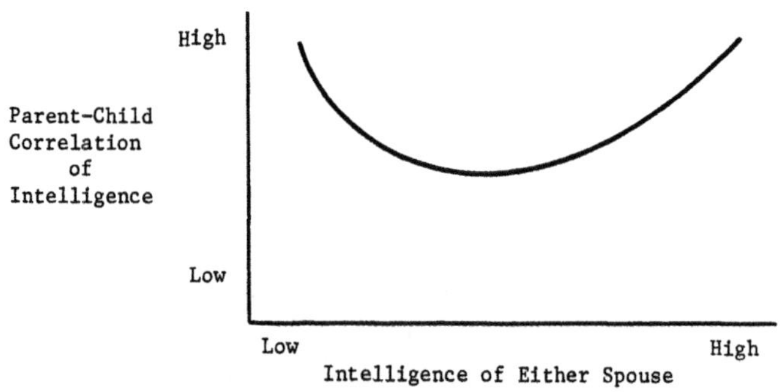

Fig. 8. Parent-Child Correlations by Level of Parent's Intelligence Under Conditions of Assortative Mating.

these same regions of the distribution, then assortative mating should strengthen the pressure for natural selection. The progeny of the adults in these regions will inherit more than the average number of genes from their parents. To the extent that these same adults also produce more children, their contribution to the gene pool increases to a point considerably in excess of their contributions under a random mating model. In essence, we are saying tha the rate of natural selection for intelligence due to differential fertility or other factors indirectly depends upon the genetic variance of the trait which, in turn, depends upon assortative mating.

Thus, assortative mating tends to amplify the genetic effects of fertility and other selection pressures. I know of no polygenic model of inheritance which has been developed to account for the *joint* effects of assortative mating and differential fertility on human evolution.

No particular laws or customs govern assortative mating for intelligence, except to the extent, I suppose, that females sometimes believe that they should marry a man just a bit smarter than themselves. Moreover, probably because assortative mating for intelligence actually occurs largely as a result of educational selection and social endogamy (and not because of deliberate choice), most people are hardly aware that it occurs at all. And certainly few recognize its effects on heredity. But considering that the potential effects in some ways are similar to cloning (i.e., copying by asexual reproduction), the subject doubtlessly warrants careful study.

REFERENCES

Bajema, C. J. 1963. Estimation of the direction and intensity of natural selection in relation to human intelligence by means of the intrinsic rate of natural increase. Eugen. Quart. 10:175-187.

——1966. Relation of fertility to educational attainment in a Kalamazoo public school population: a follow-up study. Eugen. Quart., 13:306-315.

——1971. Natural selection and intelligence: relationship between intelligence and completed fertility among third Harvard Growth Study participants. Paper presented at the annual meetings of the American Association of Physical Anthropologists, Boston.

Blau, P. M., and O. D. Duncan. 1967. The American Occupational Structure, New York, John Wiley and Sons.

Crow, J. F., and J. Felsenstein. 1968. The effects of assortative mating on the genetic composition of a population. Eugen. Quart., 15:85-97.

Davis, K. 1965. Sociological aspects of genetic control. *In* Roslansky, J. D., ed., Genetics and the Future of Man, pp. 173-204. Amsterdam, North-Holland Publishing Co., New York, Appleton-Century-Crofts.

Eckland, B. K. 1967. Genetics and sociology: a reconsideration. Amer. Sociol. Rev., 32:173-194.

——1970. New mating boundaries in educational selection. Social Biol., 17:269-277.

Jacobson, P. H. 1964. Cohort survival for generations since 1840. Milbank Mem. Fund Quart., 42:44-45.

Jensen, A. R. 1969. How much can we boost IQ and scholastic achievement. Harvard Educ. Rev., 39:1-123.

Kiser, C. V. 1968. Assortative mating by educational attainment in relation to fertility. Eugen. Quart., 15:98-112.

LaBarre, W. 1955. The Human Animal. Chicago, University of Chicago Press.

Neel, J. V., and W. J. Schull. 1972. Differential fertility and human evolution. Evol. Biol., 6 (In press).

Osborn, F. 1968. The Future of Human Heredity. New York, Weybright and Talley.

Reed, E. W., and S. C. Reed. 1965. Mental Retardation: A Family Study. Philadelphia, W. B. Saunders.

Author Index

Subject Index

GPSR Compliance

The European Union's (EU) General Product Safety Regulation (GPSR) is a set of rules that requires consumer products to be safe and our obligations to ensure this.

If you have any concerns about our products, you can contact us on ProductSafety@springernature.com

In case Publisher is established outside the EU, the EU authorized representative is:

Springer Nature Customer Service Center GmbH
Europaplatz 3
69115 Heidelberg, Germany

Batch number: 10682411

Printed by Printforce, the Netherlands